PETERSON FIELD GUIDE TO

MEDICINAL PLANTS AND HERBS

of Eastern and Central North America

PETERSON FIELD GUIDE TO

MEDICINAL PLANTS AND HERBS

of Eastern and Central North America

THIRD EDITION

STEVEN FOSTER

AND

JAMES A. DUKE

PHOTOGRAPHS BY STEVEN FOSTER

HOUGHTON MIFFLIN HARCOURT
BOSTON NEW YORK 2014

Sponsored by the National Audubon Society,
the Roger Tory Peterson Institute,
and the National Wildlife Federation

www.hmhco.com

Library of Congress Cataloging-in-Publication Data is available.

ISBN 978-0-547-94398-5

Printed in China

SCP 10 9 8 7 6 5 4 3 2 1

READERS SHOULD PAY CAREFUL ATTENTION TO THE SECTION ON HOW TO USE THIS BOOK. THE PUBLISHER AND THE AUTHORS DISCLAIM RESPONSIBILITY FOR ANY ADVERSE EFFECTS RESULTING DIRECTLY OR INDIRECTLY FROM INFORMATION CONTAINED IN THIS BOOK.

To Herb and Hope Foster
—SF

To Martha and Ed Duke
—JD

The legacy of America's greatest naturalist and the creator of the Peterson Field Guide Series, Roger Tory Peterson, is kept alive through the dedicated work of the Roger Tory Peterson Institute of Natural History (RTPI). Established in 1985, RTPI is located in Peterson's hometown of Jamestown, New York, near Chautauqua Institution in the southwestern part of the state.

Today RTPI is a national center for nature education that maintains, shares, and interprets Peterson's extraordinary archive of writings, art, and photography. The Institute, housed in a landmark building designed by world-class architect Robert A. M. Stern, continues to transmit Peterson's zest for teaching about the natural world through leadership programs in teacher development as well as outstanding exhibits of contemporary nature art, natural history, and the Peterson Collection.

Your participation as a steward of the Peterson Collection and supporter of the Peterson legacy is needed. Please consider joining RTPI at an introductory rate of 50 percent of the regular membership fee for the first year. Simply call RTPI's membership department at (800) 758-6841 ext. 226, or email *membership@rtpi.org* to take advantage of this special membership offered to purchasers of this book. For more information, please visit the Peterson Institute in person or virtually at *www.rtpi.org*.

CONTENTS

PREFACE

Herbal medicines have begun to find their rightful place as an integral, modern component of health care systems worldwide. Why a field guide to American medicinal herbs? Didn't the use of herbal medicine dissolve into obscurity after the Dark Ages? Aren't folk remedies just old wives' tales, and the stuff of witches' brew?

The answer is an unequivocal no. The plant kingdom is a storehouse of active chemical compounds. Herbs—defined as any plant used for flavor, fragrant, or medicinal purposes—encompass at least 25 percent of known flowering plants, and yet fewer than 2 percent have been thoroughly scientifically investigated for their medicinal potential. Over 40 percent of prescription drugs sold in the United States contain at least one ingredient derived from nature. As many as 25 percent of prescription drugs contain an ingredient derived from higher (flowering) plants. This statistic has not changed by plus or minus 1 percent since the 1950s.

Madagascar Periwinkle (*Catharanthus roseus* or *Vinca rosea)* is a common ornamental in the United States, often planted as a ground cover outside homes or city high-rises. Few people who encounter the plant realize that preparations derived from Madagascar Periwinkle are used in chemotherapy for leukemia and more than a dozen other types of cancer. Alkaloids derived from the fungus ergot, which grows on Rye Grass and Giant Cane, are used as uterine-contracting drugs. The many cardiac glycosides from Foxglove (*Digitalis),* commonly planted as an ornamental flower, are used in a variety of products for the management of several phases of heart disease. The primary source of material for the biosynthesis of steroid hormones is the plant kingdom. The social, economic, and political impact of oral contraceptives alone illustrates the importance of this group of plant-derived drugs. The manufacture of progesterone was made commercially feasible by the use of chemicals from

Mexican yams (*Dioscorea* species), which were then converted to progesterone. With all the advances of modern medicine, there is still nothing to replace morphine, derived from the Opium Poppy (*Papaver somniferum*), as a pain reliever for major trauma and codeine for cough. Nature, in its infinite chemical factory, endows both animals and plants with biologically active compounds.

Since the first edition of *A Field Guide to Medicinal Plants: Eastern and Central North America* was published in 1990, paclitaxel (formerly known by its registered trademark name, Taxol) has emerged as an important new drug in chemotherapy for certain forms of ovarian and breast cancer. Paclitaxel, which is derived through a semi-synthetic process from the English Yew (*Taxus baccata*), is expected to be approved in the future for other difficult-to-treat hard-cell cancer types, with sales in the billions of dollars per year. The fact is, herbs—plant drugs—are a very important and integral part of modern medicine.

We believe that safer natural compounds could be found to replace the synthetic compounds that occur in about 75 percent of our prescription drugs, and that evolution has better equipped us to deal with rational doses of preparations from medicinal plants. And since it costs, in our litigious American society, hundreds of millions of dollars to prove a new drug safe and efficacious, herbal medicines may offer a better alternative.

In modern medicine in the United States, only single isolated chemical components are used in prescription drugs, rather than the complex mix of chemicals found in a single herb. This has more to do with the structure of our drug laws than with scientific advancement. To invest hundreds of millions of dollars in proving a drug safe and effective, pharmaceutical companies are interested in substances that can be patented. Therefore, no drug company is going to spend hundreds of millions of dollars to prove Echinacea can reduce the length and severity of a cold, since Echinacea cannot be patented.

Much scientific research has been conducted in developing countries where traditional medicine systems, some thousands of years old, are still an integral part of health care systems. China and India are prime examples. Although these countries may still be classified as "developing" in terms of their economic and technical systems, the more than 5,000-year-old traditions of Traditional Chinese Medicine, and of Ayurveda in India, represent highly developed medical systems that are constantly being vindicated and enhanced by modern research. Their experience and research is valuable to our study of American medical botany.

Most American medicinal plants have yet to be thoroughly investigated in terms of pharmacology and chemistry, much less through clinical trials in humans. However, that trend has begun to change in the past decade. Finally, we see progress in medicinal plant research in the United States.

Herbs in their whole form, rather than just a single chemical derived from them, have come to play an important role in health care. Since the first edition of this book was published in 1990, interest in herbs worldwide has exploded. In the past decade alone, herb product sales in the United States have, astonishingly, increased tenfold at the retail level.

A landmark legislative development in the United States fueled the shift of herb products from the niche market of health and natural food stores into the mass market. The Dietary Supplement Health and Education Act of 1994—commonly known by its acronym, DSHEA—clearly defined herb products as "dietary supplements," allowing companies to list a product's benefits on the label, recommending how the product affects the "structure and function" of the body. Statements on the label cannot directly imply prevention or treatment of a specific condition; if such claims are made, the manufacturer must notify the FDA, and the claim must carry the caveat, "This statement has not been evaluated by the Food and Drug Administration. This product is not intended to diagnose, treat, cure, or prevent any disease." Balanced, truthful, nonmisleading, scientifically based third-party literature can also be used to provide the consumer with legitimate information on the intended use of herbal products. The system is far from perfect, but the net effect has been to stimulate enormous interest by consumers, health care professionals, scientists, regulators, and lawmakers in the role that herbs play in benefiting health.

In the early 1990s, before DSHEA, herb products were relegated to the realm of health and natural food stores, with an estimated $500 million in sales. Today herb products are sold wherever Americans shop for food or drugs—supermarkets, discount department stores, chain and independent pharmacies, and the Internet—and are advertised nationally on television. Herbal dietary supplement products represent nearly $5.3 billion in sales in the United States. The use of herbs for health purposes has become part of the American mainstream.

Consumer interest in and acceptance of the benefits of herbs is a major driving force which helped to spark legislation that makes herb products more widely available. American consumers spend more than $5 billion annually on herbal products to treat colds, burns, headaches, rashes, insomnia, PMS, depression, gastrointestinal problems, and menopause, among other conditions.

An important development in the twenty-first century is significant federal funding for medicinal plant research as our culture explores the potential of herbs not only to treat disease but, perhaps even more importantly, to prevent disease and contribute to healthier lifestyles. In 1991, soon after the first edition of this book was published in 1990, the U.S. Congress provided $2 million in funding to establish an office within the National Institutes of Health (NIH) to investigate and evaluate "unconventional" medicines, including medicinal plants. In 1999, the small office was elevated to a major research center, becoming the National Center for Complementary and Alternative Medicine (NCCAM), one of 25 major research centers at NIH. For 2013, the NIH NCCAM budget, much of it going to fund grants for research scientists, is nearly $128 million.

Along with NCCAM, the Office of Dietary Supplements (ODS) at NIH funds the NIH Botanical Research Centers Program, which promotes collaborative integrated, interdisciplinary research on botanicals. Five centers are funded from 2010 to 2015 at the Pennington Biomedical Research Center, Louisiana State University; University of Illinois at Chicago; University of Illinois at Urbana-Champaign; University of Missouri; and Wake Forest University in North Carolina. In addition, other research groups have emerged, such as the National Center for Natural Products Research, School of Pharmacy, University of Mississippi, a major research center with more than 100 Ph.D. natural product scientists working in an applied multidisciplinary research environment to discover and develop natural products for use as pharmaceuticals, dietary supplements, and agrochemicals. Today, hundreds of scientists in the United States work on creating new science on medicinal plants using new, highly specific technologies, funded by government research dollars. In dramatic contrast, in the late 1970s, coauthor James A. Duke, who spent most of his career with research programs in the U.S. Department of Agriculture (USDA), was the sole U.S. government employee charged with full-time research on medicinal plants.

The expansion and maturation of the Internet has been both a boon and a bane to finding reliable information on medicinal plants. Two developments are of particular interest: (1) the continued expansion and development of the online AGRICOLA database produced by USDA's National Agricultural Library—the largest agricultural library in the world; and (2) the continued expansion and development of the online PubMed search engine of NIH's National Library of Medicine—the largest medical library in the world. For example, a simple PubMed search for scientific articles with the word "Echinacea" in the title yields 78 references published up to the year 2000—the publication date of the previous edition of this book. Today, a PubMed search from 2000 to 2012 yields 425 scientific papers with Echinacea in the title. In short there's been an exponential explosion in research on medicinal plants. In 1990, when the first edition of this book was published, it was possible for a single researcher to keep up with new research on medicinal plants in the limited number of scientific journals then devoted to the subject. Now that task is impossible, as the PubMed database alone adds up to 70 references and abstracts of newly published herbal and medicinal plant research each and every day.

In addition, it is important to understand that new scientific research is often sparked by an historical herbal medicine research lead, or observations by ethnobotanists of nuanced plant use in traditional, yet often endangered cultures. Throughout this book we've strived to present traditional or historic uses that have been verified by modern scientific research.

Another major new development in information technology is the availability of historical herbal literature. Millions of out-of-print books essential to research on medicinal herbs have been digitized and are available for free. For example, major botanical institutions such as the Missouri Botanical Garden, the New York Botanical Garden, the National Library of Medicine, the Library of Congress, and virtually all major university libraries have digitized and make freely available scores of rare books. They can be found on sites such as Google Books (books.google.com), the Internet Archive (archive.org), and the Biodiversity Heritage Library (biodiversitylibrary.org), to name a few. When the first and second editions of this book were published in 1990 and 2000, respectively, many historical reference gems were available only in the rare-book rooms of major libraries, requiring a physical trip and appointment to view the book and garner whatever information might be available in the limited time a researcher may have at a location. The other alternative was to find and purchase a copy of the book from an antiquarian or rare book dealer. In 1990 and 2000, to access the information in the seminal reference C. S. Rafinesque's *Medical Flora; or Manual of the Medical Botany of the United States of North America,* published in two volumes (1828–1830), one had to go to a major library's rare-book collection; or, if an actual physical copy could be found on the antiquarian market, one had to pay upwards of $12,000 to own the book. Today, however, an exact digital reproduction of the original work sits on the senior author's computer desktop as a searchable PDF file—downloaded for free—along with several hundred other rare antiquarian references used in researching this book.

Another major factor is increased scientific interest in herbs, resulting in dozens of controlled clinical trials, the gold standard of scientific evidence. In 1990 the number of undergraduate and graduate university programs pertaining to medicinal plant–related sciences could be counted on one's hand. Today, nearly every medical school, every school of pharmacy, and numerous biological education programs offer courses dealing with herbs, herbal medicine, and medicinal plant topics. In the United Kingdom, based on legislation that allows herbalists to practice as primary health care practitioners, starting in 2005, several universities began matriculating students who completed four years of study. Following a licensing exam, they are authorized to set up practice, hang a shingle, and offer their services to the British public as herbalists. The same holds true for Australia.

In addition, head-to-head controlled clinical trials comparing conventional antidepressant drugs with herbal alternatives such as St. John's Wort have shown that St. John's Wort preparations are just as effective and safer than prescription antidepressants for the treatment of mild to moderate forms of depression. St. John's Wort preparations outsell prescription antidepressants by as much as 20 to 1 in Germany. Saw Palmetto fruit extracts have been shown to be as effective and at least as safe as prescription drugs for the treatment of benign prostatic hyperplasia (BPH), a nonmalignant enlargement of the prostate affecting a majority of men over 50 years of age. An indigenous American medical plant, Saw Palmetto is approved for use in BPH in Germany, France, and Italy. High-quality herb products are much less expensive than prescription drugs. The European and Asian influence (or even invasion) is a crucial factor in the rise of herbal medicines in America.

Much modern research on herbal medicines, or "phytomedicines," has been stimulated by European interest. A phytomedicine is an herb product that represents the totality of chemical constituents in an herb or plant part, rather than a single isolated chemical component. Phytotherapy, the practice of herbal medicine, is most highly developed in Germany, where over 70 percent of physicians prescribe herb products, and medical students are required to pass a section on phytomedicine in licensure exams.

In Germany, phytomedicine is not viewed as "alternative medicine." Rather, phytomedicines are part of the mainstream medical establishment. Herbal preparations are simply another tool, normally available to physicians, pharmacists, and, ultimately, the consumer. The German educational and regulatory model now serves as the basis for the development of herb product regulations throughout European Union countries.

Since this book first appeared over two decades ago, one intention has been to further an awareness of the need for plant conservation by recognizing the economic or beneficial history of plants that could provide potential future economic and medicinal benefits for humans. Conservation and preservation are necessary on both the micro and macro levels. European and Chinese medical botanists who go on collecting trips to U.S. fields and forests are struck by the abundance of our wild herbs, and they caution that we should conserve them. The demand for certain species of wild-harvested native American herbs, such as Echinacea species, Black Cohosh, and Saw Palmetto, has increased more than tenfold in the past two decades. Rational conservation efforts will be necessary. We should make an effort to conserve native medicinal plants that may provide treatment or cures for cancer, heart disease, warts, the common cold, and even AIDS. The notion that American fields and forests are an endless fountain of animal, plant, and mineral resources is a nineteenth-century idea, not appropriate to the dwindling natural resources of the twenty-first century.

A number of medicinal plants are being extirpated worldwide without regard to preservation and the continued ecological success of the species. While the Endangered Species Act and Lacey Act have helped regulate the harvest of a few medicinals, notably Ginseng, the public consciousness is still swayed more toward protecting animals than plants. Commenting on dramatic declines in Kansas populations of Echinacea (Purple Coneflower) in recent years, one frustrated researcher remarked that if Echinacea had fur and cute little black eyes, it could elicit a little attention! In the past decade, attention to conservation of wild medicinal plants has led to significant commercial cultivation, not only in the United States, but especially in China, where skill in medicinal plant production coupled with a low-cost agricultural labor force has made China a leader in supplying not only Chinese medicinal plants, but American medicinal plants to world markets. It is yet another sign of the global economy.

We hope this volume will help the reader gain a deeper appreciation of the plants around us. By understanding the traditional medicinal uses of so many of our wildflowers, woody plants, and weeds, we will gain a deeper sense of our relationship to the natural world. Knowing that a wildflower was a folk remedy for cancer, and that that knowledge may eventually produce the lead that helps researchers develop a new cancer treatment, adds a new dimension, a human element, to conservation. Enjoy, and be cautious.

Steven Foster and James A. Duke

PETERSON FIELD GUIDE TO

MEDICINAL PLANTS AND HERBS

of Eastern and Central North America

This book is not a prescriptor, just a field guide to medicinal plants of the eastern and central portion of the North American continent. The purpose of this guide is to help you identify these plants safely and accurately, and to help you avoid similar-looking plants that could be dangerous or even fatally poisonous.

General Organization

SPECIES COVERED: There are more than 800 species of plants growing in the eastern United States that can be documented as having at least some medicinal use. This book includes 500 of the most significant medicinal plant species of the eastern United States with important historical uses, present use, or future potential. We have not attempted to cover all of the alien plants found here that are used medicinally in their native lands. For example, of the hundreds of ornamentals originating from East Asia in American horticulture, more than 1,000 species can be documented as being used in Traditional Chinese Medicine. Nearly all common weeds naturalized from Europe have been used as medicinal plants in their native lands. We have included many naturalized weeds, but not all of them.

AREA COVERED: This Field Guide covers all states east of, but excluding, Colorado, Montana, and New Mexico. It does not fully cover the southern half of Florida or the southern and western halves of Texas. Adjacent regions of Canadian provinces are included. Toward the southern, western, and extreme northern extensions of the range, our coverage is less comprehensive.

BOTANICAL AND MEDICAL TERMS: Since this book is intended as a guide for the layperson, we have used as few technical terms as possible.

Nevertheless, the use of some specific terms relative to plant identification or the medicinal use of plants has been inevitable. We define those terms in the Glossary. Although we have defined terms relative to medicinal effects, we have not attempted to define each disease, condition, or ailment included in this book. We have kept disease terminology as simple as possible. For further explanations, the reader is referred to any good English dictionary or medical dictionary, online or in print.

ILLUSTRATIONS: All photos are by Steven Foster except where noted otherwise. Photographs have been taken over a period of 35 years. Most photographs were taken with various Nikon camera bodies and Nikkor lenses, particularly the Micro-Nikkor series. Most photographs were taken with Fuji Velvia or Kodachrome 64 film, until 2004, when photographic media evolved to digital capture and film essentially became obsolete.

Identifying Plants

Plants are arranged by visual features, based on flower color, number of petals, habitat, leaf arrangement, and so on. These obvious similarities help the reader to thumb quickly through the pages and find an illustration that corresponds with a plant in hand. Once you have matched a plant with a photograph, read all details in the descriptive text, making sure that all characteristics—key italicized details, range, habitat, flowering time, color, and the flower or leaf structure—correspond. Never, ever ingest a plant that has not been positively identified. (Read the section on warnings on p. 9 and the "Word of Caution" on p. 12.)

The first part of the book covers wildflowers, which are arranged by flower color (white, yellow, orange, pink to red, blue-violet, green) and other visual similarities. Please note that there is tremendous variation in flower color in the plant world. If a plant may have white or blue flowers, for example, we have attempted to include the plant in both sections of this guide, usually depicted by a photograph only in the section with typical flower color. Your interpretation of "pink to red" may be viewed as "violet to blue" by others. Be aware of these subtle differences.

Flowering shrubs, trees, and woody vines are in separate sections following the wildflowers. Flowering woody plants are generally not included in the wildflower section. Ferns and related plants follow the woody plant section. Last is a section on grasses and grasslike plants.

COMMON NAMES: One or two common names are listed at the beginning of each entry. Though some plants have only one common name, others may have many. Common names of wildflowers generally conform to those used in Peterson and McKinney's *A Field Guide to Wildflowers.* Sometimes, in an herbal context, a plant is better known by another common name. In these instances we have used the most commonly known name for the herb. For trees, shrubs, vines, grasses, and ferns, we have used the names we believe are best known.

PART USED: The plant parts used for medicinal purposes are listed in boldface type opposite the common name. In descriptions of woody plants, the word "bark" almost always refers to the inner bark of the tree or shrub, not the rough outer layer. Slippery Elm, for example, has a rough, corky outer bark that is not known to be of medicinal use, though it is often sold as Slippery Elm bark. The tawny white, fibrous, highly mucilaginous (slippery) inner bark can easily be stripped from the branches once the outer bark has been rasped away. Please see the section "Conservation and Harvesting" (p. 9) for general guidelines on harvesting various plant parts.

SCIENTIFIC NAMES: Beneath the common name is the scientific or botanical name by which the plant is generally known (in our opinion). Scientific names are not set in stone. Although changes in scientific names are annoying, sometimes they must be made according to valid new information developed by botanists with a special interest in taxonomic relationships. When the first edition of this book was published in 1990, life was simpler for the plant taxonomist. We relied largely on the scientific names found in the eighth edition of *Gray's Manual of Botany,* last revised by Merritt Lyndon Fernald in 1950, long outdated by 1990. In the past two decades, plant taxonomy has become a "brave new world," no longer relying on traditional visual morphological similarities and differences to distinguish and classify one organism in relation to another. Visual resemblance has given way to evolutionary relationships based on plants' DNA and molecular clues that offer surprising genetic relationships, often having little to do with a plant's morphology. The answer to the question "What is the correct scientific name?" has become at once simpler and more difficult to answer. Gone are the days when a single reference can answer that question, as new research is published on a daily basis.

Botanical taxonomy reflects our collective understanding of the diversity of life. Classification of flowering plants (angiosperms) is constantly being reviewed by a worldwide scientific team called the Angiosperm Phylogeny Group. Phylogeny hypothesizes the evolutionary history of organisms, represented by a branching "tree of life" diagram that delineates inferred plant relationships based on similarities and differences in their physical and/or genetic characteristics. We have used several information sources to reach our conclusions on the "correct" or current scientific name. These include databases such as www.theplantlist.org, a collaborative project of the Missouri Botanical Garden and the Royal Botanic Gardens at Kew (UK), which is a dynamic online resource with over a million scientific plant names, with a ranking of accuracy, forming a list of nearly 300,000 accepted names at the species level. It is useful not only for current scientific names, but also for searching herbal literature 100 to 300 years old, as it offers clues to names used historically; hence it is an important research aid. Concurrently we used the USDA Agricultural Research Service's Germplasm Resources Information Network (GRIN) database as a primary source.

Plant taxonomy is a moving target. No single information source has all the correct answers. We have provided synonyms in brackets beneath the main botanical name to reflect alternate or obsolete scientific names that are often encountered in recent botanical manuals, popular field guides, and herbals. The synonyms also reflect the many taxonomic changes since the first edition of this guide was published in 1990.

The scientific name is followed by an abbreviation of the name of the botanist, or "species author," who named the plant. This is a useful reference tool for taxonomists and can be a flag for the layperson as well. For example, many old American herbals list the scientific name of the Slippery Elm as *Ulmus fulva* Michx., but all modern botanical works cite this elm as *Ulmus rubra* Muhl. "Michx." is the abbreviation for the name of the French botanist André Michaux (1746–1802), who assigned the name *Ulmus fulva* to this species. "Muhl." is the abbreviation for Gotthilf Henry Ernest Muhlenberg (1753–1815), who first proposed the name *Ulmus rubra*. Muhlenberg's *Ulmus rubra* has priority over Michaux's *Ulmus fulva*, according to the rules of the

International Code of Botanical Nomenclature, and *Ulmus rubra* thus becomes the name used by botanists everywhere. Research on medicinal plants requires delving into historical literature; thus it is useful to have the author citation as a reference to the scientific name. It is a point of comparison that may help you verify which species is being discussed.

FAMILY NAMES: The scientific name is followed by the common name for the family, with the technical family name in parentheses. We did not include the technical family name in previous editions of the book, but decided it was essential to include in the present edition because modern phylogenic taxonomy has exploded long-standing family concepts. Examples include the once stable broadly circumscribed Lily family (Liliaceae), which now constitutes upwards of 30 segregate plant families. Milkweeds have long sat in their own family, the Asclepidaceae. Now milkweeds are included in the Dogbane family (Apocynaceae). The Goosefoot family (Chenopodiaceae) is no longer. It is now part of the Amaranth family (Amaranthaceae). The once familiar Figwort family (Scrophulariaceae) has had many of its once stable but difficult to define members move into other families, such as the Plantain family (Plantaginaceae).

Beneath current family name listings we have listed former family associations, including both the old family common and technical names. Also, several families, such as the Mint (Lamiaceae or Labiatae), Pea (Fabaceae or Leguminosae), and Aster (Asteraceae or Compositae) families, are known by two acceptable names—newer names ending in the suffix "aceae" and their old names ending only in "ae." We have chosen to include both in the technical family name listing. Evolution has prevailed, and we hope to help you adapt to this new way of looking at plant relationships.

DESCRIPTION: A brief description of the plant follows. The descriptions are based primarily on visual characteristics, though we have sometimes included scent as an identifying feature. We regret that scratch-and-sniff features cannot be included in the text. The description begins with the growth habit of the plant (annual, perennial, twining vine, tree, shrub, and so on) and height. Characteristic details for leaves, flowers, or fruits follow. Key identifying features of each plant are italicized.

The earliest blooming date given is usually the time when the flowering period begins in the South. Blooming time in more northerly areas often begins a month or more later. Though the blooming date may be listed as "Apr.–June," the plant may bloom for only two weeks in any particular location. Take Goldenseal (*Hydrastis canadensis*), for example. Blooming time is listed as Apr.–May. In northern Arkansas the plant usually blooms around the first week of May. The flowers last only three to five days. In revising the present edition, we adjusted many flowering times, reflecting changes that have resulted from the fact of global warming, often producing earlier flowering or fruiting times than was normal just 20 years ago.

DISTRIBUTION: The habitat in which the plant grows is under the heading "Where Found." We also indicate whether the plant is an introduced alien or is native to the United States. Many alien (nonnative) plants are now naturalized or adventive—well established on their own without being cultivated. Many plants in this category often are still cultivated in herb, kitchen, or flower gardens; others are best known as weeds. Habitat is followed by the plant's range in Canada and the eastern and central United States. Each plant's range is given from northeast to southeast and from southwest to northwest.

Although many wild plants are common and widespread, others are rare and should not be overcollected. This section also includes cautions about plants that are protected by law, and those that should be protected.

MEDICINAL USES: This section presents a brief discussion of some of the most significant medicinal uses of a plant. Historic or folk uses are usually described first, generally starting with known uses of the plants by native peoples of North America. In this edition, as much as possible we have attempted to attribute use to a specific Native American group that used a plant for a specific purpose. We have chosen to avoid use of the phrase "American Indian." We sometimes list medicinal uses in India or cite work by Indian researchers, so this helps to avoid confusion. Readers are referred to Dr. Daniel Moerman's excellent work *Native American Ethnobotany* (1998) or his website (http://herb.umd.umich.edu), and Dr. Jim Duke's *Handbook of Northeastern Indian Medicinal Plants* (1986) or his website (www.ars-grin.gov/duke/). The original references cited in these works and their bibliographies have been consulted for detailed accounts of uses of native plants by specific indigenous groups of people.

The discussion often covers Native American usage, folk usage by settlers of the North American continent over the past 400 years, historical usage by medical practitioners, and vindication of medicinal uses as suggested by presence of specific chemical components or chemical groups, pharmacological studies (mostly with animals), and, if available, clinical studies or clinical applications.

Often in our discussion we have included notes on the experience of Chinese practitioners with a closely related plant species. For more than 200 years, botanists have recognized striking similarities between the floras of eastern Asia and eastern North America. Most plants involved in this pattern of "disjunctions" in plant geography are thought to be remnants of an ancient forest that covered the Northern Hemisphere more than 70 million years ago. Because many of the more than 140 genera that share ranges in eastern Asia and eastern North America are important medicinal plants on one continent or the other, the Chinese experience is relevant to North American species, and vice versa. Some plants included in this classical pattern of plant disjunctions are the various species of ginseng (in the genus *Panax*), Witch-hazel (*Hamamelis*), Sassafras, Mayapple (*Podophyllum*), Magnolia, Sweetgum (*Liquidambar*), Spicebush (*Lindera*), and other plant groups included in this book.

In a historical context, we occasionally mention a concept known as the "doctrine of signatures." This refers to an ancient idea that if a plant part was shaped like, or in some other way resembled, a human organ or disease characteristic, then that plant was useful for that particular organ or ailment. With its three-lobed leaves, Liverleaf (as herbalists referred to the *Hepatica* species) was thought to be useful in treating liver disease. If held to light, the leaves of Common St. John's Wort (*Hypericum perforatum*) appear to have numerous holes pricked through the surface. The resemblance between these holes and the pores in human skin led some people to believe that preparations made from the leaves of this plant were useful for healing cuts. Using this doctrine, one might assume that kidney beans were good for the kidneys, or that the leaves of Broad-leaved Arrowhead (*Sagittaria latifolia*) would be useful for wounds caused by the head of an arrow. This concept has no scientific basis, though uses conceived centuries ago have persisted and may even have been corroborated by scientific evidence. Look at the two halves of a walnut. Walnut halves look like the human brain, and walnuts are one of the best dietary sources of the brain food serotonin. A strong argument can be made that the "doctrine of signatures" itself is not a conceptual tool to discover uses of medicinal plants; rather it is a tool used in traditional societies or folk cultures to help the teacher convey a plant use to the student while serving as a tool to help the student remember the particular use.

Following Native American, European, or folk use by other indigenous groups, we include information on the current status of the plant. In many cases, pharmacological studies on individual chemical components or extracts of a plant or plant part confirm ethnobotanical uses. Controlled clinical studies have been conducted in the past two decades on the benefits of preparations of such herbs as St. John's Wort, Chamomile, Saw Palmetto, Echinacea, and others. In many instances, these herbs are actually approved for therapeutic use in various countries. We also relied heavily on Jim Duke's data (www.ars-grin.gov/duke/).

PREPARATIONS: In the "Uses" section we discuss ways in which each plant has traditionally or historically been used. A simple *infusion* is made by soaking an herb (usually the leaf or flower) in hot water for 10 to 20 minutes. A cold infusion may be made by soaking the plant material in cold water for a relatively long period of time (varying from two hours to overnight), or simply letting the hot infusion sit until it is cool.

A *decoction* is made by simmering the plant material—usually the root, bark, or seed—under low heat. As a general rule of thumb, the word "infusion" is reserved for leaves and flower material, and "decoction" for roots, barks, or seeds. Check the plant part used, if available, to determine whether a tea should be simmered or "decocted" or simply "infused," based on whether it is made from the bark, root, leaves, flowers, and so on.

The term *wash* is often used for the external application of a cooled tea. A wash is usually applied to the skin over the affected area.

Poultice is another commonly used term for an external application of herbs. A poultice is typically a moist paste made from the plant material, beaten to pulp in a mortar and pestle or with some other instrument if the herbs are fresh, or soaked in warm water if the herbs are dried. The poultice is spread over the affected area. Since some plants, such as Comfrey and Mullein, have leaves with irritating hairs that may adversely affect the skin, a layer of thin cloth such as muslin may be applied to the skin, with the herb material placed on top of the cloth.

A *tincture* is a plant extracted in alcohol. Most tinctures are made with dilute alcohol (100 proof or 50 percent ethanol, 50 percent distilled water). Traditionally a tincture is made simply by soaking a certain percentage of plant material (often 20 percent by weight to the menstruum, or solvent) in the ethanol and water for a period of about two weeks, shaking the material daily, then straining the liquid through cheesecloth or filter paper before bottling. Tinctures are also made by a process known as percolation, in which the menstruum is poured through the plant material, which has been finely ground and then placed in a funnel-shaped container with a receptacle at the bottom to catch the herb-fortified liquid.

DOSAGE: Since this book is intended to help the reader identify medicinal plants and appreciate their traditional uses, and not to serve as a prescriptor, we have not included dosage except in historical context. See cautions regarding individual sensitivity below.

WARNINGS: Last but not least, we include warnings. *Please be sure to read the warnings under each species account before handling the plant, even for identification purposes: some plants can cause a painful skin rash. Some medicinal plants are very similar to and can be easily confused with a poisonous plant, with unpleasant or even fatal results. Never eat or taste any part of a wild plant, or use it in any medicinal preparation, unless you are certain of its identification and safety, and that the dosage is correct, and that the plant has been properly prepared.*

Conservation and Harvesting

If you intend to harvest a plant or plant part, you must observe certain rules and values after you have properly and positively identified it.

(1) If the plant is unusual or rare in your area, leave it alone. Contact your local Audubon Society, native plant societies, botanical gardens, or state conservation agencies for a list of rare or threatened plants in the immediate vicinity. Often plants that are common in one state may be rare in another. Pale Purple Coneflower (*Echinacea pallida*) is common in eastern Kansas, but it is very rare near the extreme limits of its range. Other plants, such as Pink and Yellow

Lady's-slippers, were historically valuable as medicinal plants, but should be left alone wherever they occur. Once you dig the root, the plant is no more. These wild orchids are difficult to propagate and cultivate. Once much more abundant in the wild, they have been historically extirpated as a medicinal plant and are currently overexploited as plants for wildflower gardens. We believe their sale should be banned, where appropriate, or carefully regulated. (Account books from the 1860s reveal that one company alone was selling more than 300 pounds of dried Lady's-slipper root per month.) It takes dozens of plants to get one pound of dried root. Although a number of orchids have traditionally been used in folk medicine, none is abundant enough for harvest.

In China, only a handful of wild-harvested ginseng roots are dug each year. The Oriental Ginseng has been valued as a medicinal plant for more than 2,000 years. It has been virtually exterminated from the wild in China. One wild root of Oriental Ginseng can sell for as much as $20,000 on the Hong Kong market. Wild American Ginseng, by comparison, has only been traded as a commodity for a mere 200 years. Tons of wild-harvested American Ginseng are shipped to Asian markets each year. How long will our plant populations be able to sustain themselves?

Goldenseal (*Hydrastis canadensis*) was once one of the best-selling herbs in domestic health food markets. In recent years, supply shortages caused by alleged heavy harvesting of wild populations could threaten the plant's future. Like American Ginseng, Goldenseal is now monitored in international trade through the provisions of an international treaty, the Convention on International Trade in Endangered Species (CITES). Cultivation efforts are underway for Goldenseal. We strongly encourage the cultivation—rather than harvesting in the wild—of all native medicinal plants that enter commerce.

(2) Never collect all of the specimens of a plant in an area. One fear in publishing a volume such as this is that the interest in harvesting medicinal herbs for profit will outweigh the necessity of conservation. Take only what you need and no more than 10 percent of the individuals in a given population. If you harvest an entire population of most herbs, you will do so only once. Careful consideration and attention to detail are necessary for identification, harvest, usage, and conservation.

(3) Find out who owns the property where you intend to harvest

a medicinal plant and obtain permission to harvest it before going on the property. In Missouri and some other states, state law prohibits the harvest of plant material without the permission of the landowner. Along roadsides, the owner is often the state itself.

(4) Harvest the plant at the correct time of year. Most herbs in which the leaves or whole herb (flowers, leaves, and stem) are used are harvested just before or just as the plant comes into flower. The biologically active chemical components in a plant vary in quantity and quality according to the stage of the plant's growth, and even the time of day when it is harvested. Flowers are best harvested as they reach their peak bloom. As a general rule, it is best to harvest leaf and aboveground plant materials before noon on a sunny day, after the dew has dried off the leaves. Seeds should be harvested only when fully ripened, but before they have dispersed.

Roots are usually harvested when a plant is dormant. It is generally better to harvest most roots in autumn rather than in spring. In spring, wet weather results in higher moisture content in the roots, which makes them more difficult to dry. Autumn-harvested roots should be dug after the plant's seeds have matured. Federal law prohibits the harvesting of Ginseng until after the fruits have ripened. Unfortunately, the practice of harvesting Ginseng roots before the seeds have had a chance to develop is still common. Such practices should be discouraged. Goldenseal is often harvested in the spring, as the plant emerges from the ground and begins flowering, but it should be harvested only after it sets seed in late summer. Roots, generally speaking, should not be harvested when the leaves and stem are still growing, in order to avoid affecting both the quality of the plant material and the plant's ability to set seeds.

(5) If you are harvesting bark, always take it from the lateral branches—do not strip it from the main trunk. Harvest only from one side of the branch. Avoid girdling the branches. Bark serves as a protective covering for plants. The rough outer bark consists mainly of corky cellular tissue that later develops wood cells, especially on the inner surface. The inner bark—the part usually gathered for medicinal purposes—consists mainly of long wood cells, often forming fibers of great strength and toughness. Bark is most easily removed when the sap rises, in spring to early summer. Of course, the bark is the tree's lifeline, from its roots to its top. Complete girdling (stripping a complete circle of bark around the trunk or branches) will usually kill a tree.

A Word of Caution

We cautiously advise our readers that this Field Guide is just that—a key to the recognition of medicinal plants. Perhaps we have been overly careful in our warnings, but in fact there are people who are allergic to any given species of plant. All food plants, like medicinal plants, contain greater or lesser amounts of minerals, vitamins, carcinogens, anticarcinogens, oxidants and antioxidants, enzyme agonists and enzyme antagonists, toxins and antitoxins, and other biologically active compounds.

In our warnings we have mentioned that dozens of these plants can cause contact dermatitis, although we have handled most of the species treated in this book and have experienced dermatitis only from Stinging Nettles and Poison Ivy. However, as our bodies change with age or other variables, our sensitivity to toxins may inexplicably change as well. Jim Duke used to imbibe a sip or two of Poke berry juice; now he is allergic to it. Steven Foster had not had dermatitis from Poison Ivy for 30 years, but exposure to Poison Ivy in 2012 frequently caused him to break out in Poison Ivy rash. Further, we suspect that the pollen of most species, if gathered and forcibly inserted in the nostrils, would induce sneezing and perhaps even allergic rhinitis in some people. And, of course, all plants included in this book contain substances that are poisonous in excess.

Dosage and proper preparation are very important. Everyone should be cautious about ingesting any new material, food, or medicine. The reaction of one individual may differ from that of another. One person's past experience may not be indicative of how one may react to a plant exposure in the present.

Unfortunately, some very poisonous plants, such as Poison Hemlock, can closely resemble some very innocuous medicinal plants, such as Wild Carrot (Queen Anne's Lace). And poisonous Water-Hemlock might be mistaken for some of the Angelicas and Skirret. We would not trust all botanists, much less all amateurs, to identify them accurately. Even professional botanists have died after misidentifying mushrooms or plants in the unrelated Parsley family. In spring, it is easy to grab a wild Iris among the new Cattail shoots. Results might not be fatal, but unpleasant. One elderly couple confused Foxglove for Comfrey and died shortly thereafter. Recently an entire Italian family was poisoned by Foxglove leaves mistaken for Borage leaves. Father or Mother Nature is not benign: He or She has produced some of our deadliest poisons. If you are imprudent, you may ingest some of nature's lethal compounds.

In spite of these perils, we lose fewer people to herbal accidents (fewer than ten per year) than we do to iatrogenic (hospital or doctor-induced accidents) or intentional ingestion of narcotics derived from plants (6,000 deaths a year in America); alcohol (100,000 deaths a year); or the smoking of the Indian gift, tobacco (300,000 deaths a year).

Basil, Comfrey, and Sassafras are some herbs that have come under fire for containing potential carcinogens. An article published in *Science* (Ames et al., "Ranking Possible Carcinogenic Hazards," 17 Apr. 1987, vol. 236, pp. 271–280) puts these carcinogens in proper perspective. A cup of Comfrey leaf tea was stated to be about $^{1}/_{100}$ as carcinogenic for its symphytine as a can of beer was for its ethanol. A gram of basil was $^{1}/_{28}$ as carcinogenic for its estragole as the beer was for its ethanol. A sassafras root beer, now banned by our FDA, was $^{1}/_{14}$ as carcinogenic for its safrole as the can of beer for its ethanol.

We cannot agree with herbalists who say that herbal medicine has no side effects. Probably all natural and synthetic compounds, good and bad, are biologically active in many ways in addition to the one we wish to harness in medication. Moreover, any medicine—herbal, natural, or synthetic—can be toxic in overdoses. We again remind our readers that this Field Guide is a key to the recognition of medicinal plants, not a prescriptor. Only your doctor or other health care professional who is licensed to do so can prescribe an herb for you. We cannot and do not prescribe herbal medication.

SYMBOLS

☠ = **POISONOUS.** Dangerous or deadly to ingest, or perhaps even to touch.

⚠ = **CAUTION.** See warning in text.

The authors would like to thank various friends and colleagues who have aided in our combined 90 years of experience in medicinal plant research. A special thanks to Les Eastman for encouragement and for tracking down photos and plants. The authors deeply appreciate the help, support, and friendship of Mark Blumenthal and staff at the American Botanical Council, Dr. Ed Croom, the late Dr. Norman Farnsworth (1930–2011), members of the Sabbathday Lake Shaker Community, the late Dr. Varro Tyler (1926–2001), and Maggie Heran and the staff of the Lloyd Library and Museum in Cincinnati (the world's largest medicinal plant library), who provided many useful and obscure research materials.

We thank Peggy Duke and Donna Foster for their behind-the-scenes contributions, without which this work could not have materialized. Additional support was provided by Colin Foster and Abbey Foster. A special thanks to the late Barbara Garsoe (1919–2013), childhood next-door neighbor, who helped instill Steven Foster's love of nature by leading a junior Audubon club.

Professor Yue Chongxi, at the Institute of Chinese Materia Medica, Academy of Traditional Chinese Medicine, Beijing, and the late Dr. Shiu-Ying Hu (1908–2012), of Arnold Arboretum, Harvard University, and the Chinese University of Hong Kong are gratefully acknowledged for information they provided on Chinese uses of closely related plant species of China and North America.

The authors are grateful for the fine work and gentle prodding of the late Harry Foster (1945–2007) and of Lisa White. Thanks to the rest of the staff of Houghton Mifflin Harcourt for all the thankless details associated with book production.

Credit must also be given to the dozens of botanists, medical

botanists, chemists, pharmacognosists, toxicologists, ethnobotanists, physicians, and researchers in all disciplines touching upon medicinal plants for their hundreds of published works, which have been frequently and repeatedly consulted. If this book cited references in the style of a scientific publication, the bibliography would be double the length of the book. A few of the most pertinent works for further reference are listed on pp. 422–425.

Tribute must be paid to the 1,100 to 5,000 generations of Native Americans whose experience and evolution with the indigenous medicinal flora ultimately made this book, and a scant two centuries of literature, possible.

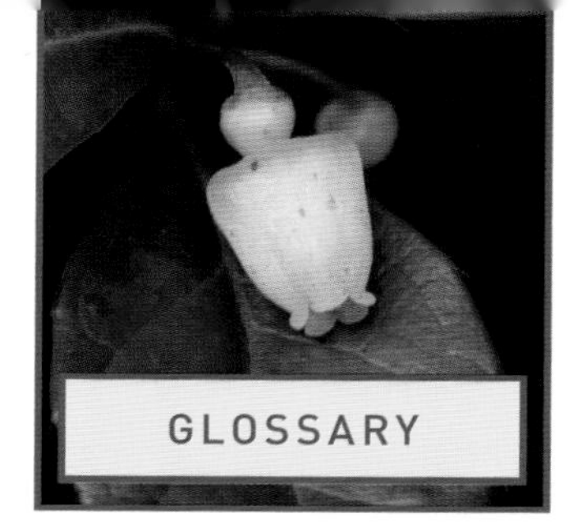

Medicinal Terms

Adaptogenic: Helping the human organism adapt to stressful conditions.

Alkaloid: A large, varied group of complex nitrogen-containing compounds, usually alkaline, that react with acids to form soluble salts, many of which have physiological effects on humans. Includes nicotine, cocaine, caffeine, etc.

Alterative: A medicinal substance that gradually restores health.

Analgesic: A pain-relieving medicine.

Anodyne: A pain-relieving medicine, milder than analgesic.

Anti-allergenic: Reducing or relieving allergies.

Anti-aphrodisiac: Suppressing sexual desire.

Antibiotic: An agent that inhibits the growth or multiplication of, or kills, a living organism; usually used in reference to bacteria or other microorganisms.

Anticonvulsant: Reducing or relieving convulsions or cramps.

Antifungal: An agent that inhibits the growth or multiplication of fungi, or kills them outright.

Antihistaminic: Neutralizing the effect or inhibiting production of histamine.

Anti-inflammatory: Reducing or neutralizing inflammation.

Antimicrobial: An agent that inhibits the growth or multiplication of microorganisms, or kills them.

Antioxidant: Preventing oxidation; a preservative.

Antiscorbutic: An agent effective against scurvy.

Antiseptic: Preventing sepsis, decay, putrification; also, an agent that kills germs, microbes.

Antispasmodic: Preventing or relieving spasms or cramps.

Antitumor: Preventing or effective against tumors (cancers).

Antitussive: Preventing or relieving cough.

Antiviral: An agent that inhibits growth or multiplication of viruses, or kills them.

Aphrodisiac: Increasing or exciting sexual desire.

Astringent: An agent that causes tissue to contract.

Bactericidal: An agent that kills bacteria.

Blood purifier: An agent that facilitates or speeds the process of transport or removal of toxins or waste products from the blood stream; in herbal traditions often by stimulating intestinal, liver, or bile function. In older medical literature often called a *depurative*.

Calmative: An agent with mild sedative or calming effects.

Cardioactive: Affecting the heart.

Carminative: An agent that relieves and removes gas from the digestive system.

Cathartic: A powerful purgative or laxative, causing severe evacuation, with or without pain.

Cholagogue: An agent that increases bile flow to the intestines.

CNS: The central nervous system.

Counterirritant: An agent that produces inflammation or irritation when applied locally to affect another, usually irritated, surface to stimulate circulation. (Example: a mustard plaster or liniment.)

Cytotoxic: An agent that is toxic to certain organs, tissues, or cells.

Decoction: A preparation made by boiling or simmering a plant part in water. Compare with *infusion*.

Demulcent: An agent that is locally soothing and softening.

Depurative: An agent that tends to purify or cleanse the blood. Compare with *blood purifier*.

Diaphoretic: An agent that induces sweating.

Digestive: An agent that promotes digestion.

Diuretic: An agent that induces urination.

Emetic: An agent that induces vomiting.

Emollient: An agent that softens and soothes the skin when applied locally.

Estrogenic: A substance that induces female hormonal activity.

Expectorant: An agent that induces the removal (coughing-up) of mucous secretions from the lungs.

Fungicidal: An agent that kills fungi.

Hemostatic: An agent that checks bleeding.

Homeopathic: Relating to homeopathy, a system of medicine founded in the late 1700s by Samuel Hahnemann. The system is based on the principle that "like cures like." Practitioners believe that a substance that produces a set of symptoms in a well person will, in minute, "potentized" doses, cure those same symptoms in a diseased individual.

Hypertensive: Causing or marking a rise in blood pressure.

Hypoglycemic: Causing a lowering of blood sugar.

Hypotensive: Causing or marking a lowering of blood pressure.

Immunomodulatory: An agent modulating immune function toward normal, healthy function.

Immunostimulant: Stimulating various functions or activities of the immune system.

Infusion: A preparation made by soaking a plant part in hot water (or cold water, for a cold infusion); in essence, a "tea." Compare with *decoction*.

Laxative: A mild purgative.

Leishmanicidal: An agent killing protozoan parasite responsible for the tropical disease leishmaniasis.

Mitogenic: An agent that affects cell division.

Moxa: A dried herb substance burned on or above the skin to stimulate an acupuncture point or serve as a counterirritant. A famous technique of Traditional Chinese Medicine, using dried pressed leaves of Mugwort (*Artemisia vulgaris*).

Mucilaginous: Pertaining to, resembling, or containing mucilage; slimy.

Nervine: An agent that affects, strengthens, or calms the nerves.

Panacea: An agent good for what ails you, or what doesn't ail you.

Poultice: A moist, usually warm or hot mass of plant material applied to the skin, or with cloth between the skin and plant material, to effect a medicinal action.

Purgative: An agent that causes cleansing or watery evacuation of the bowels, usually with griping (painful cramps).

Rubefacient: An agent that causes reddening or irritation when applied to the skin, once widely used, for example, to irritate the skin at a rheumatic joint in an effort to relieve pain by producing a "counterirritant" to the site of pain. Believed to help dilate blood capillaries and increase blood circulation at the site.

Saponin: A glycoside compound common in plants, which, when shaken with water, has a foaming or "soapy" action.

Spasmolytic: Checking spasms or cramps.

Stimulant: An agent that causes increased activity of another agent, cell, tissue, organ, or organism.

Styptic: Checking bleeding by contracting blood vessels.

Teratogen: A substance that can cause the deformity of a fetus.

Tincture: A diluted alcohol solution of plant parts.

Tonic: An ambiguous term referring to a substance thought to have an overall positive medicinal effect of an unspecified nature (see *adaptogenic*).

Tuberculostatic: Arresting the tubercle bacillus (the "germ responsible for causing tuberculosis").

Uterotonic: Having a positive effect of an unspecified nature on the uterus.

Vasoconstrictor: An agent that causes blood vessels to constrict.

Vasodilator: An agent that causes blood vessels to dilate.

Vermicidal: Having worm-killing properties; an agent that kills worms, a vermifuge.

Vulnerary: An agent used for healing wounds.

Botanical Terms

Achene: Small, dry, nonsplitting fruit with a single seed.
Anther: The top pollen-bearing structure of a stamen.
Aril: Thick, fleshy seed coat attached to, and often surrounding, a seed.
Axil: Upper angle at junction of stem and a leaf.
Basal rosette: Leaves radiating directly from the crown of the root.
Bracts: The leaflike structures of a grouping or arrangement of flowers (inflorescence).
Bulblet: Small, above-ground bulb usually at leaf axils.
Calyx: The sepals collectively; the external floral envelope.
Corolla: Collectively, all the petals of a flower.
Cyme: Flat- or round-topped flower cluster with terminal flowers blooming first.
Decompound: Divided several or many times; compound with further subdivisions.
Drupe, Drupelet: A fleshy, nonsplitting fruit with a hard casing within usually surrounding a single seed. A peach is technically a drupe.
Fascicle: A tight cluster.
Floret: A very small flower, especially one of the disk flowers of plants in the Aster family.
Glaucous: Covered with a fine, white, often waxy film, which rubs off.
Hemiparasitic: A plant that is parasitic on other plants, such as Mistletoe, but still produces its own chlorophyll.
Herbaceous: Nonwoody.
Liana: A vigorous woody vine (usually refers to tropical vines).
Obovate: Oval, but broader toward the apex; refers to leaf shape.
Ovate: Oval but broader toward the base; egg-shaped.
Palmate: With 3 or more leaflets, nerves, or lobes radiating from a central point.
Panicle: A branching flower grouping, with branches that are usually racemes (see below).
Perfect (flower): A flower that has a full complement of male and female parts as well as floral envelopes (petals and sepals).
Perfoliate: A leaf that appears to be perforated by the stem.
Petiole: A leaf stalk.
Pinnate: A featherlike arrangement; usually refers to a compound leaf with leaflets arranged on each side of a central axis.
Raceme: An unbranched, elongated flower grouping, with individual flowers on distinct stalks.
Rays (ray flowers): The straplike, often sterile flowers (commonly called "petals") surrounding the flowerhead (disk) of a plant in the Aster family. (Examples: the yellow rays of Sunflowers or the purple rays surrounding the cone of Purple Coneflower.)
Rhizome: A creeping underground stem.
Rosette (basal): Leaves radiating directly from the crown of the root.

Saprophytic: A plant (usually lacking chlorophyll) that lives on dead organic matter.

Sepals: The individual divisions of the calyx (outer floral envelope).

Sessile: Lacking a stalk; such as a leaf or flower with no obvious stalk.

Silique: A term applied to the peculiar seedpod structure of plants in the Mustard family.

Sinus: Curved or angled recess between two lobes.

Sorus, Sori: The spore clusters beneath a fern's leaf. (*Sori* is the plural.)

Spadix: A thick, fleshy flower spike (usually enveloped by a spathe) as in members of the Arum family (Skunk Cabbage, Jack-in-the-Pulpit, Dragon Arum, etc.).

Spathe: A modified, leaflike structure surrounding a spadix, as in members of the Arum family (Skunk Cabbage, Jack-in-the-Pulpit, Dragon Arum, etc.).

Spike (flower): An unbranched, elongated flower grouping in which the individual flowers are sessile (attached without stalks).

Stamens: The pollen-bearing anthers with attached filaments (sometimes without filaments).

Stipules: Appendages (resembling small or minute leaves) at the base of leaves of certain plants.

Strobile: A conelike structure with overlapping scales spiraling from a common axis as in the fruit of Hops.

Subshrub: Somewhat or slightly shrublike; usually a plant with a stem that is woody at the base, but mostly herbaceous.

Tendrils: A modified leaf or branch structure, often coiled like a spring, used for clinging in plants that climb.

Umbels: A flower grouping with individual flower stalks or floral groupings radiating from a central axis; often flat-topped and umbrella-like.

Technical Manuals

Bailey, Liberty Hyde, and Ethel Zoe Bailey. 1976. *Hortus Third: A Concise Dictionary of Plants Cultivated in the United States and Canada.* Revision by L. H. Bailey Hortorium Staff. New York: Macmillan.

Fernald, Merritt Lyndon. 1950. *Gray's Manual of Botany*. 8th ed. New York: Van Nostrand.

Flora of North America Editorial Committee, eds. 1993+. *Flora of North America North of Mexico*. 16+ vols. New York and Oxford: Oxford University Press.

Gleason, Henry A., and Arthur Cronquist. 1991. *Manual of Vascular Plants of Northeastern United States and Adjacent Canada*. 2nd ed. New York: New York Botanical Garden.

Radford, Albert E., Harry E. Ahles, and C. Ritchie Bell. 1968. *Manual of the Vascular Flora of the Carolinas*. Chapel Hill: University of North Carolina Press.

Popular Guides

Cobb, Boughton. 1963. *A Field Guide to the Ferns*. Boston: Houghton Mifflin.

Duke, James A. 1986. *Handbook of Northeastern Indian Medicinal Plants*. Lincoln, Mass.: Quarterman Publications.

———. 1997. *The Green Pharmacy*. Emmaus, Penn.: Rodale Press.

Foster, Steven. 1993. *Herbal Renaissance: Understanding, Using and Growing Herbs in the Modern World*. Layton, Utah: Gibbs Smith Publisher.

———. 1995. *Forest Pharmacy: Medicinal Plants in American Forests*. Durham, N.C.: Forest History Society.

Foster, Steven, and Christopher Hobbs. 2002. *A Field Guide to Western Medicinal Plants and Herbs*. Boston: Houghton Mifflin.

Foster, Steven, and Rebecca L. Johnson. 2006. *National Geographic's Desk Reference to Nature's Medicine*. Washington, D.C.: National Geographic.

Foster, Steven, and Varro E. Tyler. 1999. *Tyler's Honest Herbal*. 4th ed. Binghamton, N.Y.: Haworth Press.
Foster, Steven, and Yue Chongxi. 1992. *Herbal Emissaries: Bringing Chinese Herbs to the West*. Rochester, Vt.: Healing Arts Press.
Peterson, Lee Allen. 1977. *A Field Guide to Edible Wild Plants*. Boston: Houghton Mifflin.
Peterson, Roger Tory, and Margaret McKinney. 1968. *A Field Guide to Wildflowers*. Boston: Houghton Mifflin.
Petrides, George A. 1972. *A Field Guide to Trees and Shrubs*. 2nd ed. Boston: Houghton Mifflin.

Scholarly Works

Austin, Daniel F. 2004. *Florida Ethnobotany*. Baco Raton, Fla: CRC Press.
Blumenthal, Mark, ed. 1998. *The Complete German Commission E Monograph: Therapeutic Guide to Herbal Medicines*. Austin, Texas: American Botanical Council; Boston: Integrative Medicine.
Blumenthal, Mark, Alicia Goldberg, and Josef Brinckmann, eds. 2000. *Herbal Medicine: Expanded Commission E Monographs*. Austin, Texas: American Botanical Council.
Duke, James A. 1992. *CRC Handbook of Biologically Active Phytochemicals and Their Activities*. Boca Raton, Fla.: CRC Press.
———. 1992. *CRC Handbook of Phytochemical Constituents in GRAS Herbs and Other Economic Plants*. Boca Raton, Fla.: CRC Press.
Duke, James A., Mary Jo Bogenschutz-Godwin, Judi duCellier, and Peggy-Ann K. Duke. 2002. *CRC Handbook of Medicinal Herbs*. Boca Raton, Fla: CRC Press.
Duke, James A., Mary Jo Bogenschutz-Godwin, and Andrea Ottesen. 2009. *Duke's Handbook of Medicinal Plants of Latin America*. Boca Raton, Fla: CRC Press.
Hardin, James W., and Jay M. Arena. 1974. *Human Poisoning from Native and Cultivated Plants*. 2nd ed. Durham: Duke University Press.
Moerman, Daniel E. 1998. *Native American Ethnobotany*. Portland, Ore.: Timber Press.
Nelson, Lewis S., Richard D. Shih, and Michael J. Balick. 2006. *Handbook of Poisonous and Injurious Plants*. New York: New York Botanical Garden and Springer.
Schulz, Volker, Rudolf Hänsel, Mark Blumenthal, and Varro E. Tyler. 2004. *Rational Phytotherpy: A Physician's Guide to Herbal Medicine*. Berlin: Springer.
Weiss, Rudolf Fritz, and Volker Fintelmann. 2000. *Herbal Medicine*. New York: Thieme.
Witchl, M., J. A. Brinckmann, and M. P. Lindenmaier. 2004. *Herbal Drugs and Phytopharmaceuticals*. Stuttgart, Germany: Medpharm GmbH Scientific Publishers.

Historical Bibliography

Bigelow, Jacob. 1818–1820. *American Medical Botany*. 3 vols. Boston: Cummings and Hilliard.

Clapp, A. 1852. *A Synopsis: or Systematic Catalogue of the Indigenous and Naturalized, Flowering and Filicoid Medicinal Plants of the United States*. Presented to the American Medical Association, at its Session of May, 1852. Philadelphia: T. K. and P. G. Collins.

Densmore, Francis. 1926–1927. "Uses of Plants by the Chippewa Indians." In *Forty-Fourth Annual Report of the Bureau of American Ethnology*, edited by J. Walter Fewkes, 275–398. Washington, D.C.: Smithsonian Institution.

Felter, Harvey Wickes, and John Uri Lloyd. (1901) 1983. *King's American Dispensatory*. 18th ed. 2 vols. Reprint, Portland, Ore.: Eclectic Medical Publications.

Gilmore, Melvin Randolph. 1919. *Uses of Plants by the Indians of the Missouri River Region*. Washington, D.C.: Government Printing Office.

Grieve, Maude. 1931. *A Modern Herbal: The Medicinal, Culinary, Cosmetic and Economic Properties, Cultivation and Folk-lore of Herbs, Grasses, Fungi, Shrubs & Trees with All Their Modern Scientific Uses*. London: Jonathan Cape.

Griffith, R. Eglesfeld. 1847. *Medical Botany: or Descriptions of the More Important Plants Used in Medicine, with Their History, Properties, and Mode of Administration*. Philadelphia: Lea and Blanchard.

Hamel, Paul B., and Mary U. Chiltoskey. 1975. *Cherokee Plants and Their Uses: A 400 Year History*. Sylva, N.C.: Herald Publishing.

Herrick, James William. 1977. *Iroquois Medical Botany*. PhD diss., State University of New York at Albany.

Jacobs, Marion Lee, and Henry M. Burlage. 1958. *Index of Plants of North Carolina with Reputed Medicinal Uses*. Austin, Texas: Henry M. Burlage.

Lloyd, John Uri. 1921. *Origin and History of All the Pharmacopeial Vegetable Drugs, Chemicals, and Preparations*. Cincinnati: Caxton Press.

Millspaugh, Charles F. 1887. *American Medicinal Plants: An Illustrated and Descriptive Guide to the American Plants Used as Homeopathic Remedies*. New York: Boericke & Tafel.

Porcher, Francis Peyre. 1863. *Resources of the Southern Fields and Forests, Medical, Economical, and Agricultural; Being also a Medical Botany of the Confederates States: Prepared by Order of the Surgeon-General, Richmond, Va.* Charleston, S.C.: Steampower Press of Evans & Cogswell.

Rafinesque, C. S. 1828–1830. *Medical Flora: or, Manual of the Medical Botany of the United States of North America*. 2 vols. Philadelphia: Atkinson & Alexander.

Smith, Huron H. 1923. "Ethnobotany of the Menomini Indians." *Bulletin of the Public Museum of the City of Milwaukee* 4 (1):1–174.

——. 1928. "Ethnobotany of the Meskwaki Indians." *Bulletin of the Public Museum of the City of Milwaukee* 4 (2): 175–326.
——. 1932. "Ethnobotany of the Ojibwe Indians." *Bulletin of the Public Museum of the City of Milwaukee* 4 (3): 327–525.
——. 1933. "Ethnobotany of the Forest Potawatomi Indians." *Bulletin of the Public Museum of the City of Milwaukee* 7 (1): 1–230.
Tantaquidgeon, Gladys. 1942. *A Study of Delaware Indian Medicine Practice and Folk Beliefs.* Harrisburg: Pennsylvania Historical Commission.

Page numbers in **bold** refer to text photos.

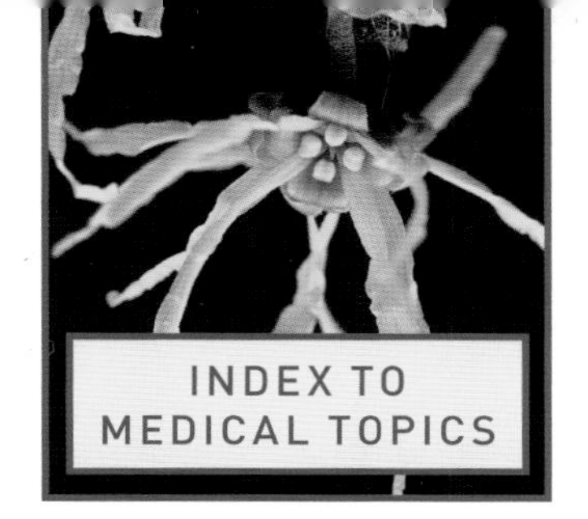

INDEX TO MEDICAL TOPICS

PETERSON FIELD GUIDES®

Roger Tory Peterson's innovative format uses accurate, detailed drawings to pinpoint key field marks for quick recognition of species and easy comparison of confusing look-alikes.

BIRDS

Birds of North America

Birds of Eastern and Central North America
SIXTH EDITION

Western Birds
FOURTH EDITION

Birds of Britain and Europe
FIFTH EDITION

Birds of Texas

Eastern Birds
LARGE FORMAT EDITION

Feeder Birds of Eastern North America
LARGE FORMAT EDITION

Hawks of North America
SECOND EDITION

Hummingbirds of North America

Warblers

Western Birds' Nests

Eastern Birds' Nests

New Birder's Guide to Birds of North America

The Young Birder's Guide to Birds of North America

PLANTS AND ECOLOGY

Eastern and Central Edible Wild Plants

Eastern and Central Medicinal Plants and Herbs

Western Medicinal Plants and Herbs

Eastern Forests

Rocky Mountain and Southwest Forests

Eastern Trees

Western Trees

Eastern Trees and Shrubs

Ferns of Northeastern and Central North America
SECOND EDITION

Mushrooms

North American Prairie

Venomous Animals and Poisonous Plants

Southwest and Texas Wildflowers

Wildflowers of Northeastern and North-Central North America

MAMMALS

Animal Tracks
THIRD EDITION

Mammals
FOURTH EDITION

INSECTS

Insects

Eastern Butterflies

Moths of Northeastern North America

REPTILES AND AMPHIBIANS

Eastern Reptiles and Amphibians
THIRD EDITION

Western Reptiles and Amphibians
THIRD EDITION

FISHES

Atlantic Coast Fishes

Freshwater Fishes
SECOND EDITION

SPACE

Stars and Planets

GEOLOGY

Geology of Eastern North America

Rocks and Minerals
FIFTH EDITION

SEASHORE

Atlantic Seashore

Shells of the Atlantic and Gulf Coasts and the West Indies

Southeastern and Caribbean Seashores

PETERSON FLASHGUIDES®

Portable and waterproof, FlashGuides are perfect for those who want to travel light. Covering 50–100 species, with brief surveys of habit and habitat, each opens to two rows with twelve full-color, laminated panels on each side.

Atlantic Coastal Birds

Birds of the Midwest

Trees

PETERSON FIRST GUIDES®

The first books the beginning naturalist needs, whether young or old. Simplified versions of the full-size guides, they make it easy to get started in the field, and feature the most commonly seen natural life.

Astronomy

Birds

Butterflies and Moths

Caterpillars

Clouds and Weather

Fishes

Insects

Mammals

Reptiles and Amphibians

Rocks and Minerals

Seashores

Shells

Trees

Urban Wildlife

Wildflowers

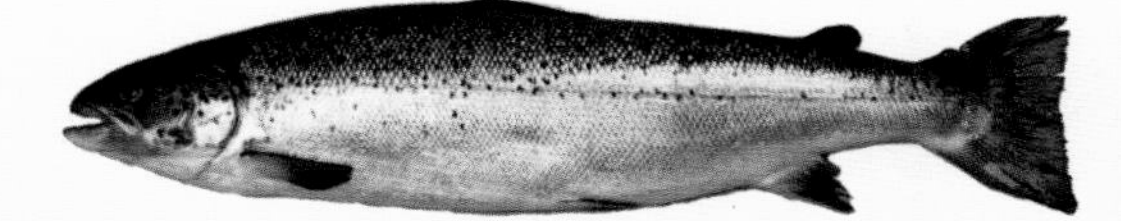

PETERSON FIELD GUIDES FOR YOUNG NATURALISTS

This series is designed with young readers ages eight to twelve in mind, featuring the original artwork of the celebrated naturalist Roger Tory Peterson.

Backyard Birds

Birds of Prey

Songbirds

Butterflies

Caterpillars

PETERSON FIELD GUIDES® COLORING BOOKS®

Fun for kids ages eight to twelve, these color-your-own field guides include color stickers and are suitable for use with pencils or paint.

Birds

Butterflies

Dinosaurs

Reptiles and Amphibians

Wildflowers

Seashores

Shells

Mammals

PETERSON REFERENCE GUIDES®

Reference Guides provide in-depth information on groups of birds and topics beyond identification.

Seawatching: Eastern Waterbirds in Flight

Gulls of the Americas

Molt in North American Birds

Behavior of North American Mammals

PETERSON AUDIO GUIDES

Birding by Ear: Western

Birding by Ear: Eastern/Central

More Birding by Ear: Eastern/Central

Bird Songs: Eastern/Central
THIRD EDITION

PETERSON FIELD GUIDE / *BIRD WATCHER'S DIGEST* BACKYARD BIRD GUIDES

Identifying and Feeding Birds

Hummingbirds and Butterflies

Bird Homes and Habitats

Finding Your Wings: A Workbook for Beginning Bird Watchers

DIGITAL

Apps available on the App Store for iPad, iPhone, and iPod Touch.

Peterson Birds of North America

Peterson Birds Pocket Edition

Peterson Feeder Birds of North America

E-books

Birds of Arizona

Birds of California

Birds of Florida

Birds of Massachusetts

Birds of Minnesota

Birds of New Jersey

Birds of New York

Birds of Ohio

Birds of Pennsylvania

Birds of Texas

SPECIES ACCOUNTS

PRICKLY POPPY; BLUESTEM PRICKLY POPPY

Stem juice, seeds, leaves

Argemone albiflora Hornem. — Poppy Family (Papaveraceae)

⚠ Bluish green annual or biennial herb; 2–3 ft. *Yellow* (or white) *juice. Thistlelike* leaves and stems with *sharp bristles.* Flowers with 4–6 petals at least 2 in. wide; May–Sept. Capsules upright, oblong or elliptical, prickly, splitting when mature. **WHERE FOUND:** Waste places; scattered. Introduced. CT to FL; TX to IA and MI. **USES:** Seed tea is emetic, purgative, demulcent. Plant infusion used for jaundice, skin ailments, colds, colic, wounds. Externally, used for headaches. Folk remedy for cancers, itching, and scabies. Little researched and, as C. F. Rafinesque advised in his 1828 *Medical Flora,* "deserving attention." **WARNING:** Contains toxic alkaloids. Seed oil causes glaucoma and edema.

TURTLEHEAD, BALMONY

Leaves

Chelone glabra L. — Plantain Family (Plantaginaceae)
[Formerly in Figwort Family (Scrophulariaceae)]

Smooth perennial; 2–3 ft. Stem somewhat 4-sided. Leaves lance-shaped to oval, toothed. Flowers white to pink, swollen; in tight clusters atop plant; July–Oct. *Flowers 2-lipped*; with swollen, *strongly arching upper lip* (resembling a turtle's head, hence the name). **WHERE FOUND:** Moist soils. NL to GA; AR to MB, QC. **USES:** Cherokee used tea of flowering tops to treat worms; Malecite used plant tea as a contraceptive to prevent pregnancy. Leaf tea said to stimulate appetite; also a folk remedy for worms, fever, jaundice; laxative. Introduced to physicians in 1820s via the Shakers for dyspepsia, liver diseases, fevers, and

LEFT: *Prickly Poppy in flower. Note the thistlelike leaves and spiny capsules.* ABOVE: *Turtlehead, Balmony in flower. Note the strongly arching upper lip.*

Lily-of-the-valley is more often found in gardens than in wild habitats.

Dutchman's-breeches flowers each have 2 inflated, "pantlike" spurs.

inflammation. Ointment used for hemorrhoids, inflamed breasts, painful ulcers, herpes. Contains a bitter resin with iridoid glycosides. Several butterfly species and other insects feed on the plant, accumulating the plant's bitter compounds to deter predators.

LILY-OF-THE-VALLEY

Root, flowers

Convallaria majalis L.

Asparagus Family (Asparagaceae)
[Formerly in Lily Family (Liliaceae)]

⚠ Perennial, spreading by root runners; 4–8 in. Leaves 2–3; basal, oblong-ovate, entire (not toothed); *veins parallel; connecting veins obvious when held to light*. Flowers bell-shaped, white; May–June. **WHERE FOUND:** Europe. Widely escaped from cultivation. **USES:** In eighteenth century France, the powdered flowers were used as a snuff to induce sneezing in an attempt to treat nervous headache and vertigo. Tea of flowers and roots traditionally used in valvular heart disease (digitalis substitute), fevers; diuretic, heart tonic, sedative, emetic. Root ointment, folk remedy for burns, to prevent scar tissue. Russians use for epilepsy. Used in European phytomedicine for the treatment of mild cardiac insufficiency, to economize cardiac efficiency, and to improve tone of the veins. Cardenolides are the active compounds. A steroidal saponin from the roots, convallamaroside, showed cancer prevention potential in a mouse experimental model. **WARNING:** Potentially toxic. Use only under a physician's supervision. Although widely prescribed in Germany, it is seldom used in the U.S. Can interact with other drugs;

is not administered in conjunction with other cardiac drugs. Leaves can be a mild skin irritant.

DUTCHMAN'S-BREECHES Leaves, root

Dicentra cucullaria (L.) Bernh. Poppy Family (Papaveraceae)

⚠ Perennial; 5–9 in. Leaves much dissected. Flowers white, yellow-tipped; appearing upside-down on an arching stalk; Apr.–May. Each flower has *2 inflated, "pantlike" spurs.* **WHERE FOUND:** Rich woods. PE to GA mtns.; AR, OK to ND. **USES:** Iroquois used leaf ointment to make athletes' legs more limber. Among the Menominee, it was the most important love charm, thrown by a suitor at his potential mate. If the root was nibbled, it was believed one's breath would attract a woman, even against her will. Leaf poultice is a folk medicine for skin ailments. Root tea is diuretic; promotes sweating. Contains alkaloid with CNS-depressant activity; used for paralysis and tremors. **WARNING:** Potentially poisonous; may also cause skin rash.

MISCELLANEOUS AQUATIC PLANTS

BUCKBEAN, BOGBEAN (NOT SHOWN) Root, leaves

Menyanthes trifoliata L. Buckbean Family (Menyanthaceae)
[Formerly in Gentian Family (Gentianaceae)]

⚠ Aquatic perennial with *cloverlike* leaves arising from the root. Flowers 5-parted, white to pinkish, on a naked raceme; Apr.–July. *Petals with fuzzy beards.* **WHERE FOUND:** Bogs, shallow water. Much of N. America, except for TX and se. U.S.; all of Northern Europe. **USES:** Dried-leaf or dried-root tea traditionally a digestive tonic; used for fevers, rheumatism, liver ailments, dropsy, worms, skin diseases; astringent; stops bleeding. Science confirms phenolic acids may be responsible for bile-secreting, digestive tonic, and bitter qualities. Folk remedy in Europe for arthritis. In Germany, once a well-regarded tonic. Leaf used in European phytomedicine for treatment of dyspeptic discomfort and loss of appetite. Stimulates flow of saliva and gastric juices. Additional compounds from the plant shown to have anti-inflammatory and immune system modulating activity. In Lapland, an ounce of leaves formerly substituted for a pound of hops in beer making. **WARNING:** Fresh plant induces vomiting.

AMERICAN LOTUS, YELLOW LOTUS Roots, leaves, seeds

Nelumbo lutea Willd. Lotus Family (Nelumbonaceae)
[Formerly in Water-Lily Family (Nymphaeaceae)]

Large aquatic perennial, with leaf stalks up to 6 ft. tall. Leaves *large, cupped in center, umbrella-like,* up to 2 ft. across. Flowers whitish to pale yellow; June–Oct. Fruit receptacle less than 1.25 times longer than wide. **WHERE FOUND:** Native species found in river flood plains, ponds, lakes,

Flowers of American Lotus are whitish to pale yellow, much smaller than those of Asian Lotus.

Lotus is sacred in Hindu and Buddhist traditions and occasionally naturalized in N. America. Flowers are white to pink-red.

marshes, and swamps from ME to FL, west to TX north to the Great Lakes. Once limited to flood plains of major rivers, as an important food source of indigenous peoples (rhizome and seed), distribution expanded as a result. **USES:** Cooked rhizome and shelled seeds an important food among various native groups of the Mississippi and Missouri River valleys and believed to have mystical powers. Contains at least seven bioactive alkaloids, similar to those in *N. nucifera* (below). Little researched, but likely to have similar biological activities to Sacred Lotus. There are only two species in the genus, the other from Asia.

LOTUS, SACRED LOTUS — Entire plant

Nelumbo nucifera Gaertn. — Lotus Family (Nelumbonaceae)
[Formerly in Water-Lily Family (Nymphaeaceae)]

Large aquatic perennial, *to 3 ft. above water surface in flower*, with large fleshy rhizomes. Leaves large, umbrella-shaped, to 2 ft. across. Flowers white to pink, large, to 8 in. across, sweetly fragrant; June–Aug. **WHERE FOUND:** Alien (Asia); escaped and naturalized in slow-moving water, ponds, lakes, ditches; NJ to FL, west to TX, MO. **USES:** The Sacred Lotus of Hindu and Buddhist traditions; seeds (hard nuts) and roots (tuberous rhizome) are widely used as food. All parts of Lotus are used in Traditional Chinese Medicine and in Ayurvedic medicine in India. The seed is used to tonify the spleen, treat diarrhea, and arrest seminal emission, heart palpitation, and insomnia. Young cotyledons from sprouting seed used similarly, to calm nerves. Dried seedpods (receptacle) used to stop bleeding. Flower stamens used to stop frequent urination. Dried leaf tea used to reduce fevers, stop bleeding (caused by "heat" in blood). Rhizomes used in prescriptions to slow abnormal uterine bleeding. The alkaloid nuciferine stimulates insulin secretion, lending credence to traditional use as an antidiabetic. Anti-inflammatory, liver-protectant, immuno-stimulating, antioxidant, anti-obesity,

Fragrant Water-lily leaves float on water's surface.

Lizard's-tail, Water-dragon produces a tail-like spike of flowers, nodding when mature.

and antidepressant effects are among activities reported from research since 2000. Includes numerous alkaloids, flavonoids, glycosides, triterpenoids, and other compounds that may contribute to pharmacological activity. See also *N. lutea* (above).

FRAGRANT WATER-LILY — Roots, leaves, flower buds

Nymphaea odorata Ait. — Water-lily Family (Nymphaeaceae)

⚠ Aquatic perennial with large, round, floating leaves; leaf notched at base. Flowers white, *floating*, to 5 in. across; sweetly fragrant; June–Sept. **WHERE FOUND:** Ponds, slow waters. NL to FL; TX to NE; most of N. America. **USES:** The large, spongy, fleshy roots (the size of a man's forearm) were traditionally used by Chippewa, Micmac, and Penobscot, particularly for lung ailments; root tea for coughs, tuberculosis, inflamed glands, mouth sores; stops bleeding; poulticed root for swellings. The Ojibwa ate the flower buds, attributed as early as 1751 for lack of scurvy among some tribes. In folk tradition, a mixture of root and lemon juice was used to remove freckles and pimples. In early nineteenth-century America, botanic physicians used the dried roots for diarrhea, dysentery, and fevers; externally as a poultice for sores, skin inflammation, and tumors. Fresh leaves considered cooling. Root tea is drunk for bowel complaints, primarily as an astringent and antiseptic for chronic diarrhea. Contains lignans and flavonol glycosides that may be responsible for historic anti-inflammatory attributes. **WARNING:** Large doses may be toxic.

LIZARD'S-TAIL, WATER-DRAGON
Root, leaves

Saururus cernuus L. Lizard's-tail Family (Saururaceae)

Perennial; 2–5 ft. Leaves large, asymmetrically heart-shaped. Flowers tiny, white, on a *showy, nodding "tail"*; May–Sept. **WHERE FOUND:** Shallow water, swamps, RI to FL, TX to MN. **USES:** The Cherokee, Choctaw, and other native groups used root poultice for wounds, inflamed breasts (the plant is also known as Breastweed), and inflammations. Tea of whole plant used as a wash for general illness, rheumatism; internally for stomach ailments. Contains several novel compounds with sedative effects. Subject of considerable chemical and pharmacological research, including potential anticancer and antioxidant activity. **RELATED SPECIES:** Only two species in the genus; the other is *S. chinensis* (Lour.) Baille., an Asian counterpart, which is also used to relieve inflammation in Traditional Chinese Medicine; strongly anti-inflammatory. It, too, contains novel sedative compounds.

AQUATIC PLANTS; 3 PETALS

WATER-PLANTAIN
Dried leaves, root

Alisma subcordatum Raf. Arrowhead Family (Alismataceae)
[*Alisma plantago-aquatica* var. *parviflorum* (Pursh) Farw.]

Erect or drooping (in deep water) perennial; 1–3 ft. in flower. Long-stemmed, nearly heart-shaped leaves. Flowers tiny (less than 1/8 in. long), in whorls on branched stalks; June–Sept. Petals same length as sepals. **WHERE FOUND:** Shallow water or mud. QC, NY, MA, to FL; west to TX, NE. **USES:** Tea diuretic; used for "gravel" (kidney stones), urinary diseases. Fresh leaves are a rubefacient—they redden and irritate skin. Native Americans used root poultice for bruises, swellings, wounds. An 1899 article by a California physician reported on the use of the root tincture (alcohol extract), mixed with equal parts water and glycerin, as a local application to nostrils to treat "nasal catarrh." **RELATED SPECIES:** The root of the closely related *A. plantago-aquatica* L. is used in China as a diuretic for dysuria, edema, distention, diarrhea, and other ailments. Chinese studies verify the plant's diuretic action. In Traditional Chinese Medicine, the herb is used to lower blood pressure, reduce blood glucose levels, and inhibit the storage of fat in the liver, with pharmacological evidence to support traditional use.

COMMON WATER-PLANTAIN
Root

Alisma triviale Pursh Arrowhead Family (Alismataceae)
[*Alisma plantago-aquatica* var. *brevipes* (Greene) Vict.]

Similar to above species, but leaves are mostly oval, though the base is often slightly heart-shaped. Flowers larger, to ¼ in.; June–Sept. Panicle often with fewer branches. **WHERE FOUND:** Shallow water, ditches. NS to MD; NE to MN and beyond. **USES:** Native Americans used root tea for lung ailments, lame back, and kidney ailments.

Four species of Water-plantain occur in our range, distinguished by technical factors.

Twenty-four species of aquatic arrowheads are found in N. America. Note the arrow-shaped leaves.

BROAD-LEAVED ARROWHEAD Roots, leaves

Sagittaria latifolia Willd. Arrowhead Family (Alismataceae)

Aquatic perennial. Leaves arrow-shaped; *lobes half as long to as long as main part of leaf.* Flowers white; 3 rounded petals; filaments of stamens smooth; June–Sept. The bracts beneath the flowers are blunt-tipped, thin, and papery. The beak of the mature fruit (achene) projects at a right angle from the main part of the fruit. **WHERE FOUND:** Ponds and lakes throughout our area. Technical details separate the 19 species in our range (most with very narrow ranges). There are 24 species in N. America. **USES:** The Chippewa and other native groups used the dried or baked, peeled tubers for food; in tea for indigestion; poulticed roots for wounds and sores. Cherokee used leaf tea for rheumatism and to wash babies with fever. Leaves were poulticed to stop milk production. The cooked corms of *S. sagittifolia* L. are used as a food source. **WARNING:** Arrowheads (not necessarily this species) may cause dermatitis. Many aquatic plants accumulate toxins from polluted waters.

RATTLESNAKE-MASTER Root, whole plant

Eryngium aquaticum L. Parsley Family (Apiaceae or Umbelliferae)

Perennial, with bluish cast; to 4 ft. Similar to *E. yuccifolium* but *leaves much narrower, linear to lanceolate, entire or with few teeth*; upper leaves spiny and often divided. Flowers a buttonlike head, to ½ in. long, with tiny whitish to green flowers; July–Oct. **WHERE FOUND:** Moist soils, bogs and marshes, often near coast from NY to ne. FL, west to AL. **USES:** Cherokee and Choctaw used root as a snakebite antidote. An infusion of the plant used to induce vomiting; root tea used as an expectorant. Often confused with or treated synonymously with *E. yuccifolium* in nineteenth-century medical literature.

RATTLESNAKE-MASTER, BUTTON-SNAKEROOT Root, whole plant

Eryngium yuccifolium Michx. Parsley Family (Apiaceae or Umbelliferae)

Perennial, also with bluish cast; 1½–4 ft. Leaves mostly basal (reduced on stem), *yuccalike* (hence the species name), *parallel-veined, spiny-edged.* Flowers white to whitish green, tiny, covered by bristly bracts; in tight heads to 1 in. across; Sept.–Nov. **WHERE FOUND:** Prairies, dry soil. S. CT to FL; TX to NE, MN. **USES:** Cherokee and Creek Indians used root as poultice for snakebites, toothaches, bladder trouble; for coughs,

Rattlesnake-master has white flowers, but bracts and upper stems are often blue-purple tinted. Found in moist habitats.

Button-snakeroot has bristly, light green leaves that superficially resemble those of yucca.

ABOVE: *Button-snakeroot. Note tiny flowers closely crowded on rounded head.* RIGHT: *Flowers of Adam's Needle Yucca grow on smooth-branched stalks.*

neuralgia; also an emetic. Traditionally, root tincture was used as a diuretic and to treat urinary disorders such as burning pain and irritation of the urethra, and frequent, scanty, and painful urination; also for female reproductive disorders, gleet, gonorrhea, hemorrhoids, and rheumatism. Chewing the root increases saliva flow. The whole plant contains several phenolic compounds and triterpenoid saponins that could be responsible for biological activities. **WARNING:** Do not confuse with False Aloe or Rattlesnake-master (*Manfreda virginica,* p. 144), which may produce strongly irritating latex.

YUCCA, ADAM'S NEEDLE Roots, leaves

Yucca filamentosa L. Asparagus Family (Asparagaceae)
[Formerly in Lily Family (Liliaceae) or Agave Family (Agavaceae)]

Perennial; to 9 ft. in flower. Leaves in a rosette; stiff, spine-tipped, oblong to lance-shaped, with *fraying, twisted threads on margins.* Flowers whitish green bells on smooth, branched stalks; June–Sept. **WHERE FOUND:** Sandy soils. S. NJ to GA. Cultivated elsewhere. **USES:** Cherokee and Catawba Indians used root in salves or poultices for sores, skin diseases, and sprains. Pounded roots were put in water to stupefy corralled fish so they would float to the surface for easy harvest. Plant yields a strong fiber; in the South during Civil War developed as "Confederate Flax," a cordage and twine source. Fruits used as food by some people. Saponins in roots of yucca species possess long-lasting soaping action and have been used in soaps and shampoos. Steroidal saponins in leaves have leishmanicidal activity (against tropical protozoan parasites). **WARNING:** Root compounds (saponins) are toxic to lower life forms.

YUCCA, SOAPWEED
Roots

Yucca glauca Nutt. Asparagus Family (Asparagaceae)

[Formerly in Lily Family (Liliaceae) or Agave Family (Agavaceae)]

Blue-green perennial; 2–4 ft. Leaves in a rosette; stiff, swordlike; narrow to ½ in. wide, rounded on back, margins rolled in. Flowers whitish bells; mostly in unbranched racemes, to 40 in. tall; May–July. **WHERE FOUND:** The most widespread of American Yuccas; dry soils, prairies. IA to TX; MO to ND. **USES:** Numerous Native American groups used plant for medicine, food, and fiber. Poulticed root on inflammations, used to stop bleeding; also in steam bath for sprains and broken limbs; hair wash for dandruff and baldness. Leaf juice used to make poison arrows. Antifungal, antitumor, and antiarthritic activity have been suggested by research. Water extracts have shown antitumor activity against B16 melanoma in mice. One human clinical study suggests that saponin extracts of yucca root were effective in the treatment of arthritis, but the findings have been disputed. Sometimes confused with or considered synonymous to *Y. arkansana* Trel. **WARNING:** Same as for *Yucca filamentosa.*

Flowers of the Soapweed Yucca grow on a single stalk. Upward of 30 species of yucca occur in N. America, most in the desert Southwest.

SHOWY BELLS OR TRUMPETS

FIELD BINDWEED
Leaf, root, flowers

Convolvulus arvensis L. Morning-glory Family (Convolvulaceae)

Aggressive, deep-rooted, intertwining vine. Leaves arrow-shaped; *lobes sharp, not blunt*; 1–2 in. Flowers white (or pink), to 1 in.; June–Sept. **WHERE FOUND:** Fields, waste places. Most of our area. Alien (Europe). **USES:** Native Americans used cold leaf tea as a wash on spider bites; internally to reduce profuse menstrual flow. In European folk use, flower, leaf, and root teas considered laxative. Flower tea used for fevers, wounds. Root most active—strongly purgative.

Field Bindweed has arrow-shaped leaves with squared lobes at base.

Hedge Bindweed has arrow-shaped leaves with blunt lobes.

HEDGE BINDWEED — Root

Calystegia sepium ssp. *sepium* (L.) R. Br. (*Convolvulus sepium* L.) — Morning-glory Family (Convolvulaceae)

⚠ Trailing vine with white or pink, morning glory–type flowers; May–Sept. Leaves arrow-shaped with *blunt lobes at base*. **WHERE FOUND:** Cosmopolitan invasive alien; thickets, roadsides. Most of our area. **USES:** Root historically used as a purgative; substitute for the Mexican jalap (*Ipomoea purga*). Traditionally used for jaundice, gallbladder ailments; thought to increase bile flow into intestines. Root a folk remedy for cancer. Contains alkaloids called calystegines and possibly other bioactive components. **WARNING:** Root and plant juice are active purgatives.

JIMSONWEED, THORN-APPLE — Leaves, root, seed

Datura stramonium L. — Nightshade Family (Solanaceae)

☠ Coarse, annual; 2–5 ft. Leaves ovate in outline, angular coarse-toothed. Flowers white to pale violet (in center); 3–5 in., trumpet-shaped, *tube segments angular, folds narrowly winged*; May–Sept. Seedpods upright, spiny, chambered, 4-parted; seeds black. **WHERE FOUND:** Originating in tropical areas of N. America; waste places. Throughout our area. **USES:** Whole plant contains atropine, scopolamine, and other alkaloids; used in eye diseases (atropine dilates pupils); causes dry mouth, depresses bladder muscles, impedes action of parasympathetic nerves, used in Parkinson's disease; scopolamine used in patches behind ear for vertigo. Leaves were once smoked as an antispasmodic for asthma and spasmodic cough; also sedative and pain-relieving. Folk remedy for cancer. **WARNING:** Violently toxic, especially the seeds. Causes severe hallucinations. Many fatalities recorded and not rare. Those who collect this plant may end up with swollen eyelids.

Jimsonweed is one of our most toxic wildflowers.

Jimsonweed. Note upright, spiny seedpod.

SACRED THORN-APPLE, INDIAN THORN-APPLE Leaves, root, seed

Datura wrightii Regel Nightshade Family (Solanaceae)

Coarse, soft, hairy, gray, strongly scented perennial (often grown as annual); 2–7 ft. Leaves ovate in outline, entire, *velvety beneath*; base of leaves offset. Flowers usually white to pale violet; 5–8 in., trumpet-shaped; May–Oct. Seedpods pendulous, spiny, chambered, 5-parted; seeds black. **WHERE FOUND:** Originating in Mex. and Southwest; dry areas; waste places, increasingly cultivated for beautiful flowers and escaped. Much of our area. **USES:** Same as Jimsonweed, above. **RE-**

Note rounded, drooping seedpod of Sacred Thorn-apple.

Beautiful flower of Sacred Thorn-apple.

Sacred Thorn-apple flowers open at sunset, opening in less than a minute. Fun to watch.

Wild Potato-vine produces heart-shaped leaves.

LATED SPECIES: Often confused with *D. inoxia* Mill. and *D. metel* L.; uses and warnings the same. **WARNING:** Violently toxic, especially the seeds. Causes severe hallucinations. Many fatalities recorded and not rare. Those who collect this plant may end up with swollen eyelids.

WILD POTATO-VINE — Root, whole plant

Ipomoea pandurata (L.) G. Mey. — Morning-glory Family (Convolvulaceae)

Twining or climbing, often purple-stemmed vine with a very large tuberous root. Leaves *heart- to pear-shaped.* Flowers large (2–3 in.), white with pink stripes from center; May–Sept. **WHERE FOUND:** Dry soils. NY, CT to FL; TX to NE, IA to MI. **USES:** The Cherokee poulticed root for rheumatism, "hard tumors." Cherokee, Creek, and Iroquois used root decoction as a diuretic, laxative, expectorant; for coughs, asthma, beginning stages of tuberculosis; blood purifier; powdered plant used in tea for headaches, indigestion. Root extract shown to have strong antimicrobial activity. The large, deep-set, starchy edible rootstock can weigh 20 lbs. or more, but is difficult to dig.

MISCELLANEOUS NONWOODY VINES

VIRGIN'S BOWER — Whole flowering plant, root

Clematis virginiana L. — Buttercup Family (Ranunculaceae)

⚠ Clambering perennial vine, apparently herbaceous, but woody at base, 8–15 ft. Leaves divided into 3 sharp-toothed leaflets. Flowers white, with *4 petal-like sepals*; in clusters; July–Sept. *Feathery plumes* attached to seeds. **WHERE FOUND:** Rich thickets, wood edges. NS to GA, LA, e. KS, NE, north to Canada. **USES:** Cherokee used root tea for stomach ailments, kidney ailments, and nervous conditions. Iroquois used dried powered root as a wash for sore penis, also kidney trouble.

One Iroquois name means "make you dream," achieved by a wash of weak stem tea on hands and face. Liniment once used by physicians for skin eruptions, itching; weak leaf tea used for insomnia, nervous headaches, nervous twitching, and uterine diseases. Surprisingly little researched given its abundance. **WARNING:** Toxic. When fresh, highly irritating to skin and mucous membranes. Ingestion may cause bloody vomiting, severe diarrhea, and convulsions.

WILD CUCUMBER, BALSAM-APPLE

Root

Echinocystis lobata (Michx.) Torr. & Gray

Cucumber Family (Cucurbitaceae)

Climbing vine with tendrils. Leaves *maple-shaped, 5-lobed*, toothed along edges. Flowers *6-petaled*, in clusters of leaf axils; June–Oct. Fruits *solitary*, egg-shaped; fleshy, covered with weak bristles. **WHERE FOUND:** Thickets. NB to FL; TX to MN; westward; most of Canada. **USES:** Ojibwa used extremely bitter root tea as a tonic for stomach troubles; also purgative. Cherokee used tea for kidney ailments, rheumatism, chills, fevers, and obstructed menses. Menominee used in love potions and as a general tonic. Pulverized root poulticed for headaches. **WARNING:** Do not confuse this plant with *Momordica balsamina* L., a tropical member of the Cucumber family also known as Balsam-apple. Its root is purgative and considered toxic. The two plants have been confused in the literature because authors did not carefully compare scientific names of "balsam-apples." Both, as members of the Cucumber family, may contain cucurbitacins, which are extremely active as antitumor and cytotoxic agents at levels of less than one part per million.

ABOVE: *Virgin's Bower, usually blooming in late summer, is easily seen from a distance because of its abundant display of flowers.*
RIGHT: *Wild Cucumber in flower.*

ABOVE: *Passion-flower is one of the few temperate species of this mainly tropical plant group. Usually violet-blue, rarely completely white as shown here.* RIGHT: *Bottle Gourd has wavy-edged flowers.*

PASSION-FLOWER, MAYPOP — Whole flowering plant

Passiflora incarnata L. — Passion-flower Family (Passifloraceae)

Climbing vine, to 30 ft.; tendrils springlike. Leaves *cleft*, with 2–3 slightly toothed lobes. Flowers large, showy, unique, whitish to purplish, with *numerous threads* radiating from center; July–Oct. Fruits fleshy, egg-shaped. **WHERE FOUND:** Sandy soil. PA to FL; e. TX to e. NE, s. MO. **USES:** Cherokee poulticed root for boils, cuts, earaches, inflammation, and wounds. Traditionally used as an antispasmodic, and as a sedative for neuralgia, epilepsy, restlessness, painful menses, insomnia, and tension headaches. Research shows extracts are mildly sedative, slightly reduce blood pressure, increase respiratory rate, and decrease motor activity. Controlled clinical studies show a tea provides short-term benefits in adults with mild sleep quality difficulties. Passion-flower is used in modern phytotherapy for treating anxiety, nervousness, generalized anxiety disorder, insomnia, neuralgia, spasmodic asthma, ADHD, hypertension, and sexual dysfunction, among others. Mechanism of actions is still poorly understood. Fruits edible, delicious; a simple syrup of fruits is cooling in fevers.

BOTTLE GOURD, CALABASH, WHITE-FLOWERED GOURD — Leaves, fruits

Lagenaria siceraria (Molina) Standl. — Gourd Family (Cucurbitaceae)

Taprooted annual vine with tendrils and musky smell when crushed; to 15 ft. or more. Plant with long, densely hairy hairs gland-tipped and sticky. Leaves with long stalks, blades oval to heart-shaped, unlobed, or with 3–5 irregular, shallow lobes. Flowers white, solitary or in pairs, *often crinkled edges*; opening at night; summer. Fruits highly variable; 3–36 in. long, globe-shaped, crook-necked, or dumbbell-shaped.

WHERE FOUND: Originating in Old World tropics; MA, NY, FL; TX to IL. **USES:** With hard durable walls, the fruits have been used for containers since prehistory. Thought to have originated in Africa; New World archeological remains date back at least 13,000 years, predating the cultivation of corn and the development of clay pottery. New evidence suggests prehistoric East Asian origin of genetic material in the Americas arriving via the Bering Strait. Whatever the origin, this species and its variations are among the most ancient of human-utilized plants; especially as vessels, implements, and containers. The Houma used a poultice of fresh leaves for headaches. Cherokee used seeds soaked in water as a poultice for boils. Used in traditional cultures worldwide for a wide range of purposes, such as treatment of jaundice, diabetes, ulcers, colitis, hypertension, congestive heart failure, rheumatism, insomnia, and various skin disorders. Fruit pulp is considered emetic, purgative, diuretic, expectorant, analgesic, anti-inflammatory, and antibacterial, among others. A fruit extract found to reduce body weight, blood glucose, and total cholesterol in laboratory animals suggests directions for future research. Chemistry highly variable depending upon origin. **WARNING:** Bitter bottle gourd may cause gastrointestinal toxicity and hypotension. May cause contact dermatitis.

WHITE ORCHIDS

PINK LADY'S-SLIPPER, MOCCASIN-FLOWER, AMERICAN VALERIAN — Root

Cypripedium acaule Aiton — Orchid Family (Orchidaceae)

Usually pink; *rarely white* in some individual plants or populations. Perennial; 6–15 in. Leaves 2, *basal*. Flower a strongly veined pouch with a deep furrow on a leafless scape; May–June. **WHERE FOUND:** Acid woods. NL to GA; AL, TN to MN, and adjacent Canada. Too rare to harvest. **USES:** C. S. Rafinesque, in his 1828 *Medical Flora,* claimed to have introduced this Indian remedy to physicians; also widely used by Thomsonian practitioners (followers of the botanic medicine system of Samuel Thomson [1769–1843]). Along with other Lady's-slippers, called American Valerian and used as a substitute for Valerian (*Valeriana officinalis* L.), widely used in nineteenth-century America as a sedative for nervous headaches, hysteria, insomnia, nervous irritability, mental depression from sexual abuse, and menstrual irregularities accompanied by despondency (possibly referring to premenstrual syndrome); claimed to relieve pain. The Pink Lady's-slipper was considered a good substitute for the more commonly used Yellow Lady's-slipper (p. 134). Both were harvested in significant tonnage in the nineteenth century, contributing to scarcity in the twentieth century. Orchids often have swollen, ball-shaped tubers, suggesting testicles; these roots are widely regarded as aphrodisiacs, perhaps reflecting the doctrine of signatures (see p. 7). Despite its widespread use historically, the chemistry and biological effects of the genus are little studied. **WARNING:** May cause dermatitis.

Pink Lady's-slipper flowers are usually pink, sometimes pale or white.

TOP RIGHT AND RIGHT: *Downy Rattlesnake-plantain blooms in late summer. Note netted-veined leaves.*

DOWNY RATTLESNAKE-PLANTAIN

Leaves

Goodyera pubescens (Willd.) R. Br. Orchid Family (Orchidaceae)

Perennial, essentially evergreen; to 16 in. (in flower). Leaves mostly basal, oval, with *white veins in a checkered pattern*, and white spreading hairs. Small whitish flowers on a dense, woolly raceme; July–Sept. **WHERE FOUND:** Moist, humus-rich soils in woods. NS, ME to FL; AR, MO to WI, w. QC. Rare—do not harvest. **USES:** The distinctive bluish green leaves with prominent white veins in a "rattlesnake" pattern earned this plant its common name, as did the use of the leaf for snakebites, based on the doctrine of signatures (see p. 7). Cherokee used leaf tea for colds, kidney ailments; leaf tea taken (with whiskey) to improve appetite, blood tonic; tea swished in mouth for treating toothaches. Externally, leaf poultice used to "cool" burns, treat skin ulcers. Exudate of leaf used for sore eyes. Physicians once used fresh leaves steeped in milk as poultice for tuberculous swelling of lymph nodes (scrofula). Fresh leaves were applied every 3 hours, while the patient drank a tea of the leaves at the same time. Famously noted as a snakebite remedy by Jonathan Carver in 1778 (*Travels through the Interior Parts of North America, in the Years 1766, 1767, and 1768*); chewed leaves, swallowing part of the juice considered the most effective rattlesnake bite remedy among the Potawatomi. Goodyerin, a flavone glycoside from various *Goodyera* species, was found to have a significant, dose-dependent sedative and anticonvulsant effect. Related compounds called goodyerosides have potential liver-protectant effects. **REMARKS:** Of historical interest only. Too scarce to harvest, which is true for the other three North American species, *G. oblongifolia* Raf., *G. repens* (L.) R. Br., and *G. tesselata* Lodd. Used similarly and interchangeably.

NODDING LADIES' TRESSES

Whole plant

Spiranthes lacera (Raf.) Raf.

Orchid Family (Orchidaceae)

Delicate, variable, fleshy-rooted orchid; 4–20 in. Basal leaves (absent at flowering) are firm, thick, pale green; leaves much reduced or absent on flowering stalk. Small, white, *cut-edged* lipped flowers in a double spiral; Aug.–Nov. **WHERE FOUND:** Bogs, meadows. Most of our area. **USES:** Native Americans used various species of *Spiranthes*; plant tea as a diuretic for urinary disorders, venereal disease, blood purifier, and as a wash to strengthen weak infants. Ojibwa used as talisman to attract game. **RELATED SPECIES:** A genus of 43 species found primarily in Asia, Europe, N. America, and S. America (20 species in our range). Ladies' tresses also used as a diuretic and aphrodisiac. Contains various flavonoids and coumarins with potential antimicrobial and anticancer activity.

Ladies' Tresses have a distinct double-spiraling flower spike. Most orchids are relatively rare and should not be harvested. Some are protected by law.

LOW-GROWING, CREEPING EVERGREENS

BEARBERRY, UVA-URSI, KINNIKKINNIK — Leaves

Arctostaphylos uva-ursi (L.) Spreng. — Heath Family (Ericaceae)

⚠ Trailing shrub; bark fine-hairy; to 1 ft. tall. Leaves *shiny-leathery, spatula-shaped.* Flowers white, urn-shaped; May–July. Fruit is a dry red berry. **WHERE FOUND:** Sandy soil, by rocks. Arctic to n. U.S. A trailing shrub found in sandy soils and near exposed rock from the Arctic south to the northern tier of American states, and mountains of Europe. The only *Arctostaphylos* species found outside N. America. **USES:** Native Americans throughout the plant's range used leaves for a wide variety of ailments; often to relieve pain. Dried-leaf tea diuretic, strongly astringent; urinary-tract antiseptic for cystitis (when urine is alkaline, achieved by adding a teaspoonful of baking soda to a glass of water), nephritis, urethritis, kidney stones, and gallstones. Also used in bronchitis, gonorrhea, diarrhea, and to stop bleeding. Leaves contain more than a dozen anti-inflammatory and antiseptic compounds. **WARNING:** Contains arbutin, which hydrolyzes to the toxic urinary antiseptic hydroquinone; therefore, use is limited to less than one week in German herbal practice. Leaves are also high in tannins (up to 8 percent). Recently implicated in potential retinal toxicity of the eye in a woman who ingested the tea for 3 years.

TRAILING ARBUTUS, MAYFLOWER (NOT SHOWN) — Leaves

Epigaea repens L. — Heath Family (Ericaceae)

⚠ Trailing perennial; to 6 in. Leaves *oval, leathery,* alternate (often closely spaced), 1–3 in. long, 1–2 in. wide. Flowers in clusters, white (or pink),

LEFT: *Bearberry's evergreen, spatula-shaped leaves with red berries.*
ABOVE: *Bearberry has white, often red-tinged flowers.*

Wintergreen's white, urn-shaped flower.

The leaves of the Wintergreen have a strong wintergreen fragrance; also known as Teaberry.

tubular, flared, 5-lobed; Feb.–July. **WHERE FOUND:** Open, sandy woods. NL to FL; MS, TN, OH to MI, and adjacent Canada. Protected in some states. **USES:** Algonquin, Cherokee, and Iroquois peoples used leaf tea for kidney disorders, stomachaches, blood purifier. Leaf tea is a folk remedy for bladder, urethra, and kidney disorders, gravel (kidney stones). Shakers sold this plant as "gravel-plant." The tribal flower of the Potawatomi, who believed the flowers came directly from the hands of their divinity. **WARNING:** Contains arbutin; although it is effective as a urinary antiseptic, it hydrolyzes to hydroquinone, which is toxic.

WINTERGREEN, TEABERRY, CHECKERBERRY — Leaves

Gaultheria procumbens L. — Heath Family (Ericaceae)

⚠ *Wintergreen-scented*, mat-forming, creeping, subshrub; to 6 in. Leaves oval, glossy. Flowers waxy, drooping bells; July–Aug. Fruit a dry, shiny, bright red berry. **WHERE FOUND:** Woods, openings. Canada to GA; AL to WI, MN, adjacent Canada. **USES:** Wintergreen refers to the evergreen nature of the leaves, but to most minds it has become associated with a specific flavor produced by the compound methyl salicylate. This plant and Black Birch were once commercial sources of wintergreen flavor, now largely replaced by synthetic methyl salicylate. Traditionally, leaf tea used for colds, headaches, stomachaches, fevers, kidney ailments; externally, wash for rheumatism, sore muscles, lumbago. The chemical behind the aroma, methyl salicylate, with anti-inflammatory and painkilling activities, has recently been shown to enable plants to communicate with one another. Experimentally analgesic, carminative, anti-inflammatory, antiseptic. In experiments, small amounts have delayed the onset of tumors. **WARNING:** Essential oil is highly toxic; absorbed through skin, harms liver and kidneys; fatalities from topical application of the oil (or its synthetic form) have been reported.

Note the 4-parted, bearded flowers of Partridgeberry.

Partridgeberry has prominent red berries.

PARTRIDGEBERRY, SQUAW-VINE Leaves, berries

Mitchella repens L. Madder Family (Rubiaceae)

Leaves opposite; rounded. Flowers white (or pink); 4-parted, terminal, *paired*; May–July. Fruit is a single dry red berry lasting over the winter. **WHERE FOUND:** Woods, sandy soils. NL to FL; TX to MN. **USES:** Widely used by Native American groups throughout its range. Historically, those uses adopted by European settlers with dried- or fresh-leaf or berry tea used for delayed, irregular, or painful menses and childbirth pain; astringent used for hemorrhoids, dysentery, and as a mild diuretic. Red berries once a popular folk remedy for diarrhea. Externally, used as a wash for swellings, hives, arthritis, rheumatism, and sore nipples, for which it was highly esteemed. The plant is still collected and sold in the herb trade. Also called Squaw-vine because the leaves have historically been used for menstrual difficulties, and to facilitate childbirth when taken as a tea 2–3 weeks before delivery, the safety of which is questionable. Despite mention in numerous herbals, there is virtually no research on this plant.

CURIOUS FLESHY PLANTS WITH SPECIALIZED GROWTH HABITS

ROUND-LEAVED SUNDEW Whole plant

Drosera rotundifolia L. Sundew Family (Droseraceae)

An *insectivorous* (insect-eating) denizen of bogs, the Sundew has unusual leaves that are barely 2 in. tall. Reddish leaves are covered with hairs exuding a sticky, dewlike secretion that catches and holds unwary insects. Once the insect is captured, the Sundew's leaves fold over it to digest it. Small (2–9 in.) perennial. Leaves tiny, to ½ in. across; rounded;

ABOVE: *The Round-leaved Sundew is usually found in acidic bog habitats.* RIGHT: *Round-leaved Sundew's tiny white flowers.*

blade mostly wider than long; covered with *reddish, glandular-tipped hairs exuding sticky "dewdrops."* Flowers white or pinkish, on a 1-sided raceme, opening one at a time; June–Aug. **WHERE FOUND:** Nutrient-poor, sunny, wet acid soil; wet meadows; bogs. Circumboreal; all of Canada south to FL; IL, MN, n. and cen. Europe, Asia. **USES:** The leaves of Sundew have traditionally been used for lung ailments, perhaps reflecting the concept of the doctrine of signatures. Famous in Europe to treat whooping cough since the seventeenth century. Since the plant somewhat resembles the bronchioles of the lungs, it was used for lung ailments. Traditionally, tea or tincture used for dry, spasmodic coughs; asthma; whooping cough; arteriosclerosis; and chronic bronchitis; also as an aphrodisiac; poultice or plant juice used on corns and warts. Europeans regard the extracts and tinctures as antitussive and spasmolytic. Contains plumbagin, which is immunostimulating in small doses; antibacterial, antifungal, and antiviral. Active constituents include naphthoquinones, flavonoids, and ellagic acid derivatives. Exudates from the leaves have been used to treat warts. **REMARKS:** Conservation concerns have arisen over the harvesting of the herb for use in European herbal medicine.

INDIAN-PIPE

Whole plant, root

Monotropa uniflora L.

Heath Family (Ericaceae)
[Formerly in Indian-pipe Family (Monotropaceae)]

Once called Ice Plant because it resembles frozen jelly and "melts" when handled. Also called Bird's Nest, in reference to the shape of the entangled root fibers. The species name *uniflora* means "one flower," referring to the single flower atop each stalk. Saprophytic perennial,

Indian-pipe mostly lacks chlorophyll and is easy to recognize by its single flower.

without chlorophyll; 6–8 in. *Whole plant is translucent white.* Scalelike leaves nearly absent. Flower a *single nodding bell*; May–Oct. **WHERE FOUND:** Woods. Native of our range; Asia. Too scarce to harvest. **USES:** Cherokee used pulverized root for fits, epilepsy, and convulsions; plant juice for inflamed eyes, bunions, warts. The Mohegan drank tea for aches and pains due to colds. Physicians once used tea as antispasmodic, nervine, sedative for restlessness, pains, nervous irritability. As a folk remedy for sore eyes, the plant was soaked in rose water, then a cloth was soaked in the mixture and applied to the eyes. Water extracts are bactericidal. **WARNING:** Safety undetermined; possibly toxic—contains several glycosides.

3–6 PETALS; LEAVES GARLIC-SCENTED

GARLIC

Bulb

Allium sativum L.

Allium Family (Alliaceae)

[Formerly in Lily Family (Liliaceae) or Amaryllis Family (Amaryllidaceae)]

⚠ To 3 ft. Leaves extend *almost to middle of stem.* Note the 2- to 4-in.-long *narrow, papery green spathe* around flowers. **WHERE FOUND:** Fields, roadsides. Alien. Planted and occasionally escaped from cultivation. NY to TN, KY, MO; north to IN. Garlic has evolved through cultivation by humans over 7,000 years. Believed to have originated in the Asian steppes, Garlic's existence is entirely dependent on humans. It is not found in a wild state except when escaped from cultivation. **USES:** Peeled cloves have been eaten or made into tea, syrup, and tincture to treat colds,

The familiar bulbs of Garlic don't produce their characteristic odor (or health benefits) until cut or crushed, which causes a chemical reaction.

fevers, coughs, earaches, bronchitis, shortness of breath, sinus congestion, headaches, stomachaches, high blood pressure, arteriosclerosis, diarrhea, dysentery, gout, and rheumatism. For external uses, end of Garlic clove is cut, then juice is applied to ringworm or acne (see warning below); folk remedy for cancer. Cough syrup traditionally made by simmering 10 Garlic cloves in 1 pint of milk, adding honey to taste; syrup taken in 1-tablespoon doses as needed. In China, Garlic is used for digestive difficulties, diarrhea, dysentery, colds, whooping cough, pinworms, old ulcers, swellings, and snakebites. Experimentally, it lowers blood pressure and serum cholesterol; antibacterial, antifungal, antioxidant, diuretic. Clinical studies suggest efficacy in gastrointestinal disorders, hypertension, heart disease prevention, and arteriosclerosis. According to demographic studies, Garlic is thought responsible for the low incidence of arteriosclerosis in parts of Italy and Spain, where Garlic consumption is heavy. Allicin, the substance responsible for Garlic's characteristic odor, is thought to be responsible for some of the plant's pharmacological qualities. Allicin is a chemical byproduct of an enzymatic reaction produced when Garlic is cut or crushed. In experiments with mice, Garlic extracts had an inhibitory effect on cancer cells. Garlic has been the subject of more than 3,000 scientific studies over the past 35 years, including more than 18 clinical studies that involved more than 3,000 patients and evaluated Garlic's effects in lowering blood lipids (cholesterol). Results are mixed, but generally positive. **WARNING:** The essential oil and juice extracted from the bulbs is extremely concentrated and can be irritating.

WILD LEEK, RAMP

Leaves, root

Allium tricoccum Ait. Allium Family (Alliaceae)
[Formerly in Lily Family (Liliaceae) or Amaryllis Family (Amaryllidaceae)]

Perennial; 6–18 in. Leaves 2–3, smooth, to 2½ in. wide; fleshy, strongly onion- or leek-scented; *leaves wither in late spring before whitish to creamy yellow ¼-in.-long flowers on globe-shaped, loose heads bloom.* Flowers June–July. **WHERE FOUND:** Rich, moist woods. Localized, but often in abundant populations. NB south to GA mtns., west through

Wild Leek, Ramp is abundant in local populations.

Field Garlic has a rank, skunklike garlic odor, often producing bulblets in place of seeds.

TN, MO, IL, IA, MN. **USES:** Cherokee ate leaves for colds, croup, and as spring tonic. Warm juice of leaves and bulbs used for earaches. Strong root decoction emetic. Similar to but less potent than Garlic. The wide range of effects attributed to Garlic (see above) probably accrue to Ramp or Wild Leek as well. Contains thiosulfinates as major flavor and fragrance components, one of which correlates with the typical flavor of Garlic.

FIELD GARLIC, WILD GARLIC

Bulb, roots

Allium vineale L.

Allium Family (Alliaceae)
[Formerly in Lily Family (Liliaceae) or Amaryllis Family (Amaryllidaceae)]

Perennial weed, 8–24 in. tall; with 5–20 clustered, hard-shelled bulbs. Leaves, 2–4, persistent, green at flowering, sheathing about half the height of flowering stalk, *hollow within, cylindrical or threadlike with rank garlic odor*. Flowers pink or white (if present), *mixed with or replaced by bulblets* (with small tail-like spouts); May–Sept. **WHERE FOUND:** Alien weed from Continental Europe. Fields, disturbed areas. Much of our range. **USES:** Also known in England as Crow Garlick or Stag's Garlick, the species name *vineale*, denoting of the vineyard, is a European species that has become a serious weed in agricultural lands. Propagated both from seeds and spreading of the vegetative bulblets. If intermixed during wheat harvest, bread or flour becomes strongly garlic-odored. If milk cows graze on the plant, milk or butter is tainted with a rank garlic odor. It became a vile, troublesome weed in the U.S. in the 1890s. Used similarly to Garlic, but a poor substitute. Shown to possess antioxidant activity. **WARNING:** May cause contact dermatitis or irritation.

3–6 PETALS; STARS, TUBES, OR FLARED TUBES

COLIC-ROOT, STARGRASS

Root

Aletris farinosa L.

Narthecium Family (Nartheciaceae)
[Formerly in Lily Family (Liliaceae)]

Perennial; 1½–3 ft. Leaves in a basal rosette, linear to lance-shaped. Flowers white or cream, tubular, tightly hugging a tall, leafless stalk; May–Aug. Flowers *swollen at base; surface mealy.* **WHERE FOUND:** Open dry sites, prairies, upland woods, moist bogs, sand. S. ME to FL; west to e. TX; north to WI, MI. **USES:** Cherokee used root tea as tonic for fevers; to strengthen stomach and womb; for coughs, lung conditions, rheumatism, jaundice, colic, and painful urination. Micmac and Rappahannock peoples used root or plant infusions to treat menstrual disorders. Root decoction used as bitter tonic for indigestion. Promotes appetite; treats diarrhea and labor pains. Used for colic, but small doses may cause hypogastric colic. Tincture used for rheumatism. Contains diosgenin, which has both anti-inflammatory and estrogenic properties. Intensely bitter resinous root more soluble in alcohol than in water. **REMARKS:** Of possible conservation concern due to overharvesting.

SNOWDROP

Dried bulbs

Galanthus nivalis L.

Amaryllis Family (Amaryllidaceae)

☠ Perennial; up to 1 ft. Leaves 2, appearing with flowers, narrow, lance-shaped, to 4 in. long, ½ in. wide. Flowers white, drooping, bell-like in outline, 1 in. long, *with 3 inner segments with green blotch at apex*, and 3 outer white segments; Feb–May. **WHERE FOUND:** Alien; Europe; naturalized from NB, NL, south to MD, VA, west to MI, and spreading. **USES:** Little used historically; active alkaloid galanthamine first discovered in the early 1950s, initially used to treat nerve pain and symptoms of polio. Galanthamine is approved in many countries (including the U.S.) as a long-acting, selective, reversible, and competitive acetylcholinesterase (AChE) inhibitor for the systematic treatment of mild to moderate cognitive impairment of early-stage Alzheimer's disease. Until recently the alkaloid was extracted from *Galanthus* species; now synthesized on an industrial scale. **WARNING:** Crude fresh and dried bulb highly toxic and not used; only the purified alkaloid found in the plant is used in small doses by intravenous injection or orally.

Colic-root, Stargrass. Note the mealy texture of the flowers.

Snowdrop is an alien spread from cultivation.

FALSE LILY-OF-THE-VALLEY, CANADA MAYFLOWER

Root, whole flowering plant

Maianthemum canadense Desf.

Asparagus Family (Asparagaceae)
[Formerly in Lily Family (Liliaceae)]

Perennial; 3–6 in., often forming large colonies. *Usually 2 leaves, 1 on sterile shoots*; oval, to 3 in. long, base strongly cleft (heart-shaped). Tiny, 4-pointed flowers in 12- to 15-flowered clusters; Apr.–June. Berries whitish, turning pinkish; speckled. **WHERE FOUND:** Woods. NL to GA mtns.; TN to IA, MB. **USES:** Ojibwa used plant tea for headaches and to "keep kidneys open during pregnancy." Also used as a gargle for sore throats. Potawatomi used root as a good luck charm to win games. Folk expectorant for coughs; soothes sore throats.

FALSE SOLOMON'S-SEAL

Root, leaves

Maianthemum racemosum (L.) Link
[*Smilacina racemosa* (L.) Desf.]

Asparagus Family (Asparagaceae)
[Formerly in Lily Family (Liliaceae)]

Perennial; 1–2 ft. Zigzag stem arched. Leaves oval. Flowers in terminal clusters; May–July. *Stamens longer than petals.* **WHERE FOUND:** Rich woods. NS to FL; AL, AR to MN and westward. **USES:** Native Americans used root tea for constipation, rheumatism, stomach tonic, and as "female tonic"; root smoke inhaled for insanity, and to quiet a crying child. Leaf tea used as a contraceptive and for coughs; externally, for bleeding, rashes, and itch. Herbalists use to induce sweating and urination; blood purifier.

False Lily-of-the-valley, Canada Mayflower in a wooded habitat.

The flowers of False Solomon's-seal grow in terminal clusters.

SOLOMON'S-SEAL — Root

Polygonatum biflorum (Walt.) Ell. — Asparagus Family (Asparagaceae) [Formerly in Lily Family (Liliaceae)]

[*P. biflorum* var. *commutatum* (Schult. & Schult. f.) Morong, *P. commutatum* (Schult. & Schult. f.) A. Dietr., *P. giganteum* A. Dietr.]

Perennial, unbranched, smooth, erect, and arching; 1–5 ft. Leaves 10–25, oval to elliptical, 3–7 in. long; alternate. Flowers tubular, about an inch long, *drooping in pairs (or 3 or more) from leaf axils*; May–June. **WHERE FOUND:** Rich woods. CT to FL; TX, NE to MN. **USES:** The Cherokee used the root tea as a mild tonic to treat "general debility." Tea used to treat diseases of the breast or lungs; profuse menstruation, dysentery; stomach ailments; poulticed for carbuncles. Chippewa used root tea as a sleep aid. Also used to treat coughs and as a laxative; fresh root poulticed (or root tea used externally as a wash) for sharp pains, cuts, bruises, and sores. Root tea a folk remedy to relieve irritation of mucous membranes, hemorrhoids, rheumatism, arthritis, lung ailments; used externally as a wash for skin irritation, including poison ivy rash. Contains steroidal saponins. Considered anti-inflammatory and astringent. **REMARKS:** Confusing taxonomy has been applied to this plant, a polyploidy complex, which, depending upon chromosome number, is highly variable in plant size, number of flowers, and habitat; larger plants south and west, smaller plants typically in northern part of range. Of the 57 species in the genus, 3 occur in N. America, 5 in Europe, and most are from East Asia. Several species used in Traditional Chinese Medicine, especially *P. odoratum* (Mill.) Druce (*Yu-zhu*) and its cultivars and hybrids, which contain steroidal saponins; used for sore throat, dry cough, and to promote production of body fluids. *P. sibericum* F. Delaroche and *P. kingianum* Collett & Hemsl. (*huang jing*) also used for dry cough (to "moisten" lungs), to support the blood, and for diabetes. Most chemical and pharmacological studies have involved Asian species; American species surprisingly little researched.

LEFT: *The flowers of Solomon's-seal droop from leaf axils.* ABOVE: *White Wakerobin is our largest* Trillium *species.*

WHITE WAKEROBIN, WHITE TRILLIUM — Root

Trillium grandiflorum (Michx.) Salisb. Melanthium Family (Melanthiaceae)
[Formerly placed in the Trillium Family (Trilliaceae) or the Lily Family (Liliaceae)]

Perennial; largest trillium to 18 in.; variable. Leaves 3, broadly oval, to 5 in. long or more, tip sharp-pointed. Flowers on *3-in.-long stalks, nodding under weight of large white flowers* (fading to pink-red) to 3 in. across; petals erect, flaring and wavy above middle; often green striped; Apr.–June. **WHERE FOUND:** Rich woods, deciduous forest. ON, ME, to mtns. of w. SC, e. TN, Ohio Valley to MN. **USES:** Chippewa used root as an anti-rheumatic; Menominee used raw root as a poultice to treat swelling of the eye; tea for cramps, menstrual disorders; local stimulant; expectorant. In his *Medical Flora* (vol. 2, 1830), C. S. Rafinesque suggested that almost all trilliums are used without regard to species; claims to have introduced it into medical practice and says that Native Americans called those species with red blossoms "male" and those with white blossoms "female," the latter being best for female conditions. Roots are considered astringent, expectorant, tonic, antiseptic; used for bloody urine, menstrual discharge, fevers, asthma, coughs. Externally for tumors, ulcers, etc. Science confirms antifungal activity.

4 PETALS; LEAVES TOOTHED, MUSTARD FAMILY

GARLIC MUSTARD, HEDGE GARLIC, SAUCE ALONE — Leaves, root

Alliaria petiolata (M. Bieb.) Cavara & Grande (*A. officinalis* Andrz. ex DC.)
Mustard Family (Cruciferae or Brassicaceae)

Erect *garlic-odored* annual or short-lived perennial, in N. America an obligate biennial, stems simple or branched above; normally 1–3 ft. tall. Leaves alternate, somewhat triangular on stem, often heart-shaped at base of plant, 2–6 in. long, ½–3 ½ in. wide, *margins with distinct, rounded teeth*. Flowers white in loose raceme, 4-petaled, about ¼ in. long; Apr.–May. **WHERE FOUND:** Alien. Eurasia. Exceptionally adaptable in shaded woodlands, waste places, fields, roadsides, and other habitat from ME to SC, westward to OK north to ON. A noxious weed, banned or subject to quarantine in some states. Spreads rapidly, blanketing forest floors or fields. **USES:** Crushed leaves strongly garlic-odored; a poor man's garlic. Root with radishlike flavor (if harvested before bolting). A decoction of the fresh herb used in European traditions as a diuretic, expectorant, and to induce sweating in colds and flu; also to stimulate appetite and digestion. In British folk medicine bruised leaves chewed for mouth ulcers, sore gums; externally applied for sore throat. In eighteenth century, fresh leaves bruised in wine applied to gangrenous tissue as an antiseptic. In England, called "sauce alone"; used as a wild herb in sauces, with bread and butter, salted meat, salads, or as a potherb. Eaten by cows and goats, but not horses, sheep, or swine. Imparts a bad taste to fowl that have eaten it; also to cow's milk. Contains the cancer-preventive isothiocyanates (of Mustard family) and the allyl sulfides of Garlic (and its relatives). Early leaf clumps (before flowering) have double the beta-carotene of spinach. **WARNING:** May cause contact dermatitis or irritation; also bad breath, simply from handling the plant.

HORSERADISH — Root

Armoracia rusticana P. G. Gaertn., B. Mey. & Scherb Mustard Family
(*A. lapathifolia* Gilib.) (Cruciferae or Brassicaceae)

Large-rooted herb; 1–4 ft. Leaves large, broad, lance-shaped (sometimes jagged); *long-stalked*. Leaves much reduced on flowering stalks. Flowers white, tiny, 4-petaled; May–July. Pods tiny, egg-shaped. **WHERE FOUND:** Moist fields. Throughout our area. Alien (Europe), persists after cultivation. **USES:** Root used as a condiment. Root tea weakly diuretic, antiseptic, and expectorant; used for bronchitis, coughs, bronchial catarrh, calculus (dental plaque). Root poultice used for rheumatism, respiratory congestion. Few things are better at opening the sinuses than too large a bite of pungent horseradish sauce. Science confirms plant is antibiotic against gram-negative and gram-positive bacteria, pathogenic fungi. Experimentally, it has antitumor activity, as science has come to expect from the Mustard family. In European phytomedicine, the root is used for treatment of inflammation of the respiratory tract and supportive treatment of urinary tract infections. Externally, prep-

ABOVE: *Garlic Mustard is increasing in our range as a noxious weed.* RIGHT: *Horseradish is often found near gardens and old fields.*

arations of root approved to treat lung congestion and minor muscle aches. **WARNING:** Large amounts may irritate digestive system. Plant tops are a fatal poison to livestock. External use may cause skin blisters.

SHEPHERD'S PURSE

Whole plant in fruit

Capsella bursa-pastoris (L.) Medik.

Mustard Family (Cruciferae or Brassicaceae)

Annual or biennial; 4–23 in. Basal leaves in rosette, narrow *dandelion-like* in outline. Stem leaves much reduced, clasping. Flowers tiny, white, 4-petaled. Seedpods *upside-down heart-shaped.* (Pods of *C. rubella,* once a separate species, now treated as synonymous with *C. bursa-pastoris,* have concave sides.) Fruiting and flowering: Jan.–Oct. **WHERE FOUND:** Waste places. Throughout N. America, Europe, and n. Asia. Alien (Europe). **USES:** Dried- or fresh-herb tea (made from seeds and leaves) stops bleeding, allays profuse menstrual bleeding; diuretic. Has proven uterine-contracting properties; traditionally used during childbirth. Dried herb a useful styptic against hemorrhage. Tea also used for diarrhea, dysentery; also externally, as a wash for bruises. Cottonball dipped in plant juice once used to stop nosebleed. Science confirms anti-inflammatory, diuretic, and anti-ulcer activity; decreases blood pressure in laboratory animals. Both water and methanolic extracts accelerate blood coagulation; a peptide possibly responsible for hemostatic activity. Aboveground parts (in fruit) used in European phytotherapy for symptomatic treatment of excessive or irregular menstrual bleeding and to stop nosebleeds; externally to allay bleeding from injuries. **WARNING:** Seeds are known to cause rare cases of blistering of skin.

LEFT: *Shepherd's Purse. Note the small white flowers and purse-shaped seedpods.* ABOVE: *Toothwort. One of our earliest wildflowers, usually white or pale pink.*

TOOTHWORT, TOOTHACHE ROOT, PEPPER ROOT — Root

Cardamine concatenata (Michx.) O. Schwarz. — Mustard Family (Cruciferae or Brassicaceae)

[*C. laciniata* (Muhl. ex Willd.) Wood, A. W.; *D. concatenata* Michx.]

Sparsely hairy perennial; showy spring wildflower; 6–14 in., with somewhat tuberous fleshy rootstock. Stem leaves *whorled or opposite*; divided into 3 deeply round-toothed or incised leaflets. Flowers 4-petaled, in terminal clusters; Feb.–May. **WHERE FOUND:** Moist woods. NB, QC, ON to FL, west to TX, ND. **USES:** Iroquois used plant tea for colds, to stimulate appetite, and for stomach complaints. Roots (fresh or dried) placed in mouth or pocket to attract women; rubbed on hunting implement or fishing hooks to act as a luck charm. Root peppery; used as a folk remedy for toothaches. **RELATED SPECIES:** Sometimes confused in literature with and used interchangeably with *Cardamine diphylla* (Michx.) Alph. Wood, Crinkle Root.

POOR-MAN'S-PEPPER, PEPPERGRASS — Seedpods, leaves

Lepidium virginicum L. — Mustard Family (Cruciferae or Brassicaceae)

⚠ Smooth or minutely hairy annual or biennial; 6–24 in. Leaves lance-shaped, sharp-toothed; *stalked at base*. Flowers white, inconspicuous, in raceme; Mar.–Nov. Petals as long as or longer than sepals. Seedpods roundish, widest at middle. **WHERE FOUND:** Fields, roadsides, waste places. Throughout our area. **USES:** Externally, bruised fresh-plant or fresh-leaf tea for Poison Ivy rash, scurvy; used as a substitute for Shep-

Poor-man's-pepper, Peppergrass.

Watercress, one of our earliest spring flowers, often grows in running water.

herd's Purse. Leaves poulticed on chest for croup. Used as a peppery flavoring for early spring greens. **WARNING:** Application may cause skin irritation, blisters.

WATERCRESS

Leaves

Nasturtium officinale R. Br. Mustard Family (Cruciferae or Brassicaceae)
[*Rorippa nasturtium-aquaticum* (L.) Hayek]

Watercress, well known for its edible, mustardlike leaves, forms large colonies in cool running water. It begins its life cycle in early spring. Creeping perennial; 4–36 in. Mustard-flavored leaves divided into *3–9 leaflets, or strongly divided.* Small white flowers in terminal clusters; Mar.–June. **WHERE FOUND:** Widespread in cool running water. **USES:** Fresh leaves are high in vitamins A and C and iodine when harvested before flowering. Traditionally used as a diuretic, blood purifier; also used for lethargy, rheumatism, heart trouble, bronchitis, scurvy, and goiter. Leaf extracts are used clinically in India to correct vitamin deficiency. Used in European phytomedicine for treatment of inflammation of the respiratory tract. Experimentally, plant juice may prevent DNA damage; leaf extract has strong antioxidant activity; leaf isothiocyanates with potential cancer-preventing activity. **WARNING:** Do not harvest leaves from polluted waters. Poisoning has resulted from eating leaves from polluted waters in which the plant has absorbed heavy metals and toxins.

BUNCHBERRY Leaves, roots, fruits
Cornus canadensis L. Dogwood Family (Cornaceae)

Perennial; 3–8 in. Leaves in whorls of 6. Flowers white, in clusters; surrounded by *4 showy, petal-like bracts*; flowers open explosively in 0.5 millisecond (fastest recorded movement in a plant), releasing stored elastic energy, catapulting pollen; May–July. Fruits scarlet. **WHERE FOUND:** Cool woods. Northern N. America south to WV mtns., also in n. CA. **USES:** Iroquois used decoction of whole plant for colds and fevers. Ojibwa used root to treat infant colic. Micmac used roots and leaves to treat "fits." Leaf tea a folk remedy for aches and pains, kidney and lung ailments, and as an eyewash; poulticed for sores. Fruits eaten as a snack food; gathered, dried, and stored for winter use. The bitter dried fruits (formerly called Bitter Redberry) were once used to stimulate appetite and poor digestion.

Bunchberry. Note beautiful Dogwood-like flowers.

Bunchberry in fruit.

The hooked prickles of Cleavers serve to support the plant.

CLEAVERS

Whole plant

Galium aparine L.

Madder Family (Rubiaceae)

Annual; stem *raspy*, 4-angled, with prominent hooked prickles. Plant weak, apparently drooping and clambering; but actually a climber, using differentiated hooked hairs on the top and bottom side of leaves to cling to host plants, then ratchet into position to expose leaf for optimum sunlight exposure; 1–3 ft. Leaves lance-shaped; *usually 8, in whorls*. Inconspicuous whitish flowers on stalks from leaf axils; Apr.–Sept. **WHERE FOUND:** Thickets. Throughout N. America. Alien; w. Eurasia, now worldwide. **USES:** Used since ancient times; herbal tea traditionally used as a diuretic and blood purifier; used for bladder and kidney inflammation, dropsy, "gravel" (kidney stones), fevers. Juice of fresh herb used for scurvy. Herb tea used internally and externally as a folk remedy for cancer. Juice contains citric acid, reported to have antitumor activity. Experimentally, extracts are hypotensive (lower blood pressure). Also contains asperuloside, which is anti-inflammatory. **WARNING:** Juice may cause contact dermatitis.

OPIUM POPPY

Latex from flower capsule

Papaver somniferum L.

Poppy Family (Papaveraceae)

⚠ Flowers usually 4-petaled, white, pink, red, to purple. See p. 189.

Opium Poppy flower petals vary in color; white with red shown here.

Opium Poppy's dried seedpods are the source of commercial culinary poppy seeds.

GOLDTHREAD, CANKER ROOT

Root

Coptis trifolia (L.) Salib. Buttercup Family (Ranunculaceae)
[*Coptis groenlandica* (Oeder) Fern.]

Mat-forming perennial; to 3 in., with *bright yellow, threadlike roots*. Leaves shiny, evergreen, 3-parted; leaflets wedge-shaped; resemble miniature strawberry leaves. Flowers with 5 small white showy sepals; May–Aug. **WHERE FOUND:** Cool forests. Canada to WV mtns.; north to n. OH, IN, MN, BC coastal forest to AK. **USES:** The Iroquois and other Native American groups used root tea as a wash or gargle for mouth sores; dried root applied as powder to mouth sores; tea for jaundice, indigestion, and infections; wash for irritated eyes. Root is highly astringent, chewed for canker sores. Contains berberine, which has many properties, including anti-inflammatory and antibacterial effects; also contains the alkaloid coptisine. Root traditionally used for dyspepsia, thrush, alcoholism, nausea, conjunctivitis, jaundice, nausea, sore throat, stomach cramps, and other ailments. Widely used in nineteenth-century America. Today it is seldom used and rarely present in the herb trade, probably because the root is literally a thread, hence difficult to harvest in quantity. Chinese species such as *Coptis chinensis* Franch. (*huang lian*) have entered the American market as a substitute for Goldenseal. Conversely, evaluated as a substitute for *C. chinensis, C. trifolia* lacks some of the alkaloids responsible for the former's medicinal value.

WOOD STRAWBERRY

Leaves, root, fruits

Fragaria vesca L. Rose Family (Rosaceae)

Perennial, with runners; 3–6 in. Leaves pointed, not rounded, at tip, tooth at tip of middle leaflet *usually longer than adjacent teeth*. Flowers white; calyx lobes spreading or recurved; May–Aug. Fruits with *seeds on*

Goldthread, Canker Root, with threadlike, bright yellow rhizomes.

Seeds grow on the surface of the fruit of the Wood Strawberry, naturalized from Europe.

surface. **WHERE FOUND:** Woods. Canada to VA; MO to ND. Alien (Europe). **USES:** Native Americans used root tea for stomach ailments, jaundice, profuse menses. Leaf tea for diarrhea and dysentery. In European folk medicine, leaf tea used as a blood purifier and as a diuretic for "gravel" (kidney stones). Tea also used as an external wash on sunburn. Root tea diuretic. Root used as "chewing stick" (toothbrush). Strawberries (both wild and domesticated) are rich in anticancer ellagitannins, as well as rutin, hyperoside, and chlorogenic acid. Fruit extract is anti-inflammatory. An extract of the leaves found to have a vasodilation action comparable to that of hawthorn extracts. Herbalists consider fruit to be cooling; promotes perspiration and relieves obstructions to the bladder and kidneys. Linnaeus, the eighteenth-century Swedish physician who created modern biological taxonomy, extolled its virtues in gout and pulmonary tuberculosis.

COMMON OR VIRGINIA STRAWBERRY — Leaves, root, fruits

Fragaria virginiana Duchesne — Rose Family (Rosaceae)

Generally larger than Wood Strawberry (above); leaves more rounded, tooth at tip of middle leaflet *usually shorter than adjacent teeth.* Seeds embedded within fruits; Apr.–July. **WHERE FOUND:** Fields, openings. Most of our area. Native. **USES:** Cherokee held fruit in mouth to help reduce dental tartar. Native Americans and early settlers used leaf tea as a nerve tonic. Also used to treat bladder and kidney ailments, jaundice, scurvy, diarrhea, stomachaches, and gout. Considered slightly astringent. Fresh-leaf tea used for sore throats. Berries eaten for scurvy, gout. Root tea traditionally used to treat gonorrhea, stomach and lung ailments, irregular menses; as a diuretic.

Common or Virginia Strawberry. Seeds are embedded in the fruits.

BOWMAN'S ROOT, INDIAN PHYSIC

Whole plant

Porteranthus trifoliatus (L.) Britt.
[*Gillenia trifoliata* (L.) Moench]

Rose Family (Rosaceae)

⚠ Smooth, slender perennial; 2–3 ft. Leaves alternate; divided into 3 nearly stalkless, sharp, unequal, toothed leaflets. Flowers terminal, in a loose panicle; May–July. Flowers white with a reddish tinge; petals scraggly. **WHERE FOUND:** Rich woods. ON to GA; AL to MI. **USES:** Traditionally, plant tea is strongly laxative and emetic; minute doses used for indigestion, colds, asthma, hepatitis. Poultice or wash used for rheumatism, bee stings, swellings. **RELATED SPECIES:** *P. stipulatus* (Muhl.) Britt. [*G. stipulata* (Muhl. ex Willd.) Nutt.] has prominent leaflike stipules and is used similarly and historically considered the stronger of the two. **WARNING:** Potentially toxic.

Bowman's Root, Indian Physic has long, twisted petals.

5–7 "PETALS" (SEPALS), NUMEROUS STAMENS; ANEMONES

CANADA ANEMONE

Roots, leaves

Anemone canadensis L.

Buttercup Family (Ranunculaceae)

⚠ Perennial; 1–2 ft. Basal leaves on long stalks. Stem leaves stalkless, *tightly hugging stem*. Leaves deeply divided with 5–7 lobes. Flowers white; "petals" (actually 5 showy sepals) 1–1½ in. long; May–Aug. **WHERE FOUND:** Damp meadows, thickets. NS south through New England to WV; west to IL, MO, KS; BC south to NM. **USES:** Astringent, styptic. The Chippewa used root or leaf tea (as a wash or poultice) for wounds, sores, nosebleeds. Meskwaki used root tea as eyewash to treat twitching and to cure cross-eyes. In Ojibwa traditions, the root was chewed to clear the throat to improve singing at ceremonies. Among Plains Indian groups, such as the Omaha and Ponca, the root was highly esteemed as an external medicine for many ailments, and mystical qualities were attributed to the plant. **WARNING:** Probably all our anemones contain the caustic irritants so prevalent in the Buttercup family.

LEFT: *Canada Anemone has stalkless leaves that tightly hug the stem.* ABOVE: *Pasqueflower typically has pale violet or white petals.*

PASQUEFLOWER

Whole plant

Anemone patens L. ssp. *multifida* (L.) Pritzel [*Pulsatilla patens* L. (Mill.)]

Buttercup Family (Ranunculaceae)

⚠ Perennial; 2–16 in. Silky leaves arising from root. Leaves dissected into linear segments. Showy flowers, 1–1½ in. wide; "petals" (sepals) purple or white, in a cup-shaped receptacle; Mar.–Aug. See p. 235.

THIMBLEWEED

Roots, seeds

Anemone virginiana L.

Buttercup Family (Ranunculaceae)

Perennial; 2–4 ft. Leaves *strongly veined*, with distinct stalks (not sessile). Flowers 2 or more, with greenish white, petal-like sepals (no true petals); late May–Aug. Fruit thimblelike. Seeds covered in cottony fluff. **WHERE FOUND:** Dry, open woods. ME to GA; AR to KS; north to ND. **USES:** Peter Kalm (1716–1779), who traveled to America to collect seeds and plants for Linnaeus, reported that the fluffy seeds, dipped in alcohol, were put into

Thimbleweed has strongly veined leaves with distinct stalks.

a hollowed tooth to ease pain. Roots considered expectorant, astringent, emetic. Cherokee used root decoction for whooping cough, tuberculosis, diarrhea. The Menominee poulticed root for boils. In Meskwaki traditions, in order to revive an unconscious patient, the smoke of the seeds was blown into the nostrils. To divine the truth about acts of a "crooked wife," Iroquois men slept with a piece of the root under their pillow to induce dreams.

MISCELLANEOUS FLOWERS WITH 5 PETALS

BACOPA, BRAHMI, WATER-HYSSOP, HERB-OF-GRACE — Whole plant

Bacopa monnieri (L.) Wettst. — Plantain Family (Plantaginaceae)
[*Bramia monnieri* (L.) Drake; *Lysimachia monnieri* L.] — [Formerly in the Figwort Family (Scrophulariaceae)]

Small, creeping, mat-forming perennial of wet soils. Flowers solitary at nodes, mostly purple, pale violet, or white; June–Sept. See p. 232.

MADAGASCAR PERIWINKLE — Leaves

Catharanthus roseus (L.) G. Don — Dogbane Family (Apocynaceae)
(*Vinca rosea* L.)

Short-lived perennial to 2 ft.; branching or spreading from base. Leaves opposite, short-stalked, oblong to lance-shaped, glossy or with minute hairs. Flowers typically rose-pink, white or mauve; dark color toward center; 5 showy petals spread from narrow funnel-like tube, *with small hairs surrounding tube opening.* **WHERE FOUND:** Alien (Madagascar); occasionally naturalized; a cultivated annual throughout our range. **USES:** Within 50 years of its discovery in Madagascar in 1757, Madagascar Periwinkle seeds hitched a ride with European explorers and spread throughout tropical regions of the world; it was locally adopted as a

Bacopa has small 5-petaled flowers and mat-forming, semi-succulent leaves.

Madagascar Periwinkle has white, pink, rose, or violet flowers.

medicinal plant. In Cen. and S. America, the bitter (and toxic) leaf tea was used to induce nausea for ritual cleansing; also a gargle for sore throat and laryngitis. In India, fresh juice squeezed from the leaves was used topically to treat wasp stings. Today the plant is the source of important chemical compounds used in chemotherapy. In 1950s studies on traditional use of the leaves to treat diabetes led two research groups to surprising findings—the discovery of anticancer activity from toxic alkaloids in the leaves. In 1961 vinblastine was approved to treat Hodgkin's disease. Two years later, another alkaloid, vincristine, was approved as a treatment for childhood leukemia. Since the early 1960s, tens of thousands of cancer patients, especially with leukemia or lymphomas, have benefited from these important nature-derived chemotherapeutic anticancer agents. **WARNING:** Contains toxic alkaloids; may cause skin irritation.

SPRING BEAUTY, VIRGINIA SPRING BEAUTY — Roots

Claytonia virginica L. — Montia Family (Montiaceae)
[Formerly in Purslane Family (Portulacaceae)]

Perennial with globe-shaped tubers, 6–12 in. high; leaves linear, mostly basal, *with pair of smooth, linear leaves halfway up the stem.* Flowers, 5-petaled whitish to pink, usually pinkish, rose, or pink-lavender candy-striped; Mar.–Apr. **WHERE FOUND:** Woodlands, lawns, open forests, prairies; ME, ON, south to GA; TX to MN. A common and widespread spring wildflower. **USES:** Iroquois used powdered root infused in water as an anticonvulsant in children; raw plant eaten as a contraceptive. Dried root or sliced root in water to ward off evil in the presence of witches; and to help the deceased move to other realms. Nineteenth-century physicians referred to tuberous root as "Pigroot" as it was relished by pigs. Externally used for scrofula (tuberculosis of lymphatic glands). Root eaten as a survival food; safety unknown.

Spring Beauty, with white to pink-striped flowers.

ABOVE: *Bouncing Bet, Soapwort ranges from white to rose.*
TOP RIGHT: *The leaves of the winter annual Chickweed, here with hoar frost, hold up to cold weather.*
RIGHT: *Chickweed flowers have deeply split petals.*

BOUNCING BET, SOAPWORT

Leaves, roots

Saponaria officinalis L. — Pink Family (Caryophyllaceae)

⚠ Flowers white or rose, 1 in. across; July–Sept. Petals reflexed, notched, see p. 198.

CHICKWEED

Whole plant

Stellaria media (L.) Villars — Pink Family (Caryophyllaceae)

Highly variable winter annual, prostrate, sprawling weed; 6–15 in. Leaves oval, smooth (long leafstalks are hairy). Flowers small, white; Mar.–Sept. *Petals 2-parted, shorter than sepals.* **WHERE FOUND:** Waste places. Throughout N. America. Abundant alien cosmopolitan weed. **USES:** Tea of this common Eurasian herb is traditionally used as a cooling demulcent and expectorant to relieve coughs, rheumatism; also used externally for skin eruptions and to allay itching, and as an anti-inflammatory. Still much used. Said to curb obesity. Science confirms that a methanolic (but not alcoholic) extract of the plant, which contains beta-sitosterol (similar in structure to dietary fat), produces physical competition in the gastrointestinal tract to reduce fat absorption. The extract was tested in a progesterone-induced obese laboratory model, showing promise for further research. **REMARKS:** Often used as an edible winter to early spring wild green, eaten like spinach. In folk tradition a natural barometer, with flowers remaining closed if rain is approaching, and in dry weather open from early morning to noon.

CANADA VIOLET

Root, leaves

Viola canadensis L.

Violet Family (Violaceae)

Slightly hairy to smooth perennial; to 10 in. Leaves oval to heart-shaped. Flowers white; Apr.–July. Petals yellowish at base, becoming violet-tinged at base, especially when older. **WHERE FOUND:** Mostly in northern deciduous woods. NL, NH to SC mtns., AL, AR, IA to ND, north and westward. **USES:** The Ojibwa used root tea for pain in bladder region. Root traditionally used to induce vomiting, purgative; blood purifier for skin diseases; leaves poulticed for skin abrasions, boils.

Viola *species are often used generically, rather than by the specific species.*

5-PARTED FLOWERS; NIGHTSHADE FAMILY

CAYENNE, RED PEPPER

Fruits

Capsicum annuum L.

Nightshade Family (Solanaceae)

⚠ Extremely variable smooth-stemmed or sometimes slightly hairy annual (or perennial in frost-free climates), to 30 in. Leaves lance-shaped, 1–5 in. long. Flowers mostly bright to dull white, *solitary, or rarely in pairs at leaf nodes*; 5-lobed to ½ in. across; anthers purplish; June–Nov. Fruits highly variable in size, shape, and color (red, orange, yellow, green, or purple); pungency or hotness determined by amount of capsaicin in fruits. Genetically diverse with thousands of cultivated forms available. **WHERE FOUND:** Mostly escaped from cultivation, NY, CT, to FL; LA to TX. Believed native to s. AZ and NM, cultivated varieties originating in ancient times throughout tropical America, dating back at least 9,500 years. **USES:** Worldwide one of the most commonly used spices, with an estimated 25 percent of the world's population consuming red peppers daily. The ancient Mayans used Cayenne for oral sores and inflammation of the gums. First introduced to Europe by Christopher Columbus. By 1594, in his *English Herball*, John Gerard noted that "it warmeth the stomacke, and helpeth greatly, the digestion

Cayenne, with its thousands of varieties, is the most widely consumed spice in the world.

Horse-nettle. Shown with white flowers; often pale violet.

of meates." Fruits used as a digestive stimulant, to induce sweating; hence herbalists consider the herb to be "cooling" by virtue of dispersing heat from the body. Use as a medicinal herb popularized in nineteenth-century America by Samuel Thomson (1769–1843), who used it as a stimulant to "produce a strong heat in the body" and "restore digestive powers." Pure capsaicin or its analogs are used in over-the-counter topical drug products for the treatment of pain associated with postherpetic neuralgia (persistent pain at the site of an infection that has apparently healed) and diabetic neuropathy. It is also used in the treatment of herpes zoster (shingles). Capsaicin is the subject of clinical studies for a wide range of topical applications, including psoriasis, vitiligo, intractable pruritus, post-mastectomy pain, phantom pain syndrome, postsurgical pain, sciatica, and arthritis pain, among others. **WARNING:** Cayenne is highly irritating. In manufacturing facilities where pure capsaicin is manufactured into drugs, workers wear full-body spacesuitlike protective clothing.

HORSE-NETTLE — Leaves

Solanum carolinense L. — Nightshade Family (Solanaceae)

☠ Perennial; 1–4 ft. *Stems sharp-spined.* Leaves oval to elliptical; lobed to coarse-toothed. Flowers white to pale violet stars; May–Oct. Fruits orange-yellow; Aug.–Sept. **WHERE FOUND:** Sandy soil. Old fields, farmlands, waste places. New England to FL; TX to s. SD. **USES:** Cherokee gargled wilted-leaf tea for sore throats; poulticed leaves for Poison Ivy rash; drank tea for worms. Berries fried in grease were used as an ointment for dog's mange. Properly administered, berries were historically used for epilepsy and as a diuretic, painkiller, antispasmodic, aphrodisiac. Contains a steroidal glycoside, carolinoside. **WARNING:** Toxic. Fatalities reported in children from ingesting berries.

Common Nightshade, a highly variable cosmopolitan weed, occurs throughout the world.

COMMON NIGHTSHADE Leaves, berries

Solanum nigrum L. Nightshade Family (Solanaceae)

Highly variable, smooth-stemmed to hairy branching annual; 1–2½ ft. Leaves broadly triangular or oval, irregularly toothed. Flowers are white stars with protruding yellow stamens; petals often curved back; May–Sept. Fruits are shiny black berries. **WHERE FOUND:** Waste places; cosmopolitan weed. NS to FL; local westward and spreading. Alien (Europe). **USES:** Externally, leaf-juice preparations have been used as a folk remedy for tumors, cancer. Berries formerly used as a diuretic; used for eye diseases, fever, rabies. Extracts used in tea in India, China, Europe, Japan, Africa, etc.; experimental anticancer, CNS-depressant, estrogenic, and antiviral activity. **WARNING:** Some varieties contain solanine, steroids; deaths have been reported from use. In India, some varieties are eaten as vegetables, but similar varieties may be violently toxic. Highly variable in appearance, chemistry, and quantity of toxic components. Taxonomy is unsettled. Some varieties emerging as a food source, especially in Africa.

NODDING WAXY FLOWERS WITH 5 PETALS

SPOTTED PIPSISSEWA Leaves

Chimaphila maculata (L.) Pursh Heath Family (Ericaceae)

Perennial; 4–10 in. Leaves lance-shaped, in whorls, *midrib broadly white-marked*. Flowers whitish pink, drooping, waxy; June–Aug. **WHERE FOUND:** Rich woods. ME to GA; AL, TN to MI. **USES:** Substitute for *C. umbellata* (below). Cherokee used an infusion of the leaves for colds, fevers, urinary problems; externally, a wash for ulcers, skin cancers, and rheumatism. **WARNING:** Said to be a skin irritant.

PIPSISSEWA

Leaves

Chimaphila umbellata (L.) Nutt.

Heath Family (Ericaceae)

⚠ Perennial; 6–12 in. Leaves in *whorls*; lance-shaped, toothed, shiny. Flowers whitish pink, with a ring of red anthers; drooping, waxy; June–Aug. **WHERE FOUND:** Dry woods. NS to GA; OH to MN. **USES:** Native Americans used leaf tea for backaches, coughs, bladder inflammations, stomachaches, kidney ailments; blood purifier, diuretic, astringent; drops used for sore eyes. Leaves were smoked as a tobacco substitute. A popular folk remedy of colonial America by herbalists, "Indian doctors," and the Shakers. Physicians formerly used leaf tea for bladder stones, kidney inflammation (nephritis), prostatitis, and related ailments. In Appalachia, a folk remedy for rheumatism. Science confirms antimicrobial, anti-inflammatory, antioxidant, diuretic, astringent, and urinary antiseptic activity. Loaded with biologically active compounds—arbutin,

ABOVE: *Spotted Pipsissewa has dark green leaves with distinct white veins.* TOP RIGHT: *Pipsissewa flower close-up. Note waxy surface.* BOTTOM RIGHT: *A Pipsissewa plant in flower.*

sitosterol, ursolic acid. Chimaphilin, found in the leaves, has antiseptic, anti-yeast, antibacterial, and blood-thinning activities. The constituents may be responsible for inducing skin irritation. Ursolic acid proves to be one of those "miracle aspirin" compounds, the so-called COX-2 inhibitors. A multiherbal formula in Japan with Pipsissewa as primary ingredient used for benign prostatic hyperplasia in men. **WARNING:** Leaves poulticed on skin may induce redness, blisters, and peeling. Arbutin hydrolyzes to the toxic urinary antiseptic hydroquinone.

FLOWERS WITH 6–9 PETALS; UMBRELLA-LIKE LEAVES

UMBRELLA-LEAF (NOT SHOWN) — Root

Diphylleia cymosa Michx. — Barberry Family (Berberidaceae)

⚠ Smooth perennial; 8–36 in. Two leaves on a stout stalk. Leaves *cleft, umbrella-like; each division with 5–7 toothed lobes.* Flowers white, in clusters; May–Aug. **WHERE FOUND:** Rich woods (rare); mountains. VA to GA. Too rare to harvest. **USES:** A related Chinese species (see below) is used in Traditional Chinese Medicine, but the scarcity and narrow range of the American species probably have limited the interest in medicinal use of this plant. The Cherokee used the root tea of Umbrella-leaf to induce sweating. It was considered diuretic, antiseptic, and useful for smallpox. Physicians thought its effects might be similar to those of Mayapple (see below). **RELATED SPECIES:** A closely related Chinese species, *D. sinensis* H. L. Li (not shown), contains the toxic anticancer compound podophyllotoxin. In Traditional Chinese Medicine *D. sinensis* is used for coughs, malaria, cancerous sores, snakebites, and jaundice, and is considered antiseptic. Historical and modern uses of *D. sinensis* indicate parallels in chemistry and use with Mayapple (see below). **WARNING:** Probably toxic.

TWINLEAF — Whole plant, root

Jeffersonia diphylla (L.) Pers. — Barberry Family (Berberidaceae)

Perennial; 8–16 in. Leaf broadly rounded in outline, but with 2 distinct *sinuses (notches).* Flowers white, 8-petaled; seed capsule *leathery, with distinct lidlike apex,* Apr.–May. **WHERE FOUND:** Rich woods. W. NY to MD, WV, VA, NC; TN, KY, OH to WI, IA. Too rare to harvest. **USES:** Cherokee and Iroquois used root tea for cramps, spasms, nervous excitability, diarrhea; diuretic for "gravel" (kidney stones), dropsy, urinary infections, gargle for sore throats; externally, used as a wash for rheumatism, sores, ulcers, inflammation, and cancerous sores. Used by colonial herbalists as a diuretic for edema; externally applied to sores and ulcers. **WARNING:** Probably toxic. Surprisingly little researched.

Twinleaf flower as the leaves first emerge.

Twinleaf has winglike leaves.

Note the flowers beneath leaves in fork of stem of Mayapple.

Mayapple grows in large patches in a variety of habitats.

MAYAPPLE, AMERICAN MANDRAKE — Root

Podophyllum peltatum L. — Barberry Family (Berberidaceae)

⚠ Perennial; 12–18 in. Leaves smooth, paired, *umbrella-like; distinctive.* A single waxy white flower, to 1 in. across, *droops from crotch of leaves*; Apr.–June. Fruit is globe-shaped to egg-shaped, about 2 in. long; edible but relished by rodents who collect as soon as ripe, robbing humans of a wild food; July–Aug. **WHERE FOUND:** Woods, clearings. S. ME to FL; TX to MN. **USES:** One of the most important, medical practitioner–administered American medicinal plants, both historically and today.

Native Americans, such as the Cherokee and Cree, and early settlers used roots as a strong purgative, "liver cleanser," emetic, worm expellent; for jaundice, constipation, hepatitis, fevers, and syphilis. Resin from root, podophyllin (highly allergenic), used to treat venereal warts. Extract active against herpes, influenza, and vaccinia viruses. Podophyllotoxin, an important lignan from the root, has anticancer and antimalarial activity. Two semisynthetic derivatives, etoposide and teniposide, are used in chemotherapy against several cancer types. Etoposide is used in combination with other compounds for refractory testicular cancers and as a primary treatment for small-cell lung cancer; also used for various forms of leukemia. Teniposide is used to fight childhood acute lymphocytic leukemia. Fruits edible; all other plant parts toxic. **WARNING:** Tiny amounts of root or leaves are poisonous. Powdered root and resin can cause skin and eye problems.

LOW, SHOWY SPRING FLOWERS WITH 6–10 PETALS

WINDFLOWER, RUE ANEMONE

Root

Anemonella thalictroides (L.) Spach
[*Thalictrum thalictroides* (L.) Boivin.]

Buttercup Family (Ranunculaceae)

⚠ Delicate perennial; 4–8 in. Leaves in *whorls*; small, 3-lobed. Flowers white (or pink), with 5–11 "petals" (sepals); early spring wildflower; Mar.–May. **WHERE FOUND:** Rich woods. ME to FL; AR and e. OK to MN. **USES:** Cherokee used root tea for diarrhea and vomiting. Tuberous roots considered edible. Historically, root preparation used by physicians as an experimental application to treat hemorrhoids. **WARNING:** Possibly toxic.

Windflower, Rue Anemone is a common spring wildflower.

BLOODROOT

Root

Sanguinaria canadensis L.

Poppy Family (Papaveraceae)

⚠ Perennial; 6–12 in. Juice is *orange*-red. Leaves distinctly round-lobed. Flowers white, to 2 in., with 8–10 petals; flowers appear before or with leaves; Mar.–June. Roots, a thick rhizome, *bright red within*. **WHERE FOUND:** Rich woods. NS to FL, e. TX to MB. **USES:** The blood-red fresh root was used in minute doses as an appetite stimulant, in larger doses as an arterial sedative. Formerly, root was used as an ingredient in cough medicines. Cherokee used root tea for rheumatism, asthma, bronchitis, lung ailments, laryngitis, fevers; also as an emetic. Most

TOP LEFT: *Bloodroot gets its name from the blood red color of the roots.* BOTTOM LEFT: *Bloodroot leaves are distinctly lobed.* ABOVE: *Bloodroot's poppylike flowers appear before the leaves in early spring.*

native tribes within its range used Bloodroot in combination with other herbs. Root juice is applied to warts; also used as a dye and as a decorative skin stain, also to color baskets, clothing, etc. A bachelor of the Ponca tribe would rub a piece of the root as a love charm on the palm of his hand, then scheme to shake hands with the woman he desired to marry. After shaking hands, the girl would be found willing to marry him in 5–6 days. One of the most widely used American medicinal plants in colonial America, which C. S. Rafinesque aptly describes in his *Medical Flora* (1830, vol. 2, p. 80): "Few medicinal plants unite so many useful properties; but it requires to be administered with skillful hands, and may become dangerous in empirical [amateur] hands." Experimentally, the alkaloid sanguinarine has shown anti-inflammatory, antioxidant, antiseptic, anesthetic, anticancer, and immunomodulatory activity. It was used commercially as a plaque-inhibiting agent in toothpaste, mouthwashes, and rinses from the 1990s to early 2000s, but has since disappeared from world markets as a dentifrice agent, possibly because of its potential to produce a precancerous oral lesion. Ingestion of the root can cause a positive opiate test in urinalysis. Salves containing Bloodroot are widely touted on the Internet for treating topical cancers, but are not approved for such use. **WARNING:** Toxic. Do not ingest. Jim Duke has experienced tunnel vision from nibbling the root. Sanguinarine said to cause glaucoma.

The Sharp-lobed Hepatica flowers range from white to deep blue-purple.

Sharp-lobed Hepatica has pointed leaf tips.

SHARP-LOBED HEPATICA, LIVERLEAF

Leaves

Hepatica nobilis var. *acuta* (Pursh) Styerm.
[*Anemone acutiloba* (DC.) G. Lawson; *A. hepatica* var. *acuta* (Pursh) Pritz.; *Hepatica acutiloba* DC.]

Buttercup Family (Ranunculaceae)

Flowers usually bluish lavender or pinkish, though often whitish; Feb.–early June. See p. 236.

ROUND-LOBED HEPATICA

Leaves

Hepatica nobilis var. *obtusa* (Pursh) Styerm.
[*Hepatica americana* (DC.) Ker.; *H. acutiloba* DC.; *Anemone americana* (DC.) H. Hara; *A. hepatica* var. *acuta* (Pursh) Pritz.]

Buttercup Family (Ranunculaceae)

Similar to Sharp-lobed Hepatica, but leaf lobes are rounded. Flowers are mostly white; Mar.–June. **WHERE FOUND:** Dry woods. NS to GA, AL; MO to MB. **USES:** Same as for *H. acutiloba* (see above). See p. 236. **REMARKS:** No modern taxonomic work is in agreement on the current taxonomic position and placement of *Hepatica*. Therefore, we continue to use the taxonomy from the second edition of this work.

Round-lobed Hepatica usually has white flowers.

The leaves of the Round-lobed Hepatica usually have rounded tips.

FLOWERS IN SINGLE, GLOBE-SHAPED CLUSTERS; FRUITS RED

RED BANEBERRY

Root

Actaea rubra (Ait.) Willd.
(*Actaea spicata* L. var. *rubra* Ait.)

Buttercup Family (Ranunculaceae)

⚠ Perennial; 2–3 ft. Similar to White Baneberry (see p. 70), though the flowerhead is pyramidal, usually as wide as long, and the berries are *red* (often white) and on less stout stalks. Flowering April–May. Fruits; July–Oct. **WHERE FOUND:** Deciduous forest and mixed coniferous forests. S. Canada to n. NJ, WV; west through OH and IA to SD, CO, UT, and OR, northward. **USES:** Among the Algonquin, Ojibwa, Cheyenne, root tea used for stomach pain. The Blackfoot used a root tea for coughs and colds. Many native tribes used the root decoction for menstrual irregu-

ABOVE: *Red Baneberry has white globular clusters of flowers.* RIGHT: *Red Baneberry produces fleshy red fruits.*

larity, postpartum pains, and as a purgative after childbirth; also widely used to treat rheumatism. Sometimes root appears as an adulterant to Black Cohosh root. Chemical analytical methods have been established to distinguish this and other Black Cohosh adulterants, including other American and Asian species of *Actaea.* Red Baneberry contains 11 triterpene compounds, possibly responsible for traditional uses. **WARNING:** Plant is potentially poisonous—may cause vomiting, gastroenteritis, irregular breathing, and delirium.

GOLDENSEAL

Root

Hydrastis canadensis L. Buttercup Family (Ranunculaceae)

Hairy perennial; 6–12 in. Usually *2 leaves on a forked branch; one leaf larger than the other*; each is rounded with 5–7 lobes; double-toothed. Flowers are single with greenish white stamens in clusters; Apr.–May. Berries like those of raspberry. **WHERE FOUND:** Rich woods. VT to GA; AL, AR to MN. Goldenseal is becoming less common in our eastern deciduous forests because of overcollection, hence it is now monitored in international trade. **USES:** Root traditionally used in tea or tincture to treat inflamed mucous membranes of mouth, throat, digestive system, uterus; also used for jaundice, bronchitis, pharyngitis, gonorrhea. Tea (wash) a folk remedy for eye infections. Until the 1980s, components of the root were used commercially in eyewash preparations. Contains berberine, an antibacterial agent that increases bile secretion and acts as an anticonvulsant. Experimentally, berberine lowers blood pressure and acts as mild sedative. The root also contains the alkaloids hydrastine and canadine. Recent research suggests a synergistic action of the major alkaloids in reducing muscle-spasm activity. When any single alkaloid was removed from the mix, the effect was greatly lessened.

LEFT: *Goldenseal leaves in a deep, wooded ravine.* ABOVE: *Goldenseal in flower.*

Goldenseal in fruit.

Goldenseal root, showing bright yellow interior.

A 2011 study found that flavonoids in the plant synergistically act with alkaloids to enhance antimicrobial activity. More of the flavonoids are found in the leaf than the root, suggesting that a whole plant extract (including leaf and root) may be optimal, contrary to conventional beliefs. One study suggests that Goldenseal may help fight drug-resistant tuberculosis, one of today's most lethal infectious bacterial diseases. Science confirms liver-protectant activity. **WARNING:** Avoid during pregnancy. Based on the plot of a 1901 novel, *Stringtown on the Pike,* by the pharmacist John Uri Lloyd, Goldenseal's popularity is in part stimulated by the notion that it masks detection of illicit drugs in urinalysis. This has prompted drug-testing labs to test for Goldenseal alkaloids in the urine, prior to screening for drugs, to avoid the possibility of false negative findings. Scientists have disproved rumor that Goldenseal masks morphine in urine tests.

FLOWERS IN RADIATING CLUSTERS; RIPE FRUITS NOT RED

WHITE BANEBERRY, DOLL'S EYES — Root

Actaea pachypoda Elliot. — Buttercup Family (Ranunculaceae)
[*Actaea alba* (L.) Mill.]

⚠ Perennial; 1–2 ft. Leaves twice-divided; leaflets oblong, sharp-toothed. Flowers in oblong clusters on thick red stalks. *Fleshy white berries with a dark dot at tip*; July–Oct. The dark dot at the tip of the white berry earns this plant the common name Doll's Eyes. **WHERE FOUND:** Rich woods. S. Canada to GA, LA; OK to MN. **USES:** Menominee used small amounts of root tea to relieve pain of childbirth, and headaches due to eye strain. Cherokee used root tea as gargle for sore throat; root tea to attempt to revive patient near death. Once used for coughs, menstrual irregularities, colds, and chronic constipation; thought to be beneficial to circulation. Appearing in commercial supplies of Black Cohosh as an adulterant or mistakenly harvested. Both plants occur in the same hab-

LEFT: *White Baneberry, with elongated white flower clusters.*
ABOVE: *White Baneberry produces distinct white, dry, fleshy fruit on thick red stalks.*

itat, often with numerous sterile (nonflowering) plants. This and other *Actaea* species can be distinguished by their chemical differences. **WARNING:** Potentially toxic; all parts may cause gastrointestinal inflammation and dermatitis; stem and leaf juice can be toxic.

INDIAN HEMP

Root, stems, latex

Apocynum cannabinum L.

Dogbane Family (Apocynaceae)

⚠ Shrublike herbaceous perennial; 1–4 ft., stems often reddish. Leaves (except lowermost ones) with definite stalks, to ½ in. long, broadly oval to elliptical, variable in size and shape. Flowers terminal, whitish green, bell-like, *5-sided, petals and sepals of nearly equal length,* less than ¼ in. long; June–Aug. Seedpods paired; 4–8 in. long. **WHERE FOUND:** Wide-ranging habitats from bottomlands to upland forest, fencerows, ditches, disturbed areas. Much of our area and beyond. **USES:** Used as in *A. androsaemifolium* (p. 204); also,

Indian Hemp flowers in late summer.

stems used for fiber, particularly for paper; cordage; harvested at fruiting, it makes a fine strong thread. Milky sap is a folk remedy for venereal warts. Cherokee used root for uterine obstructions, rheumatism; as a diuretic, also for coughs, asthma, and whooping cough. Many native tribes used root for cardiac edema, and as cardiac stimulant. Kiowa squeezed latex from plant as a chewing gum, allowing it to harden overnight. **WARNING:** Considered toxic. Contains toxic cardioactive (heart-affecting) glycosides. Cymarin and apocymarin have shown antitumor activity; the latter also raises blood pressure. Latex can be caustic; was once used for ringworm, leprosy, and to remove hair from animal hides. Treat with caution.

WHORLED MILKWEED — Root, leaves

Asclepias verticillata L. — Milkweed Family (Asclepiadaceae)
[Modern works place in Dogbane Family (Apocynaceae)]

⚠ Slender, sparsely branched perennial with milky sap, to 2½ ft. tall. Soft-hairy, threadlike leaves, about 2 in. long, *paler beneath, margins curving beneath*, in whorls of 5–6, along the stems. Flowers center in 2–14 loose clusters, with fewer than 20 flowers per cluster; petals strongly reflexed greenish white, with 5 white hoods; May–Sept. **WHERE FOUND:** Fields, pastures, dry prairies, dry open woods, glades, and roadsides; s. VT, MA to FL, west to AZ, north to SK. **USES:** Choctaw considered the root their most valuable snakebite remedy; used in South (Civil War era) as a domestic remedy for snakebites and venomous insect stings. Root tea used to induce sweating, and as a stimulant. Root also chewed,

Whorled Milkweed has small flowerheads in loose clusters.

Virginia Waterleaf has white to violet flowers.

swallowing saliva to deliver benefits. Lakota women used plant tea to induce milk flow. Thirteen pregnane glycosides are reported from the plant; tested against breast cell cancer test system for potential anticancer activity with negative results. **WARNING:** Leaves as forage reportedly potentially toxic to sheep and cows.

VIRGINIA WATERLEAF
Whole plant

Hydrophyllum virginianum L. Waterleaf Family (Hydrophyllaceae)
[Modern works place in Borage Family (Boraginaceae)]

Flowers whitish to violet. See p. 238.

TINY FLOWERS IN GLOBE-SHAPED CLUSTERS OR PANICLES; GINSENG FAMILY; MOSTLY HERBACEOUS; FRUITS RED OR BLACK

HAIRY SARSAPARILLA
Root, leaves

Aralia hispida Vent. Ginseng Family (Araliaceae)

Hairy Sarsaparilla occurs in dry woods.

Shrubby; 1–3 ft. Foul-smelling. Stem with *sharp, stiff bristles*. Leaves twice-compound; leaflets oval, cut-toothed, *wedge-shaped at base*. Small greenish white flowers in *globe-shaped umbels*; June–Aug. Fruits are dark, foul-smelling berries. **WHERE FOUND:** Sandy open woods. E. Canada, New England south to VA, WV, west to IL, MN. **USES:** Potawatomi used as tonic and blood purifier. Leaf tea promotes sweating, increasing circulation. Bark (root bark especially) diuretic, "tonic"; allays kidney irritation and associated lower back pain, increases secretions in dropsy and edema. **RELATED SPECIES:** Genetically similar to the East Asian species *A. cordata* Thunb.

WILD SARSAPARILLA
Root, berries, leaves

Aralia nudicaulis L. Ginseng Family (Araliaceae)

Smooth perennial, forming large colonies; to 2 ft. Leaves twice-divided; each of the 3 divisions has 3–5 toothed, oval leaflets. Flowers in a single umbel on a separate stalk, *below leaves*; Apr.–July. Root is long-running, horizontal, fleshy, and sweetly aromatic. **WHERE FOUND:** Moist woods. NL to GA; west to n. MO, IL, ND to CO and ID. **USES:** Widely used by Native American groups throughout its range; a pleasant-flavored root as beverage tea, blood purifier, tonic; used for lassitude, general

ABOVE: *Wild Sarsaparilla is one of the most common understory plants in New England forests.*
RIGHT: *Spikenard is usually a relatively large, spreading woodland plant.*

debility, stomachaches, and coughs; blood strengthener. Externally, the fresh root, bruised, chewed, or poulticed on sores, burns, itching, ulcers, boils, and carbuncles; also used to reduce swelling and cure infections. Roots eaten on treks for war expeditions; roots and berries fermented as a beer. Roots and leaves chewed and applied to heal wounds in colonial America. In folk tradition, root tea or tincture was used as a diuretic and blood purifier; promotes sweating; used for stomachaches, fevers, coughs. Root poultice used for wounds, ulcers, boils, carbuncles, swelling, infection, rheumatism. Former substitute for true (tropical *Smilax*) sarsaparilla. Berries were used as a cordial for the treatment of gout. Little used today.

SPIKENARD — Root

Aralia racemosa L. — Ginseng Family (Araliaceae)

Perennial; 3–5 ft. Stem *smooth; dark green or reddish*. Leaves compound, with 6–21 toothed, *weakly heart-shaped* leaflets. Flowers whitish, in small umbels on branching racemes; May–Aug. Root spicy-aromatic. **WHERE FOUND:** Rich woods. QC to GA; west to KS; north to MN. **USES:** Same as for *A. nudicaulis* (above); also used for coughs, asthma, lung ailments, and rheumatism, syphilis, and kidney troubles. Formerly used in cough syrups. Root tea widely used by Native Americans for menstrual irregularities, for lung ailments accompanied by coughs, and to improve the flavor of other medicines. Externally, root poulticed on boils, infections, swellings, and wounds. Contains saponins, possibly with expectorant action.

AMERICAN GINSENG Root

Panax quinquefolius L. Ginseng Family (Araliaceae)
(*Panax quinquefolium* L., a common misspelling)

Perennial; 1–2 ft. Root fleshy, sometimes *resembling human form.* Leaves palmately divided into 4–5 (occasionally 3–7) sharp-toothed, oblong-lance-shaped leaflets. Flowers whitish, in round umbels; May–July. Fruits are 2-seeded red berries. The rhizome at the top of the root, often referred to as the neck, reveals annual scars left by the year's leaf stem. The age of the root, which affects quality and price, is determined by counting the leaf scars. Only roots five years old or older are desirable (active components—ginsenosides—increase significantly between the fourth and fifth year of growth). **WHERE FOUND:** Rich woods. ME to GA; OK to MN. Wild root is becoming increasingly scarce because of dramatic price increases in the past 200 years. Interstate commerce of the root is regulated by the federal government. It is unethical (and illegal) to harvest the roots before the red berries ripen and set seed; this occurs in late summer or early autumn. In a 5-year transplant population, Duke found that approximately 10 percent of plants did not emerge aboveground each growing season, but remained dormant. **USES:** Widely used by Native Americans: Cherokee used root as tonic, chewed for colic, tea for nervous conditions, vertigo, headache. Iroquois used root tea to induce childbirth, improve appetite, stop vomiting, fainting, and as a tonic. Meskwaki used to make other medicine more powerful. Menominee used as a tonic and to increase mental powers. Penobscot used to increase female fertility. Root considered demulcent, tonic, adaptogenic. Research suggests it may increase mental efficiency and physical performance and aid in adapting

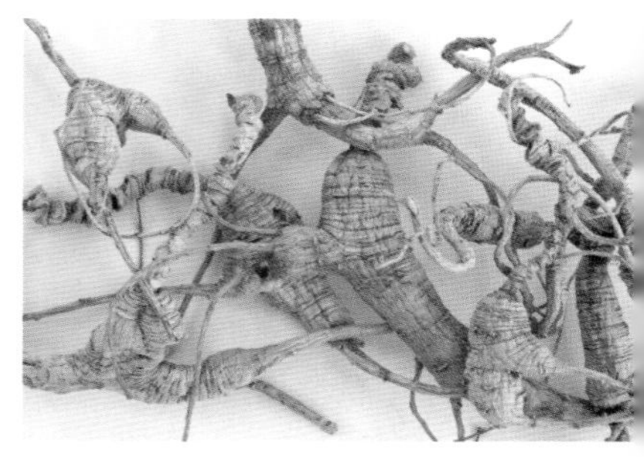

LEFT: *The bright red berries of American Ginseng appear in late summer.* ABOVE: *American Ginseng's root is one of the highest-priced products from American forests.*

to high or low temperatures and stress (when taken over an extended period). Ginseng's effect is called "adaptogenic"—tending to return the body to normal while increasing resistance to adverse influences on the body. Ginseng has long been considered an aphrodisiac. A recent study explains this effect via nitric oxide release in the corpus cavernosum. May also possess cancer-preventive, antidiabetic, and immunostimulant (cold prevention) activity. Asian ginseng (*Panax ginseng* C. A. Meyer) is widely recognized as a tonic in times of fatigue, debility, declining work capacity and concentration, and during convalescence.

DWARF GINSENG — Leaves, root

Panax trifolius L. — Ginseng Family (Araliaceae)
(*Panax trifolium* L., a common misspelling)

Globe-rooted perennial; 2–8 in. Leaves divided into 3 (occasionally 5) *toothed, oblong to lance-shaped* leaflets. Flowers white to yellow (or pinkish), in small umbels; Mar.–May. Fruits green or yellow. **WHERE FOUND:** Rich woods. NS to PA, GA mtns., IN, IA to MN. In a transplant experiment, Duke has confirmed that plants may have male flowers one year and female flowers the next year and vice versa. **USES:** Cherokee chewed root for shortness of breath, coughs, colic; tea for rheumatism, nervous debility, gout. The Iroquois rubbed root tea on the arms and legs of lacrosse players in ritual to force opposing players to lose the ball. Sliced root placed in cup of water with fishing tackle overnight; root chewed while fishing to attract fish. Little used or researched. Aboveground for only 6–8 weeks. Contains at least four ginsenosides, though not thoroughly characterized.

Dwarf Ginseng is a seldom-used ginseng.

NUMEROUS SMALL WHITE FLOWERS IN LOOSE CLUSTERS

GOAT'S BEARD — Root

Aruncus dioicus (Walter) Fern. — Rose Family (Rosaceae)

⚠ Shrublike; 4–6 ft. Leaves mainly basal; divided into large, serrated, oval leaflets. Tiny, white to yellowish flowers. See p. 150.

ABOVE: *Goat's Beard occurs in woods throughout much of the Northern Hemisphere.* RIGHT: *Buckwheat is a weak-stemmed, white-flowered annual.*

BUCKWHEAT

Leaves, seeds

Fagopyrum esculentum Moench — Buckwheat Family (Polygonaceae)

Annual; 1–4 ft.; stems weak, somewhat succulent. Leaves broadly triangular to arrow-shaped, to 3 in. long. Flowers white or pink, small, loosely crowded together in drooping terminal inflorescence; May–Sept. Fruits are the familiar triangular buckwheat of commerce, a smooth, shining achene; July–Oct. **WHERE FOUND:** Near cultivated ground. Alien. Escapes from cultivation. Occasional throughout. **USES:** Tea of the leaves used to treat erysipelas. Poultice of the powdered seeds used in buttermilk as an English folk remedy to stimulate milk flow. Plant is high in rutin, which reduces capillary fragility. In European traditions, the herb has been used to treat venous and capillary problems, as it increases tone of veins, and is taken to prevent hardening of the arteries. A controlled clinical study evaluated a buckwheat herb tea in the treatment of chronic venous insufficiency; results were that it reduced edema (water retention), improved blood flow through the femoral vein, and enhanced capillary resistance. Buckwheat is an antioxidant. **RELATED SPECIES:** *F. tataricum* (L.) Gaertn., also a commercial source of buckwheat, has about 50 times more rutin in its seeds and higher nutritional value. Both have been cultivated as food plants for at least 6,500 years.

JAPANESE KNOTWEED, MEXICAN BAMBOO, GIANT KNOTWEED — Roots, leaves

Fallopia japonica Houtt. (Ronse Decr.) — Buckwheat Family (Polygonaceae)
(*Polygonum cuspidatum* Siebold & Zucc.; *Reynoutria japonica* Houtt.)

⚠ Tall, weedy perennial, with *hollow stems and swollen leaf nodes* to 8 ft. Leaves alternate, entire, broadly elliptical to round to 5 in. long, with abrupt tip. Flowers small, whitish green, in terminal often-branched clusters; June–Sept. Fruits distinctly winged. **WHERE FOUND:** Widespread in e. Asia. Introduced from Japan as a garden ornamental in the nineteenth century, now naturalized and an aggressive noxious weed in some states. Throughout our range. **USES:** Cherokee found the leaves edible. In rural China, spring shoots are eaten like asparagus. The rhizomes and root, called *hu zhang* in Traditional Chinese Medicine, are used in prescriptions to build and detoxify the blood and reduce swelling in cases of dysentery, leukorrhea, headache; externally for mastitis and wounds. Science confirms skin collagen promotion, and anti-inflammatory activity relative to wound healing. Components in the root have antibacterial, antiviral, antioxidant, analgesic, and anti-inflammatory activity; cardiovascular protective effects, anticancer potential, among others. Rich in anti-oxidant resveratrol (best known from grape seeds and skins). **WARNING:** May interfere with enzyme system that affects the absorption of some prescription drugs (notably carbamazepine, used in the treatment of epilepsy and bipolar disorder), thereby increasing the effective dose of a drug with a narrow therapeutic window.

MEADOWSWEET, QUEEN OF THE MEADOW — Flowers

Filipendula ulmaria (L.) Maxim. — Rose Family (Rosaceae)
(*Spiraea ulmaria* L.)

⚠ Perennial; 3–7 ft. tall. Leaves with 2–5 pairs of lateral leaflets, deeply divided into 3–5 ovate to oblong, double, sharply serrated leaflets in 2–5 pairs; ½–2 in. long; the terminal leaflet somewhat rounded. Flowers white, small, mostly 5-parted in showy large panicles, *both terminal and lateral*; June–Aug. **WHERE FOUND:** Occasional in fields, near homes; escaped from cultivation. PE to NJ, WV; KY, IL to ON, QC. Alien. Eurasia; widespread in Europe, sw. Asia, Russia to n. China. **USES:** Traditionally in European folk medicine, the fresh-gathered flowering tops were used to promote sweating and as an astringent for fevers, diuretic; externally poulticed for wounds, tea also taken internally to allay bleeding. The flowers infused in alcohol impart a pleasant slight wintergreen-like flavor. In European phytomedicine, used for dyspepsia, heartburn with hyperacidity, rheumatic pains, and prevention and treatment of peptic ulcers. Contains chemical forerunners of aspirin. Polyphenols in the plant have anti-inflammatory, antioxidant, and antibacterial activity. Salicin, the popular analgesic derived from poplars and willows, probably decomposes in the digestive tract to salicylic acid, a compound first isolated from Meadowsweet flower buds in 1839. The semisynthetic acetyl-salicylic acid (aspirin) is said to have fewer side effects than the natural compound from which it is derived. Still,

ABOVE: *Japanese Knotweed is a large, weedy plant from Asia with succulent hollow stems and swollen nodes, earning it the common name "bamboo" locally in n. New England.* RIGHT: *Meadowsweet has flowers in both terminal and lateral panicles.*

nonsteroidal anti-inflammatory drugs, including aspirin, account for 10,000–20,000 deaths per year. Probably all medicines, natural and synthetic, have side effects. **WARNING:** Like aspirin, salicin-containing herbal medicines can thin the blood and cause internal bleeding.

DROPWORT, COMMON MEADOWSWEET — Leaves

Filipendula vulgaris Moench — Rose Family (Rosaceae)

⚠ Similar to *F. ulmaria* above, generally a much smaller plant to less than 3 ft. Leaves divided into 10–25 pairs of leaflets, to ½ in. long. Flowers white, usually with 6 petals *in flatter terminal clusters*; May–July. **WHERE FOUND:** Fields, near homes; escaped from cultivation mostly in the Northeast; New England, northward. Alien. Eurasia. **USES:** Similar to *F. ulmaria.* Used since ancient times in Europe, though neglected by the nineteenth century. Distilled water from plant was once believed an antidote for the plague. Used in folk medicine for fevers, coughs, colds, sore throat, headache, and externally as a wash for itching and burning of eyes, relieving sunburn, and reducing freckles. Polyphenols have been found to be antibacterial against *E. coli.* A folk remedy for nervousness; experimentally, extracts have been shown to have anxiolytic (anti-anxiety) activity surpassing that of Valerian extracts. Leaf essential oil high in salicylaldehyde (68 percent) and has strong antimicrobial activity. **WARNING:** Like aspirin, salicylate-containing herbal medicines can thin the blood and cause internal bleeding.

Dropwort has flowers in flattened terminal clusters.

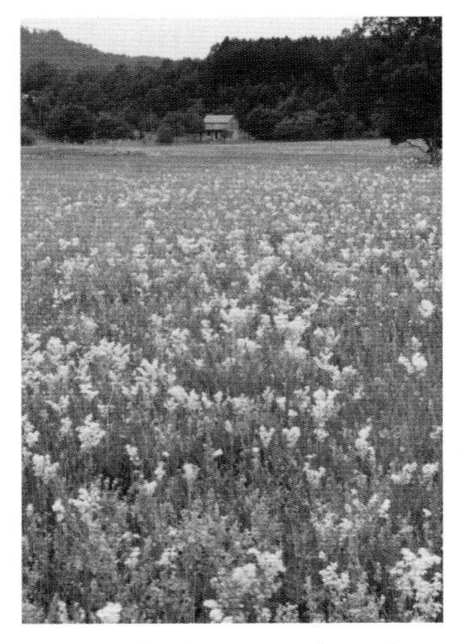

A field with a large, typical population of Dropwort.

FLOWERS IN LONG, SLENDER, TAPERING CLUSTERS

BLACK COHOSH Root

Actaea racemosa L. Buttercup Family (Ranunculaceae)

[*Cimicifuga racemosa* (L.) Nutt.]

Perennial; 3–8 ft. Leaves thrice-divided; sharply toothed; terminal leaflet 3-lobed, middle lobe largest. Flowers white, in wandlike long spikes; May–Sept. *Tufts of stamens conspicuous.* **WHERE FOUND:** Rich woods. MA, NY to GA; AR, MO to WI. Increased modern use could pressure wild populations. **USES:** Cherokee used the root (infused in ethanol) for the treatment of rheumatism and to stimulate menstruation; tea for colds, coughs, rheumatism, to relieve back pain, and as a sleep aid; effects considered pain-relieving and diuretic. Cree used root externally as liniment for rheumatism. Iroquois used root tea to stimulate milk flow; relieve pain of backache and rheumatism; steeped in cider for arthritis. Also a snakebite remedy (called Black Snakeroot). As early as 1801 Benjamin Smith Barton observed that Native Americans placed high value on the plant (called "squawroot") used for female maladies. By 1832 it emerged as an important remedy of physicians under obscure names, including Macrotys, Macrotrys, and Blacksnake Root; valued to improve appetite and digestion, as a mild sedative, and a primary treatment for rheumatism and neuralgia, muscular pain, dull headache, and a wide range of female conditions. By the mid-1950s in Germany, extracts emerged as an alternative to hormone replacement therapy

LEFT: *Black Cohosh flowers are in tall, showy wands.* ABOVE: *Black Cohosh leaves are difficult to distinguish from those of Baneberry.*

for menopause symptoms. Now one of the most widely researched American plants, most studies have focused on its benefits, chemistry (triterpene glycosides), and possible mechanism of action for menopausal symptoms, often with mixed results. Its complex chemistry defies specific attribution as it relates to clinical results. Is emerging as one of the most important American medicinal plants for the twenty-first century: several clinical studies have confirmed that root extracts are useful in the treatment of menopausal symptoms, particularly hot flashes. It is approved in Germany for treatment of premenstrual symptoms, painful or difficult menstruation, and to reduce severity of menopausal symptoms. Prescribed by European gynecologists for more than 50 years, with clinical experience in millions of patients. A recent study suggests that evidence from the historical literature coupled with recent studies showing analgesic activity points to the need for more research on the root's potential to relieve pain. Most modern research has assumed a hormonal action, based on use for PMS, menstrual pain, and menopausal symptoms. Those assumptions should be revisited in light of ethnobotanical and historical literature, along with new evidence pointing to its potential analgesic and antinociceptive effects. **WARNING:** Avoid during pregnancy.

POKEWEED, POKE

Root, fruits, leaves

Phytolacca americana L. — Pokeweed Family (Phytolaccaceae)

Coarse, large-rooted perennial; 5–10 ft. *Stem often red at base.* Leaves large, entire (toothless), oval. Flowers with greenish white, petal-like sepals; July–Sept. Fruits *purple-black*, in drooping clusters; year-round depending upon location; spring in South; through Oct. northward.

Pokeweed, Poke is usually considered poisonous.

Pokeweed has a large taproot; stem is dark red.

WHERE FOUND: Waste places. Much of our area (except ND, SD). **USES:** Cherokee used the berries in tea, or infused in wine to treat rheumatism and arthritis; greens cooked in two waters were used to build the blood; root tea used for kidney conditions and to build blood; externally in salve for cancer and ulcerated sores. Iroquois poulticed root for sprains, bruises, and swollen joints; externally, root salve used for bunions. Berries used as a dye. Old-timers in the Ozarks still eat one Poke berry a year as a preventive or to treat arthritis. Root is a folk remedy poulticed for rheumatism, neuralgic pains, bruises; wash used for sprains, swellings; leaf preparations once used as an expectorant, emetic, cathartic; poulticed for bleeding, pimples, blackheads. Widely used by nineteenth-century medical practitioners; root considered emetic and

Pokeweed's fruits are purple-black, in drooping clusters.

ABOVE: *Note the curved spikes of flowers, giving it the name Lizard's-tail.* RIGHT: *The root of Culver's-root was formerly used as a strong laxative.*

purgative; various preparations used for rheumatism and arthritis. French and Portuguese added berries to wine, until prohibited by royal ordinance of King Louis XIV "on pain of death, as it injured the flavor!" Dr. Benjamin Rush (1746–1813), a colonian physician and signer of the Declaration of Independence, reported that Yale students experienced severe diarrhea after eating pigeons that had fed on Poke berries. Plant contains a highly toxic Pokeweed mitogen (a mixture of glycoprotein lectins), investigated for anticancer, antiviral, and anti-HIV potential, as powerful immunostimulant and antifungal. Other toxins found in all plant parts. **WARNING:** All parts are poisonous, though leaves are eaten as a spring green, after cooking through 2 changes of water, still causing unpredictable toxic reactions. (Do not confuse with American White Hellebore, *Veratrum viride*, pp. 144 and 278, which is highly toxic.) Plant juice of Pokeweed can cause dermatitis, even damage chromosomes.

LIZARD'S-TAIL, WATER-DRAGON — Root, leaves

Saururus cernuus L. — Lizard's-tail Family (Saururaceae)

Aquatic or wet-ground–loving perennial. See p. 22.

CULVER'S-ROOT — Root

Veronicastrum virginicum (L.) Farw. — Plantain Family (Plantaginaceae)
(*Leptandra virginica* (L.) Nutt.) [formerly Figwort Family (Scrophulariaceae)]

Perennial; 2–5 ft. Leaves lance-shaped, toothed; in *whorls of 3–7*. Flowers tiny white (or purple) tubes with 2 projecting stamens; on showy spikes; June–Sept. **WHERE FOUND:** Moist fields, upland woods, prairies. VT to FL; e. TX to MB. **USES:** Cherokee, Iroquois, and other native groups used controlled dosages of root tea as a strong laxative, to induce sweating, to stimulate liver, to induce vomiting; diuretic. Used similarly by physicians. **RELATED SPECIES:** A Chinese species pair, *Veronicastrum sibericum* (L.) Pennell, contains immunosuppressive diterpene compounds, along with anti-inflammatory and analgesic compounds. **WARNING:** Traditionally, dried root is used; fresh root violently laxative. Potentially toxic.

FLOWERS IN UMBRELLA-LIKE CLUSTERS (UMBELS); 5 TINY PETALS; LEAVES FINELY DISSECTED

CARAWAY

Seeds

Carum carvi L. Parsley Family (Apiaceae or Umbelliferae)

⚠ Smooth biennial; 1–2 ft. Stem hollow. Leaves finely divided, *carrotlike*. Flowers tiny, white (or pink); May–July. Seeds slightly curved, ribbed, caraway-scented. **WHERE FOUND:** Fields. Scattered throughout our area; generally absent from South. Alien. **USES:** Seed tea carminative, expectorant. Relieves gas in digestive system, soothing to upset stomach; also used for coughs, pleurisy. Thought to relieve menstrual pain, promote milk secretion. Externally, used as a wash for rheumatism. Oil is antibacterial. Caraway seed, the seed on most rye breads in this country, is one of the best sources of limonene, with potential for preventing and treating breast cancer. Use of seeds is approved in European phytomedicine for mild spasms of the gastrointestinal tract accompanied by bloating and a feeling of fullness. **WARNING:** Young leaves are very similar to those of Fool's Parsley and Poison Hemlock.

Caraway fruits have a familiar caraway fragrance. Well known as a culinary herb whose fruits (seeds) are used to flavor breads and baked goods. It is less well known as a medicinal plant.

Poison Hemlock, with white umbels. It is highly toxic.

Poison Hemlock has spotted stems. It is highly toxic.

POISON HEMLOCK

Poison—Identify to avoid

Conium maculatum L.

Parsley Family (Apiaceae or Umbelliferae)

☠ Branched perennial; 2–9 ft. Stems hollow, *grooved, purple-spotted*. Leaves carrotlike, but in overall outline are more like an *equilateral triangle*, and with more divisions. Leaves *ill-scented when bruised*. Leafstalks hairless. Flowers white, in umbels; Apr.–Aug. **WHERE FOUND:** Waste ground. Widespread and common, most of our area. Invasive alien. **USES:** A famous mortal poison from remote antiquity, well known as the fatal poison of Socrates; once used in Athens, Greece, as a mode of execution. Whole plant a traditional folk remedy for cancer, narcotic, sedative, analgesic, spasmolytic, anti-aphrodisiac. **WARNING:** Deadly poison. Ingestion can be lethal. Contact can cause dermatitis. Juice is highly toxic. Young Poison Hemlock plant closely resembles the western Osha root, Wild Carrot, and various members of the Parsley family. In recent years, most deaths attributed to Poison Hemlock result from mistaken identity of the plant with other members of the Parsley family.

QUEEN ANNE'S LACE, WILD CARROT

Root, seeds

Daucus carota L.

Parsley Family (Apiaceae or Umbelliferae)

Bristly stemmed biennial; 2–4 ft. Leaves finely dissected. Flowers in a flat cluster, with 1 *small deep purple floret at center*; 3-forked bracts beneath; Apr.–Oct. **WHERE FOUND:** Waste places, roadsides. Common throughout our area. Alien (Europe). **USES:** Root tea traditionally used as a diuretic, to prevent and eliminate urinary stones and worms.

Queen Anne's Lace has a purple floret in the center of the umbel.

Valerian has white to pale lavender flowers. It is widely naturalized in the Northeast.

Science confirms its antioxidant, bactericidal, diuretic, hypotensive, and worm-expelling properties. Seeds a folk "morning after" contraceptive. Experiments with mice indicate that seed extracts may be useful in preventing implantation of fertilized egg. (Not recommended for such use.) Recent studies indicate a cancer-preventive effect associated with the root, and anticancer activity is attributed to components in the seeds. The familiar garden carrot is a cultivated race of this species. Those harvesting wild carrot roots expecting a vegetable akin to garden carrots will meet disappointment. **WARNING:** May cause dermatitis and blisters. Proper identification is essential so that it is not confused with one of the deadly poisonous members of the Parsley family (Apiaceae or Umbelliferae), such as Poison Hemlock.

VALERIAN — Root

Valeriana officinalis L. — Honeysuckle Family (Caprifoliaceae) [Formerly in Valerian Family (Valerianaceae)]

See p. 192.

UMBRELLA-LIKE CLUSTERS; STEMS PURPLISH OR PURPLE-TINGED; LEAF SEGMENTS NOT FERNLIKE

ANGELICA — Leaves, root, seeds

Angelica atropurpurea L. — Parsley Family (Apiaceae or Umbelliferae)

⚠ Smooth biennial or short-lived perennial (dying after fruiting); stem is smooth *purple or purple-tinged*, hence the species name *atropurpurea* (dark purple); 4–9 ft. Leaves with *3 leaflets, each divided again 3–5 times*. Upper leafstalks have *inflated sheaths*. Flowers in large, semiround,

Angelica has dark purple stems.

or globe-shaped heads, up to 6 in. across with numerous tiny white flowers radiating from 25–25 small stalks; May–Aug. **WHERE FOUND:** Rich, wet soil. NL to DE, WV; IL to WI. **USES:** Iroquois used a root tea to revive patient with exposure to freezing temperatures; smashed root boiled for rheumatism, tonic for female weakness, colds, fever, flu, pneumonia, chills, rheumatism, and blood purifier. Cherokee used root as tonic for obstructed menses, colds, fever, and colic. Leaf tea used for stomachaches, indigestion, gas, anorexia, obstructed menses, fevers, colds, colic, flu, coughs, neuralgia, rheumatism. Roots, seeds strongest; leaves weaker. Stimulates secretion of gastric juices and alleviates smooth-muscle spasms. **RELATED SPECIES:** Other *Angelica*s, including *A. sinensis* (Oliv.) Diels and *A. dahurica* (Hoffm.) Benth. & Hook.f. ex Franch. & Sav., are famous Chinese drugs for "female ailments." The European *A. archangelica* L., commonly grown in northern herb gardens, sometimes naturalized in eastern N. America, reportedly contains 15 calcium blockers, one as potent as the synthetic verapamil, used for the treatment of angina. Root approved in European phytomedicine for treatment of loss of appetite, stomach discomfort due to mild gastrointestinal tract spasms, flatulence, and a feeling of fullness in the abdomen. **WARNING:** Contains furanocoumarins, which can cause photosensitivity and photodermatitis. Because it grows in the same habitat as the Poison Hemlock, and the untrained botanical collector may confuse the two, harvest should be avoided unless positively identified by a trained botanist.

WATER-HEMLOCK

Poison—Identify to avoid

Cicuta maculata L. Parsley Family (Apiaceae or Umbelliferae)

Biennial; 1–7 ft. Stems smooth; *purple streaked or spotted.* Leaves divided 2–3 times; leaflets lance-shaped, coarsely toothed. Flowers in loose umbels; May–Sept. Strong odor. **WHERE FOUND:** Wet meadows, swamps. Most of our area. **USES:** Too lethal for use; contains compounds similar to those found in Poison Hemlock (below). **WARNING:** Highly poisonous. Do not confuse with harmless members of the Parsley family.

Water-hemlock occurs in wet habitats.

Cow-parsnip is one of our largest Parsley family members.

POISON HEMLOCK

Poison—Identify to avoid

Conium maculatum L. Parsley Family (Apiaceae or Umbelliferae)

☠ Stems *purple-spotted*. **WARNING:** Deadly poison. Ingestion can be lethal. Contact can cause dermatitis. See p. 85.

COW-PARSNIP

Root, leaves, tops

Heracleum lanatum Michx. Parsley Family (Apiaceae or Umbelliferae)
(*Heracleum maximum* Bartr.)

⚠ One of our largest umbelliferous plants. Large, woolly, strong-smelling biennial or perennial; to 6–9 ft. Leaves large, divided into *3 maplelike segments*; sheath *inflated*. Umbels large: 6–12 in. across; flowers with notched petals; May–Aug. **WHERE FOUND:** Moist soils. NL to GA mtns.; west to MO, KS, ND, and westward. **USES:** Root tea was widely used by Native Americans for colic, cramps, headaches, sore throats, colds, coughs, flu; externally poulticed on sores, bruises, swellings, rheumatic joints, boils. In folk use, root tea was used for indigestion, gas, asthma, epilepsy. Used similarly by early European settlers, especially with a reputation for treating epilepsy. Powdered root (1 teaspoon per day over a long period) was taken, with a strong tea of the leaves and tops, for epilepsy. Root contains psoralen, which is being researched and tested in the treatment of psoriasis, leukemia, and AIDS. A compound isolated from the root of interest as an antitumor compound for certain skin cancers. **WARNING:** Foliage is poisonous to livestock; roots contain phototoxic compounds, including psoralen. Acrid sap can cause blisters on contact. The authors know an herbalist who badly blistered his forearms and wrists after handling the plant. Fresh plant (and root) potentially more toxic than dried root.

FLOWERS IN UNEVEN OR SPARSE UMBELS; LEAVES PARTED AND TOOTHED

SWEET CICELY

Root

Osmorhiza claytonii (Michx.) C. B. Clarke

Parsley Family (Apiaceae or Umbelliferae)

⚠ Soft-hairy perennial; 1–3 ft. Root *slightly anise-scented*. Leaves fern-like; thrice-compound. Flowers tiny, white, *style shorter than petals*; Apr.–June. **WHERE FOUND:** Moist woods. NS to GA mtns.; AL, AR, e. KS to SK. **USES:** The Chippewa chewed the root or gargled root tea for sore throats; poulticed root on boils, cuts, sores, wounds. Ojibwa and Menominee used tea as a wash for sore red eyes, drank it for coughs, and consumed root as a tonic. **WARNING:** Do not confuse with Poison Hemlock (*Conium*). See p. 85.

SWEET CICELY, ANISE ROOT

Root

Osmorhiza longistylis (Torr.) DC. Parsley Family (Apiaceae or Umbelliferae)

⚠ Similar to *O. claytonii*, but stouter and nearly smooth. Tiny white flowers in open umbels; *style exceeding the petals, and persistent on fruits*, to 1/16 in. long (hence the species name "*longistylus*"); distribution more western. Root very sweet, aromatic, fleshy, *usually strongly anise-*

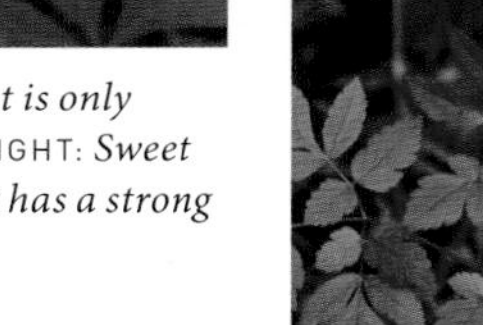

ABOVE: *Sweet Cicely root is only slightly anise-scented.* RIGHT: *Sweet Cicely, Anise Root's root has a strong anise scent.*

scented, edible. **WHERE FOUND:** NS to GA; NM north to AB. **USES:** Various tribes, including the Cheyenne, Chippewa, and others, used root tea for general debility, panacea; tonic for upset stomach, parturition (childbirth); root poulticed on boils, wounds; root tea an eye-wash. In folk medicine, used as an expectorant, tonic for coughs and stomachaches. Root eaten or soaked in brandy. The Meskwaki chewed root to regain flesh and strength. Herbalists consider it an aromatic stomachic, carminative, an expectorant useful for coughs, and a gentle stimulant for digestive problems. **RELATED SPECIES:** Both eastern N. American species are closely related to the widespread temperate Asian species *O. aristata* (Thunb.) Rydb., which also contains anethole (anise flavor) and estragole (tarragon flavor) in the root's essential oil; valued as a treatment for abdominal pain and headache. **WARNING:** Do not confuse with Poison Hemlock (*Conium*). See p. 85.

BLACK OR CANADIAN SANICLE — Leaves, root

Sanicula canadensis L. — Parsley Family (Apiaceae or Umbelliferae)

Biennial; to 36 in; roots slender and fibrous. Leaves long-stalked; palmate, with *3–5 leaflets that are double-toothed or deeply incised*; upper leaves reduced, becoming bract pairs. Flowers greenish whitish with white anthers, in uneven umbels; Apr.–July. Fruits grow on a small but distinct stalk. **WHERE FOUND:** Dry woods, openings. QC to FL; TX to SD. **USES:** The Chippewa used root to stimulate menses; abortive. The Houma used as a heart remedy. Leaves, which contain allantoin, were poulticed for bruises, inflammation. Used like *S. marilandica,* below, as the species are separated by technical details unlikely observed by collectors.

Canadian Sanicle flowers are white-green to yellowish.

BLACK SANICLE OR SNAKEROOT (NOT SHOWN) — Root

Sanicula marilandica L. — Parsley Family (Apiaceae or Umbelliferae)

Perennial; 1–4 ft. Leaves palmate, with 5–7 leaflets; 2 of the leaflets are deeply cleft, suggesting 7 leaflets. Flowers whitish, in uneven umbels with leaflike bracts beneath; Apr.–July. *Prickled fruits sessile (without stalks); base of bristles bulbous.* **WHERE FOUND:** Thickets, shores. NL to FL; to e. KS, ND, westward to BC. **USES:** The thick rhizome (root) was used by Iroquois as a wash for sore navels in children. Micmac used root tea for menstrual irregularities, pain, kidney ailments, rheumatism, fevers. Ojibwa used root poultice for poisonous snakebites. Valued as a domestic remedy for fevers, sore throat, erysipelas, and nervous conditions. **RELATED SPECIES:** Of the 40 species in the genus, several, including *S. europaeus* L., possess antioxidant and antiviral activity.

FLOWERS WITH 5 "PETALS," IN FLAT-TOPPED CLUSTERS

FLOWERING SPURGE — Leaves, root, stems

Euphorbia corollata L. — Spurge Family (Euphorbiaceae)

⚠ Deep-rooted, milky-juiced, smooth-stemmed perennial; 1–3 ft. Leaves alternate, without stalks, *oval to linear*, to 2½ in long; margins entire. Flowers white, in many forked umbels, rising from *whorl of reduced leaves*; note the 5 "petals" (actually bracts) surrounding flowers; May–Aug. **WHERE FOUND:** Fields, roadsides. ME, ON, NY to FL; TX to SD, MN. **USES:** Cherokee used tea to arrest bleeding after childbirth; root used for toothache, urinary disease such as gonorrhea; purgative. Stem juice for skin eruptions, sore nipples. Meskwaki used to treat pinworm. Suggested in nineteenth century as a substitute for ipecac as an emetic; also purgative and diaphoretic and used to induce blistering externally (not recommended). **WARNING:** Extremely strong laxative. Stem latex may cause skin chemical burns and blistering.

Flowering Spurge. Its stems produce a milky juice.

SMALL WHITISH FLOWERS IN LEAF AXILS; PARASITIC OR SEMIPARASITIC PLANTS

DODDER

Whole plant

Cuscuta species — Morning-glory Family (Convolvulaceae)

⚠ Parasitic, chlorophyll-lacking (though not entirely), leafless annual; leaves replaced by a few scales. Stems *yellow or orange*. White flowers, small to 1/8-in., 4–5-parted in clusters; June–Oct. **WHERE FOUND:** Low ground. Most of our area. Dodders clamber over other growth and cause serious damage as annual weeds. **USES:** Stems used by Cherokee as a poultice for bruises. **RELATED SPECIES:** About 15 species in our area, distinguished by flowers and fruit details. Common Dodder (*C. gronovii* Willd.) has small, whitish, waxy, 5-lobed flowers in loose or crowded clusters. In China, the stems of other *Cuscuta* species are used in lotions for inflamed eyes. The Chinese value the seeds of dodders for urinary tract ailments. The tea of Chinese Dodder (*C. chinensis* Lam.) has demonstrated anti-inflammatory, cholinergic, and CNS-depressant activity. **WARNING:** Dodders are called love vines and "vegetable spaghetti" but are not generally considered edible. Some parasitic flowering plants take up toxins from their hosts.

EYEBRIGHT

Whole plant

Euphrasia stricta D. Wolff ex J. F. Lehm. (*Euphrasia officinalis* L.) — Broom Rape Family (Orobanchaceae)

[Formerly in Figwort Family (Scrophulariaceae)]

⚠ Slender, hemiparasitic (root is attached to grasses, gaining water and mineral nutrients from host, while producing carbon from photosynthesis) annual (or perennial); 4–8 in. Leaves tiny, bristle-toothed. Flowers June–Sept. The 3 *lower lobes are notched*, with purple lines. Highly variable. **WHERE FOUND:** This complex plant group comprises more than 350 named species (mostly European) that are distinguished by botanists on minute technical details (highly variable and weakly differentiated microspecies), though herbalists treat the group as a single herb referred to as *Euphrasia officinalis* L., in Europe often designated as *E. rostkoviana* Hayne. Usually hidden among grass and difficult to find in dry or moist fields, roadsides, waste places. Subarctic south to QC, ME, MA, NY; N. Hemisphere. **USES:** Tea astringent. A folk remedy the juice of which was taken in a teaspoonful in a glass of wine, as a tea, or made into beer, as a remedy to strengthen the eyes and optic nerve, restore weak sight, and of benefit to a weak brain and memory; for eye ailments with mucous discharge; coughs, hoarseness, earaches, headaches with congestion. The small flowers, which look like eyes, evoke the medieval doctrine of signatures, suggesting use for eye ailments. Poorly researched, *Euphrasia* species contain anti-inflammatory, antibacterial, and antioxidant phenolic compounds, iridoid glucosides, and a glycoside called eukovoside, perhaps providing a basis for traditional use. A retrospective clinical analysis suggested that an Eyebright-based eye-drop product used in northern Europe improved redness, swelling,

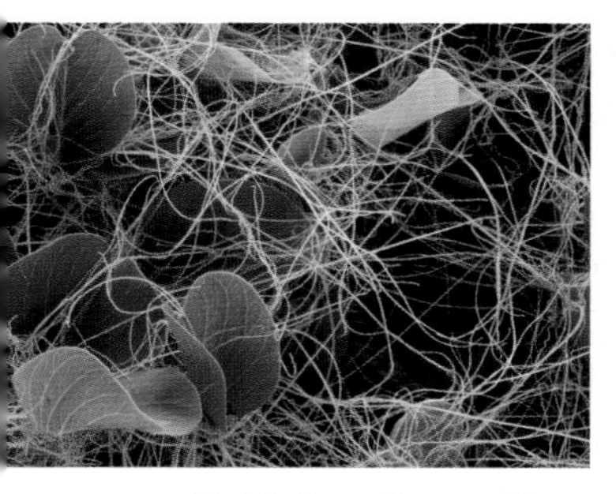

ABOVE: *Dodder is a yellow, mostly chlorophyll-lacking parasite that entangles other plants.* TOP RIGHT: *Eyebright is a low-growing plant of the far north, typically associated with European herbal medicine.* RIGHT: *Eyebright species are separated on highly technical characteristics and collectively known in herbal literature as* Euphrasia officinalis.

and secretion in inflammatory and catarrhal conjunctivitis at a dose of 1–3 drops per day. **RELATED SPECIES:** Collectively, most *Euphrasia* species have been used similarly. Some are rare and of potential conservation concern from overcollection locally. **WARNING:** Experimentally, may induce side effects, including dim vision. Use in eyes only under a licensed practitioner's advice.

SQUARE-STEMMED AROMATIC HERBS (MINTS); FLOWERS IN AXILS OR TERMINAL

LEMON BALM, MELISSA

Leaves

Melissa officinalis L. — Mint Family (Lamiaceae or Labiatae)

Perennial; 1–3 ft. Leaves opposite; oval to egg-shaped, round-toothed; strongly *lemon-scented* when fresh; dissipates on drying. Flowers white or at first yellowish, small, 1/3 to 1/2 in. long, in whorls; stamens not protruding beyond upper lip of flower; lobes of upper calyx wider than long;

LEFT: *Lemon Balm, Melissa is easily distinguished by its lemon-scented leaves.* ABOVE: *Lemon Balm has small white flowers with a hairy calyx.*

Apr.–Aug. **WHERE FOUND:** Barnyards, old house sites, open woods. Scattered over much of our area. Commonly grown in herb gardens; can become weedy. Alien (Europe). **USES:** Used in European herbal medicine from Medieval to modern times. The 1789 *Edinburgh Dispensatory* states, "So high an opinion have some of the chemists [pharmacists] entertained of balm, that they have expected to find in it a medicine which should prolong life beyond the usual period." Dried- or fresh-leaf tea is a folk remedy for fevers, painful menstruation, headaches, colds, insomnia; mild sedative, anxiolytic, carminative; leaves poulticed for sores, tumors, insect bites. Experimentally, hot water extracts have been shown strongly antiviral for Newcastle disease, herpes, mumps. Extracts show promise as mild sedative and sleep aid for anxiety; antioxidant; shows promise in reducing age-related cognitive and memory function. Strong extracts (200:1) sold in ointments in Europe to treat cold sores and genital herpes. This use is backed by clinical studies. In European phytomedicine, Lemon Balm preparations are approved for treatment of sleeplessness caused by nervous conditions and digestive tract spasms. Like extracts of Rosemary and Sage, Lemon Balm extracts slow the breakdown of acetylcholine, a messenger compound deficient in brain-cell cultures of Alzheimer's disease patients. Contains at least 8 antiviral compounds (against herpes), 8 sedative compounds, and 12 anti-inflammatory components. Leaves also have antibacterial, antihistaminic, antispasmodic, and antioxidant activity.

CANADIAN MINT — Leaves

Mentha canadensis L. — Mint Family (Lamiaceae or Labiatae)

(listed as or confused with *Mentha arvensis* L. in older texts)

Flowers whitish at first, or lilac. See p. 251.

ABOVE: *Canadian Mint occurs in moist habitats.* RIGHT: *White Horsemint occurs in woodlands.*

WHITE HORSEMINT

Leaves

Monarda bradburiana Beck — Mint Family (Lamiaceae or Labiatae)

Perennial; 1–2½ ft.; leaves with peppery scent, ovate to broad-ovate; ¾ to 2 in. broad, 2–3½ in. long, sharp-toothed. Flowers in terminal head, white or rose with prominent purple dots; *upper lip of flower cottony-hairy*; upper flower lip about equal length of corolla tube; Apr.–June. **WHERE FOUND:** Rocky wooded hills. IN to s. AL, west to TX and se. KS, IA. Our earliest blooming *Monarda* species. **USES:** Leaf tea used for fevers, upset stomach, digestive gas; cold and cough remedy; pleasant beverage tea. Probably the least used *Monarda.* Contains isopulegol in essential oil, giving it a slight pennyroyal-like fragrance. **RELATED SPECIES:** Occasionally treated as synonymous with *Monarda russeliana* Nutt. but the latter has smaller, narrower leaves; upper lip of corolla is not bearded at tip, among other differences.

VIRGINIA MOUNTAIN MINT

Leaves, root

Pycnanthemum virginianum (L.) Pers. — Mint Family (Lamiaceae or Labiatae)

Perennial; 2–4 ft. Leaves lance-shaped, without stalks; base rounded. Flowers whitish to lilac, in dense terminal clusters; July–Sept. **WHERE FOUND:** Dry thickets. ME to NC; MO, e. KS to ND. **USES:** Leaf tea of this and other *Pycnanthemum* species used for fevers, colds, coughs, colic, stomach cramps; said to induce sweating, relieve gas; also for amenorrhea, dysmenorrhea. Native Americans took the powdered root to treat stoppage of menstrual periods, interpreted to mean that it was used as an early-term abortifacient. Flowers and buds used to flavor meats and soups. **REMARKS:** Other *Pycnanthemum* species used similarly, such as

Mountain Mints are among the most aromatic wild mints. Pictured here is Pycnanthemum virginianum, *one of more than a dozen* Pycnanthemum *species in our range.*

Slenderleaf Mountain Mint has a similar fragrance to Virginia Mountain Mint.

Slenderleaf Mountain Mint, *P. tenuifolium,* which has linear aromatic leaves.

SQUARE-STEMMED PLANTS; LEAVES OPPOSITE; WEAK OR RANK SCENTS; MINT FAMILY

AMERICAN BUGLEWEED, CUT-LEAVED WATER HOREHOUND

Whole plant

Lycopus americanus Muhl. ex W.P.C. Barton

Mint Family (Lamiaceae or Labiatae)

The most abundant of the more than 7 species in our range. Perennial; 1–2 ft. Grows from edible, whitish, screwlike horizontal root. Leaves deeply cut; *lower ones suggest oak leaves.* Flowers in whorls of leaf axils; June–Sept. Stamens protruding; 5 calyx lobes sharply pointed, longer than mature nutlets. **WHERE FOUND:** Wet places. Throughout our area. **USES:** Thought to be the same as for *L. virginicus* (below). **REMARKS:** The Bugleweeds (*Lycopus* species) are also known as Water Horehound. It is likely that most N. American species of *Lycopus* are harvested and used generically in the supply of *L. virginicus* (below), without distinction to species.

American Bugleweed is our most common bugleweed.

European Bugleweed has spread as a naturalized species in e. N. America.

EUROPEAN BUGLEWEED

Whole plant

Lycopus europaeus L.

Mint Family (Lamiaceae or Labiatae)

⚠ Very similar in appearance to *L. americanus* above and hybridizing with it. Generally more hairy with leaves wider and more blunt-toothed. **WHERE FOUND:** Once limited to the St. Lawrence River Basin, it has spread in recent years, becoming naturalized in much of eastern N. America, preferring moist soils, much like its American counterparts. **USES:** Bugleweeds native to N. America, Asia, and Europe are used similarly on their respective continents, primarily as a suppressant to an overactive thyroid, along with related symptoms such as heart palpitations and nervousness; also used for coughs and insomnia. Pharmacological and chemical evidence from the mid-1950s onward suggests that extracts may block thyroid-stimulating hormone production, inhibit iodine metabolism, and lower prolactin levels in blood. Anti-allergic, antimicrobial, and antioxidant effects have also been reported. Rigorous controlled clinical trials are lacking; a recent prospective study found that treatment with European Bugleweed improved patients with mild forms of hyperthyroidism, particularly mild cardiac symptoms. Used in German phytomedicine in cases of thyroid hyperfunction and thyroid enlargement (without functional disorders). **WARNING:** In rare cases, Bugleweed has been found to enlarge the thyroid. If therapy is discontinued, symptoms may return.

BUGLEWEED

Whole plant

Lycopus virginicus L. Mint Family (Lamiaceae or Labiatae)

⚠ Perennial; 6–40 in. Leaves lance-shaped, *strongly toothed*; lower ones with long, narrow bases. Flowers in axils, *with 4 broadly triangular calyx lobes*, shorter than nutlets; July–Oct. **WHERE FOUND:** Wet places. NS to GA, AR to OK, NE to MN. **USES:** Traditionally used as a mild sedative, astringent; especially for heart diseases, chronic lung ailments, coughs, fast pulse, thyroid disease, diabetes. Bugleweed has a long history of empirical use for a variety of conditions, including treatment of heart palpitations, goiter, hyperthyroidism, and Graves' disease. Science has confirmed the potential value of this plant in treating hyperthyroidism. Bugleweed extracts inhibit iodine metabolism and thyroxine release in the thyroid. Biological effects caused by Graves' disease may be reduced by Bugleweed extracts. Leaf extracts are considered more active than the root. Used in German phytomedicine in cases of thyroid hyperfunction and thyroid enlargement (without functional disorders). **WARNING:** In rare cases, Bugleweed has been found to enlarge the thyroid. If therapy is discontinued, symptoms may return.

HOREHOUND

Leaves

Marrubium vulgare L. Mint Family (Lamiaceae or Labiatae)

White-woolly, sharp-scented perennial; 12–20 in. Leaves round-oval; toothed, *strongly wrinkled*. Flowers white *in prickly whorls*; May–Sept. White calyx with 10 bristly, curved teeth. **WHERE FOUND:** Waste places; escaped. Scattered over most of N. America. Alien. A European native common in pastures and often cultivated in herb gardens. **USES:** The sharp-fragranced, bitter leaves are a well-known ingredient in cough syrups and throat lozenges, especially in Europe. Bitter-tasting herb is a famous folk remedy as a stomachic to stimulate digestion, reduce pain and inflammation, and as an expectorant primarily used to break up phlegm, relieve coughs, soothe sore throats, and relieve bronchitis and other upper-respiratory ailments. Also used for stomach and gallbladder disorders, jaundice, hepatitis; fresh leaves poulticed on cuts, wounds. Experimentally, marrubiin is an expectorant, increases liver bile flow, and has gastroprotective activity against ulcers. Volatile oil is an expectorant, acts as a vasodilator; antibacterial and antifungal. An alcoholic extract has been confirmed to help reduce pain and reduce muscle spasms of the gastrointestinal tract. Water extract is antioxidant. Marrubic acid has liver-protectant effects. In Algeria traditionally used as an antidiabetic herb; confirmed experimentally. In various European countries, Horehound preparations are widely used for supportive treatment of coughs and colds and as a digestive aid and appetite stimulant.

CATNIP

Leaves, flowering tops

Nepeta cataria L. Mint Family (Lamiaceae or Labiatae)

Perennial; 12–24 in. Leaves *stalked*, ovate; *strongly toothed*. Flowers in crowded clusters; June–Sept. Flowers whitish, *purple-dotted*;

Bugleweed and other species are distinguished by technical characteristics. Most species are used interchangeably.

Horehound, with wrinkled gray-green leaves.

Catnip is well known for the response it elicits in cats.

Catnip has small, purple-spotted white flowers. Flowerheads are stronger in fragrance than leaves, hence loved by cats.

calyx soft-hairy. **WHERE FOUND:** Much of our area. Alien. **USES:** Tea made from leaves and flowering tops is a folk remedy for bronchitis, colds, diarrhea, fevers, chicken pox, colic, headaches, irregular menses; said to induce sleep, promote sweating, alleviate restlessness in children; leaves chewed for toothaches. Experimentally, nepetalactone, a mild sedative compound in Catnip, also possesses herbicidal and insect-repellent properties. Best known as a feline euphoric, Catnip produces the "Catnip response," which includes sniffing, licking, chewing, head-shaking, rolling, and rubbing, affecting not only housecats but large cats as well. This phenomenon is the result of an inherited autosomal dominant gene, absent in about one-third of cats. Similar effects are not experienced by humans. Recent studies suggest that catnip oil has underdeveloped potential as a safe and effective insect repellent ingredient, though it can be irritating to the skin in some individuals.

DELICATE TINY FLOWERS LESS THAN ¼ IN. LONG, IN DENSE SPIKES

DEVIL'S-BIT — Root

Chamaelirium luteum (L.) Gray — Melanthium Family (Melanthiaceae) [Formerly placed in Lily Family (Liliaceae)]

Perennial; to 3 ft. (in flower). Leaves *smooth, oblong, in basal rosettes.* Male and female flowers on separate plants. Flowers white, fading to yellow, in crowded spikes *(usually drooping at tip)*; May–July. **WHERE FOUND:** Rich woods. W. MA, NY to FL; AR to IL, MI. **USES:** Small doses of powdered root used for colic, stomach ailments, appetite stimulant, indigestion, and to expel worms. Root tea said to be a uterine tonic. Used for a wide variety of ailments associated with male and female reproductive organs. In early nineteenth-century America, the plant was considered a tonic, diuretic, and as a strong tea used as an insecticide and to kill lice. A folk notion held that if corn kernels were steeped in Devil's-bit tea for 24 hours before sowing, birds would not eat them. **WARNING:** Avoid during pregnancy.

NARROW-LEAVED PLANTAIN, ENGLISH PLANTAIN — Leaves, seeds

Plantago lanceolata L. — Plantain Family (Plantaginaceae)

Annual; 10–23 in. Leaves *lance-shaped; 3-ribbed.* Flowers tiny, whitish, in a *short cylindrical head* on a grooved stalk; Apr.–Nov. **WHERE FOUND:** Waste places. Throughout our area. Alien weed very common in temperate regions throughout the world. **USES:** Traditionally, leaf tea used for coughs, diarrhea, dysentery, bloody urine. Science confirms bronchodilation action; used in Europe for bronchitis and bronchial spasms due to colds. Approved in European countries for treatment of catarrh of the upper respiratory tract and inflamed mucous membranes of the mouth and throat. Leaves applied to blisters, sores, ulcers, swelling, insect stings; also used for earaches, eye ailments; thought to reduce

Contrary to the species name luteum, meaning yellow, Devil's-bit flowers are generally white.

Narrow-leaved Plantain is named for its narrow-ribbed leaves.

Narrow-leaved Plantain. Close-up of flowerheads.

heat and pain of inflammation. Science has vindicated utility in healing sores, with mild antibiotic, antioxidant, and anti-inflammatory activity. The mucilage from any plantain seed may lower cholesterol levels. **RELATED SPECIES:** *P. asiatica* L. is used clinically in China to reduce blood pressure (50 percent success rate). Seeds of *P. ovata* Forssk. and *P. psyllium* L. (not shown) are widely used in bulk laxatives; also reduce rate of coronary ailments. There are more than 270 *Plantago* species worldwide, many of which are cosmopolitan weeds, used locally as folk medicines or formal traditional medicine systems. **WARNING:** Some plantains may cause rare instances of contact dermatitis; recently ingestion of *P. lanceolata* leaves was associated with two cases of photodermatitis after exposure to bright sunlight.

COMMON PLANTAIN

Leaves, roots, seeds

Plantago major L. Plantain Family (Plantaginaceae)

⚠ Perennial; 6–18 in. Leaves *broad-oval; wavy-margined* or toothed, ribbed; stalk is grooved. Flowers in a slender, elongate head; Apr.–Oct. **WHERE FOUND:** Waste places. Throughout our area and beyond. Alien from Europe; widespread elsewhere. **USES:** Same as for *P. lanceolata* (above). Prominent folk remedy for cancer in Latin America. Used widely in folk medicine throughout the world. Root formerly used as an astringent to ulcers, bleeding, bloody urine, diarrhea, and for fevers. Bruised fresh leaves for slight wounds, sores, spider and insect bites, and as a folk remedy for cancer. Confirmed antimicrobial, antioxidant, liver-protectant; stimulates healing process. Main active compounds include iridoid glycosides, mucilage, and various phenolic acids. **REMARKS:** There are a half-dozen or more broad-leaved plantain species in our area that are likely confused with or used interchangeably with *P. major.* **WARNING:** Rare contact dermatitis is reported.

SENECA SNAKEROOT

Root

Polygala senega L. Milkwort Family (Polygalaceae)

Perennial; 6–18 in. Leaves alternate; lance-shaped, small, 1–3 in. long at mid-stem, much reduced at base of plant. Small to ¼ in. long white *pealike* flowers in a terminal spike; May–July. **WHERE FOUND:** Rocky woods. NB to GA; AR to SD, most of Canada. **USES:** The most famous snakebite remedy of colonial America. Dr. John Tennent, residing in Virginia, published a 100-page book in 1736, extoling use of the root by the Seneca Indians in treating rattlesnake bites, and his success in using "this root [that] will be of more extensive use than any medicine" in gout, rheumatism, dropsy, and pleurisy. Soon exported in tonnage to Europe, where it became an official drug of apothecary shops, and by the American Revolution offered by every druggist. As "Seneca Rattlesnake Root" in 1787, Thomas Jefferson listed it among the most promising native medicinal plants of Virginia. In 1818, in *American Medical Botany* (vol. 2, p. 100) Harvard professor Jacob Bigelow poignantly observes, "The first reputation of the Seneca root was one which it divides with multitude of other plants, that of curing the bite of the Rattlesnake. . . . When, however, we consider the number of cases of recovery from the bite of this serpent, under every variety of treatment, we cannot avoid the conclusion, that these injuries are not necessarily dangerous, and that spontaneous recoveries are perhaps as frequent as those of which are promoted by medicine." Soon its reputation as a snakebite cure faded except for the plant's common name. The Seneca, Cherokee, and other native groups chewed the root, swallowing juice, then applied to poisonous snakebites. The Cherokee also used the root tea as a diuretic, to induce menstruation, and for colds, pulmonary conditions, and rheumatism. Among the Cree, Micmac, and Ojibwa the root tea was used for colds, and a gargle for sore throat. Historically, physicians used root preparations similarly; also in pneumonia, chronic bronchitis, asthma; thought to "relax respiratory mucous membranes."

Common Plantain is a common lawn weed.

Seneca Snakeroot. Close-up of flowers.

Seneca Snakeroot excited seventeenth-century American Colonists as a potential snakebite remedy.

Isolated components have been shown to lower blood sugar, reduce alcohol absorption; anti-inflammatory. Methyl salicylate (see Wintergreen, p. 36) in root suggests a rationale behind use of this plant to relieve pain, rheumatism, etc. Now used more widely in modern Japan and Germany than in the U.S. The root is used in European phytomedicine as an expectorant for treatment of upper respiratory tract inflammation. Related N. American *Polygala* species used similarly at least in a historical context.

WHITE SWEET-CLOVER

Flowering plant

Melilotus albus Medik. Pea Family (Fabaceae or Leguminosae)
(*Melilotus alba* Medik, a spelling variant)

Biennial; 1–9 ft. Leaves cloverlike; leaflets elongate, slightly toothed. Small white pealike flowers in long, *tapering spikes*; Apr.–Oct. **WHERE FOUND:** Roadsides. Throughout our area. Alien (Europe). **USES:** Dried flowering plant once used in ointments for external ulcers. In animal studies, components lowered blood pressure. Contains coumarins, which, when the plant dries, present a vanilla-like sweet fragrance. In the 1920s as White Sweet-clover and Yellow Sweet-clover spread in rangelands, cattle and sheep that grazed on moldy Sweet Clover often developed fatal bleeding, called "sweet clover disease." In the 1940s, it was discovered that coumarins oxidized in moldy Sweet-clover hay, leading to the isolation of a compound called dicoumarol. Later in the 1940s, large-scale isolation of the compound was made possible from research funded by the Wisconsin Alumni Research Foundation (WARF). An analog of dicoumarol was developed into a fast-acting rodent poison (causing death by internal bleeding) named for the funding organization (WARF) and called warfarin. The modified rodenticide warfarin became the human anticoagulant drug Coumadin® (warfarin sulfate). Among the first well-known human patients to receive warfarin in 1955 was Dwight D. Eisenhower. Today warfarin is the world's most widely used anticoagulant, originating from research on the coumarins in Sweet-clover that caused sweet clover disease in livestock. **WARNING:** Coumarins in Sweet-clover may decrease blood clotting.

White Sweet-clover is an abundant late-summer roadside weed.

WHITE CLOVER

Whole plant, flowers

Trifolium repens L. Pea Family (Fabaceae or Leguminosae)

Perennial; 4–10 in. Leaves 3-parted, often *with "V" marks*. Flowers stalked, white (often pink-tinged); in *round heads*; Apr.–Sept. **WHERE FOUND:** Fields, lawns. Throughout our area. Alien (Europe). **USES:** Cherokee used leaf tea for leukorrhea, fevers, and as a diuretic. Iroquois used flower tea to improve blood, also for menopausal symptoms. In European folk medicine, flower tea is used for rheumatism and gout. Like Red Clover, and probably most clovers, White Clover contains the estrogenic isoflavone genistein, with a multitude of activities, including cancer-preventive and antioxidant effects.

White Clover is very common across the U.S.

Close-up of flowers.

STIFF-STEMMED LEGUMES WITH 15 OR MORE LEAFLETS

CANADIAN MILKVETCH — Roots

Astragalus canadensis L. — Pea Family (Fabaceae or Leguminosae)

Herbaceous, robust, erect perennial; 2–5 ft. tall. Leaves alternate, odd-pinnate; *leaflets oblong in 10–13 pairs (15–25 leaflets), entire*. Flowers whitish yellow to cream colored in racemes, at least as long as leaf beneath, 2–6 in. long; racemes *with 30–70 flowers;* May–Aug. Fruits numerous; a dry, small, 3/8–5/8 in. long, rounded, mostly smooth pod. **WHERE FOUND:** Open woods, often near creeks, streambanks, shores; moist soils. Most of N. America and Siberia. Our most common *Astragalus* spp. **USES:** Blackfeet Indians used root, or chewed for cuts. Lakota chewed root for chest and back pains; also coughs. **RELATED SPECIES:** The genus *Astragalus* is one of the largest and most diverse genera of flowering plants, with over 2,500 species, including medicinal, edible, and poisonous species (such as the locoweeds of western N. American deserts), with complex technical language to describe differences. The root of the Chinese species *A. membranaceus* Moench, called *huang qi,* is one of the most widely used herbs in Traditional Chinese Medicine, used to invigorate vital energy (qi) and in prescriptions for shortness of breath, general weakness, and lack of appetite; tonic and immunostimulant.

GROUND PLUM, PRAIRIE-PLUM, GROUNDPLUM MILKVETCH — Root

Astragalus crassicarpus Nutt. — Pea Family (Fabaceae or Leguminosae)

Robust, somewhat hairy perennial, with stems clustered and drooping at tips from a large taproot. Leaves odd-pinnate with *15–29 oblong to elliptical leaflets*. Seedpods spreading or ascending. Flowers whitish yellow to cream with a *blue-purple spot at tip of keel* atop flower, *3–25*

Canadian Milkvetch flower stalks are up to 6 in. long with 30–70 cream-colored flowers.

Ground Plum has larger and fewer flowers than Canadian Milkvetch.

flowers per raceme. Apr.–May. Pods fleshy and plumlike in appearance, hence the common names. **WHERE FOUND:** WI to LA; TX to MT, to BC. Five varieties of this species separated on technical characteristics. **USES:** The Chippewa used the root, mostly in combination with other herbs, as an anticonvulsant, stimulant, and tonic; externally for wounds to stop bleeding. Plant, fruits, and root sometimes eaten as food by various native groups. Given the importance of Asian species of *Astragalus* in traditional medicine systems, the N. American species are surprisingly little researched.

PRAIRIE MIMOSA

Leaves, seeds

Desmanthus illinoensis (Michx.) MacMill.

Pea Family (Fabaceae or Leguminosae)

Smooth-stemmed, erect perennial; 1–4 ft. Leaves twice-divided; 20–30 tiny leaflets. Flowers greenish white, *in globular heads*; June–Aug. Pods curved, in loose globular heads. **WHERE FOUND:** Prairies, fields. PA, OH to FL; TX, CO to ND. **USES:** Pawnee used leaf tea as a wash for itching. A single report states that a Paiute Indian placed 5 seeds in the eye at night (washed out in morning) for chronic conjunctivitis. The leaves are reportedly high in protein. Reported to contain DMT (N,N-dimethyltryptamine) and analogs, alkaloid identified in preparations of the South American psychotropic cocktail known as Ayahuasca. **WARNING:** Of unknown toxicity.

Prairie Mimosa, with white flowers. Its fruits turn brown in twisted clusters.

Wild Licorice produces spiny seedpods.

WILD LICORICE

Root, leaves

Glycyrrhiza lepidota Pursh — Pea Family (Fabaceae or Leguminosae)

⚠ Shrubby perennial; 5–9 ft. Leaves compound; 15–19 leaflets, oblong to lance-shaped, *glandular-dotted* (use lens). Flowers whitish, on short spikes; May–June. Fruits oblong, with *curved prickles*; June–Aug. **WHERE FOUND:** Prairies, fields. W. ON to TX, MO west to WA. Mostly in western states, but scattered east from VA to ME. **USES:** Widely used by native peoples, including the Bannock, Great Basin Indians, Montana Indians, Pawnee, and other groups; root chewed for sore throat and coughs, diarrhea, stomachache, earache, and for fevers in children. Pawnee applied a poultice of leaves infused in hot water to ears to treat earaches. Contains an antiviral compound. **RELATED SPECIES:** This American species is similar to the Eurasian Licorice Root (*G. glabra* L.) and the Chinese species (*G. uralensis* Fisch.), both extensively used in European and Asian herbal medicine. Sweet to musty-flavored roots of these related species were traditionally used for soothing irritated mucous membranes, inflamed stomach, ulcers, asthma, bronchitis, coughs, bladder infections. *G. glabra* and *G. uralensis* have been extensively investigated; considered estrogenic, anti-inflammatory, anti-allergenic, anticonvulsive, and antibacterial. Used in European phytomedicine for the supportive treatment of gastric and duodenal ulcers and for congestion of the upper respiratory tract. Chinese studies indicate antitussive effects of these plants are equal to and longer-lasting than those of codeine. Clinically useful against gastric and duodenal ulcers, bronchial asthma, coughs. Licorice root is one of the most extensively used drugs in Chinese herbal prescriptions. In combinations, the Chinese believe that it helps to detoxify potentially poisonous drugs, weakening their effects. Our Wild Licorice (little studied) contains up to 6 percent of the active component glycyrrhizin. **WARNING:** Wild Licorice can raise blood pressure. May cause water retention and hypertension, a result of sodium retention and potassium loss. Avoid use in cases of hypertension. In Germany, use is limited to 4 to 6 weeks.

COMPOSITES WITH SMALL FLOWERHEADS IN MANY-FLOWERED FLAT-TOPPED CLUSTERS: LEAVES VARIOUS

YARROW

Whole plant in flower

Achillea millefolium L. — Aster Family (Asteraceae or Compositae)

⚠ Soft-hairy, usually strongly fragrant perennial; 1–3 ft. Leaves *lacy, finely dissected*. Flowers white (less frequently pink), in flat clusters; Apr.–Oct. Each tiny flowerhead has 5 petal-like rays that are usually slightly wider than long; each ray has 3 teeth at tip. **WHERE FOUND:** Fields, roadsides. Throughout N. Hemisphere. **USES:** Medicinal plant used since ancient times, used worldwide. A chief use is as a vulnerary—an agent to stop bleeding of wounds. The Latin name *Achillea* honors Achilles; legend holds that he used a poultice of Yarrow flowers to stop the

Yarrow has very finely divided leaves; millefolium *literally means a thousand leaves.*

Yarrow flowers are in flat-topped clusters; each head is a collection of small individual ray and disk flowers.

bleeding of his wounded soldiers in battle; stops inflammation, heals wounds. Herbal tea (made from dried flowering plant) used for colds, fevers, anorexia, indigestion, gastric inflammations, insomnia, and internal bleeding. Fresh herb a styptic poultice. Expectorant, analgesic, and sweat-inducing qualities of some components may provide relief from cold and flu symptoms. Used similarly by native cultures throughout the Northern Hemisphere. Science confirms most traditional use; extracts are hemostatic, anti-inflammatory, anxiolytic, gastric antispasmodic; improves resurfacing of skin epidermis cells; vasodilating, hypotensive, cardio-depressant; may improve vascular inflammation. Contains more than 100 biologically active compounds, including more than a dozen anti-inflammatory compounds. **WARNING:** May cause dermatitis. Large or frequent doses taken over a long period may be potentially harmful. Contains thujone, a toxic compound.

WHITE SNAKEROOT

Root, leaves

Ageratina altissima var. *altissima* (L.) R. M. King & H. Rob. (*Eupatorium rugosum* Houtt.; *Ageratum altissimum* L.)

Aster Family (Asteraceae or Compositae)

⚠ Variable perennial; 2–5 ft. Leaves opposite, on slender stalks; *somewhat heart-shaped*, toothed. Flowers white, in branched clusters; July–Oct. **WHERE FOUND:** Thickets, wood edges under dappled shade. QC to GA; TX to SK. **USES:** Cherokee used root tea as a stimulant, tonic, and diuretic in the treatment of fever, diarrhea, and urinary gravel (kidney stones). Iroquois used for venereal conditions; in sweat baths to cool a patient; poultice for snakebites. Smoke of burning herb used to revive unconscious patients. A famously toxic plant. In nineteenth-century America, a serious, often fatal neurological disorder known as "trembles" or "milk sickness" was first described in a medical journal in 1809. One curious aspect of the now little-known disease is that after the first frost, the malady discontinued. The disease affected cows, sheep, and other livestock. Humans developed the disease by drinking fresh cow's milk or consuming other dairy products. The mystery was solved in 1928 when a USDA researcher determined an oily alcohol named tremetol (a mixture of compounds) from the fresh plant caused the disease. Tremetol occurs only in the fresh plant and is destroyed when heated or dried, which explains why the disease ceased to exist each year after a hard freeze. If

White Snakeroot is the culprit in "milk sickness."

cows on small family farms grazed on the fresh plant, toxic tremetol was concentrated in their milk. The disease claimed thousands of lives, often entire families. The most famous victim was Abraham Lincoln's mother, who died of milk sickness when Lincoln was seven years old. **WARNING:** Fresh plant causes milk sickness; symptoms include prostration, severe vomiting, tremors, liver failure, constipation, delirium, and death. The disease is rarely encountered today.

BONESET, THOROUGHWORT — Leaves, roots

Eupatorium perfoliatum L. — Aster Family (Asteraceae or Compositae)

Sparsely branched, hairy perennial; 1–4 ft. Easily distinguished at any stage of growth by its *wrinkled perfoliate* leaves (stem appears to pass through the joined opposite leaves), which resemble an elongated diamond shape. Flowers white to pale purple, in flat clusters; July–Oct. **WHERE FOUND:** Moist low ground, wet pastures, swamps, moist soils. Very widespread from NS to FL; LA, TX to SK. **USES:** An important medicinal plant among Native American groups throughout its range; leaf and tops, as well as roots, universally used as a sweat-inducing treatment for colds and flu; decoction bitter; induces vomiting. The leaves were used to treat "break-bone fever" (dengue fever), characterized by severe aching down to the bones. This is how the name Boneset originated. A common home remedy of colonial and nineteenth-century America, extensively employed by Native Americans and European

Boneset was an important medicinal plant used by colonial Americans.

Boneset's species name perfoliatum *refers to the perfoliate leaves, in which the bases of opposite leaves are fused together as one.*

settlers, as a folk remedy and prescribed by physicians. Widely used, reportedly with success, during flu epidemics in nineteenth and early twentieth century. Leaf tea once used to induce sweating in fevers, flu, and colds; also used for malaria, rheumatism, muscular pains, spasms, pneumonia, pleurisy, gout, etc. Recent research identified new flavonoids and sesquiterpene lactones, one of which, guaianolide, has antiprotozoal activity against *Plasmodium falciparum,* one of the mosquito vector organisms that causes malaria; also anti-inflammatory, antioxidant. Leaves poulticed onto tumors. German research suggests nonspecific immune-system-stimulating properties, perhaps vindicating historical use in flu epidemics. Anti-inflammatory activity may correlate with use for common cold, rheumatism, arthritis, and other inflammatory conditions. **WARNING:** Emetic and laxative in large doses. May contain controversial and potentially liver-harming pyrrolizidine alkaloids.

WILD QUININE, MISSOURI SNAKEROOT — Root, leaves, tops

Parthenium integrifolium L. — Aster Family (Asteraceae or Compositae)

Large-rooted perennial; 2–5 ft. Large, oval, lance-shaped leaves, *to 1 ft. long; rough,* blunt-toothed. Flowerheads to ¼ in. wide; white, in loose umbels; May–July. **WHERE FOUND:** Prairies, rock outcrops, roadsides. MA to GA; e. TX to MN. **USES:** The Catawba poulticed fresh leaves on burns. Flowering tops were once used for "intermittent fevers" (such as malaria). Root used as a diuretic for kidney and bladder ailments,

Wild Quinine is common in prairie habitats.

Wild Quinine's cut root (left) resembles that of Pale Purple Coneflower (right); hence, it is a common Echinacea *adulterant.*

gonorrhea. One study suggests Wild Quinine may stimulate the immune system. Root once a widespread and common adulterant—historically and in modern times—to commercial root supplies of Purple Coneflower (*Echinacea purpurea,* p. 267). Confusion extended to scientific literature in the mid-1980s when German researchers published on what they thought were new compounds from *Echinacea purpurea* that actually turned out to be from Wild Quinine—a scientific effort resulting in an entire restructuring of analytical methods and regulatory approvals for *Echinacea* products in Germany. It also highlights the need for authentication of plant material and retention of reference samples for scientists studying all aspects of medicinal plants. **WARNING:** May cause dermatitis or allergies.

WHITE CROWN-BEARD, FROSTWEED — Root

Verbesina virginica L. — Aster Family (Asteraceae or Compositae)

Tough, rough-hairy, *winged-stemmed* perennial up to 7 ft. tall. Leaves mostly alternative, lance-shaped to ovate-lance-shaped, with few irregular coarse teeth or sinuses. Flowerheads in flat to panicle-like clusters, white, with *3–5 irregularly spaced ray flowers* on coarse flowerheads with 8–12 disk flowers; to about an inch across; July–Oct. **WHERE FOUND:** Roadsides, wood edges, disturbed sites, bottom lands. MD to FL; TX to KS, IL. **USES:** Various Southeastern tribes, including the Chickasaw, Choctaw, and Seminole, used the root tea as a diuretic, also for fevers, an emetic for stomachache. In the 1850s physicians used the root as a powerful sweat-inducing agent (usually employed in fevers); also as a purifying agent, principally used as a diuretic. Little studied; contains triterpenoid saponins verbesinosides A–F. **REMARKS:** The name Frostweed refers to the fact that at the time of the first few autumn frosts, the dead stalks still pull moisture from the soil, which becomes extruded ribbons of ice on the stems, easily flagging the plant's location on a frosty morning.

White Crown-beard is a coarse plant with winged stems.

The name Frostweed refers to the fact that the stems produce "frost flowers" during the first few autumn frosts.

FLOWERHEADS LONG, CYLINDRICAL; COMPOSITES WITH WEEDY GROWTH HABITS

PALE INDIAN PLANTAIN

Leaves

Arnoglossum atriplicifolium (L.) H. Rob.
(*Cacalia atriplicifolia* L.)

Aster Family
(Asteraceae or Compositae)

Large perennial; 4–9 ft. Stems smooth or slightly striated. Leaves broadly rounded to triangular, with irregular rounded teeth; glaucous beneath, *palmately veined*. Flowers in flat clusters; July–Oct. Each tubular head has 5 flowers, apparently without rays (petals minute). **WHERE FOUND:** Dry woods, forest edges, openings. NY to FL; AR, OK, NE to MI, MN. **USES:** Cherokee used the leaves as a poultice for cancers, cuts, and bruises, and to draw out blood or poisonous material as a fresh-leaf poultice, changed frequently. Dried powdered leaves reportedly used as a seasoning. Generically, Indian Plantains were sometimes called Wild Cabbage and used as food like beet leaves. **RELATED SPECIES:** *Arnoglossum reniforme* (Hook.) H. Rob. [*A. muehlenbergii* (Sch.-Bip.) H. Rob.; *Cacalia muehlenbergii* (Sch. Bip.) Fern.] (not shown) is similar, but its leaves are green on both sides; stems grooved. Indian Plantain *Arnoglossum plantagineum* Raf. [*Cacalia plantaginea* (Raf.) Shinners] has distinct parallel veins (like plantain). Asian relatives have purported antimicrobial, antioxidant, and anticancer potential.

Pale Indian Plantain and related species have flowers in flat-topped clusters.

Indian Plantain leaves have parallel veins similar to those of Plantain leaves.

HORSEWEED, CANADA FLEABANE

Whole plant

Conyza canadensis (L.) Cronq. Aster Family (Asteraceae or Compositae)
(*Erigeron canadensis* L.)

Bristly annual or biennial weed; 1–7 ft. Leaves numerous, lance-shaped. Flowers greenish white. See p. 279.

PILEWORT, FIREWEED

Whole plant

Erechtites hieraclifolius (L.) Raf. Aster Family (Asteraceae or Compositae)
(Frequent spelling variants: *E. hieracifolia, E. hieraciifolius*)

Annual; 1–9 ft. Leaves lance-shaped to oblong, 2–8 in. long; toothed, often lacerated. Flowers white, with no rays; *flowers are enveloped in a swollen group of leafy bracts, somewhat like a bullet shell*; July–Oct. Whole plant with rank, nauseous odor; unpleasant bitter taste. **WHERE FOUND:** Thickets, burns, waste places. ME to FL; TX, OK, SD to MN.

LEFT: *Pilewort in full bloom does not inspire use in a flower arrangement.* BOTTOM LEFT: *Pilewort has leaves with coarse teeth, often lacerated.* BOTTOM RIGHT: *Horseweed in full bloom. This tall plant has a weedy appearance.*

USES: Tea or tincture of whole plant formerly used as an astringent and tonic in mucous-tissue ailments of lungs, bowels, stomach; also used externally for muscular rheumatism, sciatica, contusions, wounds, and ulcers. Used in diarrhea, cystitis, dropsy, etc. In large doses emetic. Essential oil once recommended in prescriptions for hemorrhoids, hence the common name Pilewort. Neglected by scientific investigators.

WHITE LETTUCE, RATTLESNAKE ROOT — Whole plant

Prenanthes alba L. — Aster Family (Asteraceae or Compositae)

Perennial; 2–5 ft. Stem smooth, purple, with whitish bloom. Leaves triangular or deeply lobed; toothed. Flowers white, in *drooping clusters*; July–Sept. "Seed" (technically fruit) fuzz a deep rust color. **WHERE FOUND:** Rich woods, thickets. ME to GA; MO to ND. **USES:** The Chippewa put powdered root in food to stimulate milk flow after childbirth. Iroquois used tea as a wash for "weakness." Ojibwa used stem latex as a diuretic in "female" diseases; boiled in milk, taken internally for snakebites. Leaves poulticed on snakebites. Roots poulticed on dog bites and snakebites. Tea drunk for dysentery. Other *Prenanthes* species seem to have been used similarly.

White Lettuce, Rattlesnake Root. Note the drooping flowers.

WOOLLY PLANTS; EVERLASTING FLOWER CLUSTERS

PEARLY EVERLASTING — Whole plant

Anaphalis margaritacea (L.) Benth. & Hook. f. — Aster Family (Asteraceae or Compositae)

Perennial; 1–3 ft. Highly variable. Stem and leaf undersides cottony. Leaves linear; gray-green above. Flowers in a cluster of globular heads; July–Sept. Heads with *several rows of dry, white petal-like bracts* (male flowers have yellow tufts in center). **WHERE FOUND:** Dry soil, fields. NL to NC; CA to AK. Alien from Europe; widespread elsewhere. **USES:** Expectorant, astringent, anodyne, sedative. Used for diarrhea, dysentery. Leaves smoked for throat and lung ailments. Cherokee used tea or chewed leaves for colds and sore throat. Dried leaves smoked for

ABOVE: *The flowers of Pearly Everlasting remain intact after drying.* RIGHT: *Plantain-leaved Pussytoes. Historically most species of pussytoes are used interchangeably.*

bronchial coughs. Poultice used for rheumatism, burns, sores, bruises, and swellings. Components isolated from leaf with potential antibacterial activity.

PLANTAIN-LEAVED PUSSYTOES — Whole plant
Antennaria plantaginifolia (L.) Hook. — Aster Family (Asteraceae or Compositae)

Highly variable, mat-forming woolly-stemmed perennial; 3–16 in. Basal leaves *spoon-shaped*, silky-gray-woolly (or smooth), with 3–5 nerves (veins); more woolly on upper-leaf surface than below. Stem leaves small, lance-shaped. Flowers white in several flowerheads; Apr.–June. **WHERE FOUND:** Dry woods, fields. NB, NS, ME to GA; TX to SK. **USES:** Cherokee used tea for excessive menstrual bleeding. Boiled in milk, this plant was a folk remedy for diarrhea, dysentery. Tea drunk for lung ailments. Leaves poulticed on bruises, sprains, boils, and swellings. One of the multitude of snakebite remedies. **RELATED SPECIES:** The genus *Antennaria* is highly variable and is polymorphic, with specific identifying characters melding from one species to another, leaving their identity to the realm of specialists in the plant group. For practical purposes, most of what occurs in the Appalachians, Atlantic Seaboard, and areas of WI and MN are assigned to *A. plantaginifolia* Hook. Much of what was previously assigned to *A. plantaginifolia* is now assigned to two subspecies of *A. parlinii* (see below). It is likely that all eastern N. American species with gray-woolly leaves were used interchangeably in a generic sense, despite historical attribution.

PARLIN'S PUSSYTOES, PLAINLEAF PUSSYTOES — Leaves

Antennaria parlinii Fern. — Aster Family (Asteraceae or Compositae)

Much of what was formerly assigned to *A. plantaginea* is now included under *A. parlinii* and its varieties. Variable, mat- or small-colony-forming, herbaceous, densely soft hairy perennial; 4–15 in. tall. Leaves persistent year-round, ovate or broadly elliptical, 1¼–2½ in. long, half as wide, with 3–5 prominent nerves; more hairy beneath. Flowers dull white in loose or tight clusters with several compact flowerheads; *female flowers in elongated head* (to about ¼ in. tall), male flowers in *compressed buttonlike head*; Apr.–June. **WHERE FOUND:** Acidic soils, sandstone on rocky dry wooded slopes. ME south to GA, west to TX, north through eastern prairies to se. SD to s. ON. **USES:** Known historically as White Plantain, Poor Robin, Rattlesnake Plantain, or Squirrel Ear, the leaves used in tea or syrup for coughs and fevers; tonic for debility. Externally for bruises, inflammation, and for rattlesnake bite. Meskwaki women drank leaf tea for two weeks after childbirth to prevent illness. Diterpenes with antibacterial activity have been isolated from *Antennaria* species.

Male and female flowers of Parlin's Pussytoes are on separate plants. Male flowers, shown here, are in flat-topped heads.

The elongated female flowerheads of Parlin's Pussytoes.

SWEET EVERLASTING, RABBIT TOBACCO (NOT SHOWN) Leaves, flowers
Pseudognaphalium obtusifolium (L.) Hilliard & Burtt Aster Family
(*Gnaphalium obtusifolium* L.) (Asteraceae or Compositae)

Soft-hairy, faintly fragrant annual or winter annual; 1–2 ft. Leaves alternate, lance-shaped, without stalks. Flowers dirty white globular heads in spreading clusters; July–Nov. Flowerheads *enclosed by dry, petal-like bracts*. **WHERE FOUND:** Dry soil, fields. Much of our area. **USES:** Leaves and flowers (chewed or in tea) traditionally used for sore throats, pneumonia, colds, fevers, upset stomach, abdominal cramps, asthma, flu, coughs, rheumatism, leukorrhea, bowel disorders, mouth ulcers, hemorrhage, tumors; mild nerve sedative, diuretic, and antispasmodic. Fresh juice once considered aphrodisiac. **RELATED SPECIES:** Other eastern N. America species of *Pseudognaphalium,* such as *P. helleri* (Britton) Anderb., confined to the Southeast, and *P. micradenium* (Weath.) G. L. Nelson, are both referred to by the same common names and are more sweetly aromatic and fragrant. These species have likely been used interchangeably.

DAISYLIKE FLOWERS

MAYWEED, DOG FENNEL Whole plant
Anthemis cotula L. Aster Family (Asteraceae or Compositae)

⚠ *Bad-smelling* annual; 8–20 in. Leaves finely (thrice-) dissected. Flowers white; May–Nov. Disk flowers studded with *stiff chaff*. **WHERE FOUND:** Waste places. A weedy Eurasian species found throughout N. America. **USES:** Tea used to induce sweating, vomiting; astringent, diuretic. Used for fevers, colds, diarrhea, dropsy, rheumatism, obstructed menses, and headaches. Leaves rubbed on insect stings. Contains a sesquiterpene lactone called anthecotulide, which is not found in German Chamomile and may be used by analytical chemists to detect the admixture of Mayweed with Chamomile supplies. Dried flowerheads have been shown to have antimicrobial activity. **WARNING:** Touching or ingesting plant may cause allergies.

ENGLISH DAISY Whole plant
Bellis perennis L. Aster Family (Asteraceae or Compositae)

Small perennial, to 6 in., with *spatula-shaped, wavy-toothed* leaves in a basal rosette. Leaves 1–2 in. long. Flowers *small, white, daisylike*, about 1 in. across; May–Sept. **WHERE FOUND:** Lawns, roadsides, waste places. Alien Europe, w. Asia. Naturalized from PE to NC, TN, north to MN, ND. **USES:** Leaves in ointments historically used in Europe for the treatment of wounds, bruises, and skin inflammation (sometimes called Bruisewort). Decoction of roots and/or leaves used for gout, fevers, and liver disorders. Science confirms wound healing activity, notably reducing scar tissue formation; antioxidant, antimicrobial; components of research interest for the treatment of metabolic syndrome. **WARNING:** May cause rare contact dermatitis.

LEFT: *Mayweed, Dog Fennel. Crushed leaves and flowers have an unpleasant fragrance.*
ABOVE: *English Daisy, a small perennial found on lawns in some areas, has small daisylike flowers.*

CHICORY

Roots, leaves

Cichorium intybus L. — Aster Family (Asteraceae or Compositae)

Flowers usually blue (rarely white or pink). See also p. 263.

ECLIPTA

Whole plant

Eclipta prostrata (L.) L. — Aster Family (Asteraceae or Compositae)
[*Eclipta alba* (L.) Hassk.]

Annual; up to 1½ ft. tall, with rough hairs; stems weak, rooting at nodes. Leaves opposite, *linear to lanceolate*, apparently entire, but with a few marginal teeth; up to 4 in. long, with weak or without leafstalks. Flowers in terminal or axillary small heads, with white ray flowers, small, to less than 3/8 in. across; mostly July–Oct., but can flower year-round. **WHERE FOUND:** Waste places, cultivated fields, damp sandy soils. Alien. A widespread colonizer in disturbed habitats, obscuring its origins. Found throughout warm temperate and tropical areas worldwide. Mostly absent north of MA. **USES:** Used in Traditional Chinese Medicine (called *li chang; mo han lian*) for bleeding, both internal and external; dysentery, premature gray hairs, bleeding gums, loosening of teeth. One of the major herbs used in Ayurveda, the traditional medical system of India, where it is called *bhringaraja.* The leaves are used as a tonic to treat liver and spleen disease; considered to be cooling. Plant juice used for jaundice and catarrh in infants; rubbed on gums for toothaches; applied externally with oil to relieve headaches. Also used to treat abscesses and snakebites. A study found that it neutralized the venom of the S. American rattlesnake (*Crotalus durissus terrificus*) in the laboratory. Relatively well studied, particularly in India, with sci-

LEFT: *Eclipta is a common tropical- and temperate-climate weed.* ABOVE: *Annual Fleabane is widespread, with smaller flowers than Daisy Fleabane.*

entific confirmation of antibacterial, anticandidal, antidiabetic, antihypertensive, antioxidant, astringent, anti-inflammatory, liver-protectant, and immunostimulating activity. A methanol extract was shown to promote hair growth in laboratory animals. It is deemed a potential nootropic herb—one that may enhance memory and cognitive functions.

DAISY FLEABANE (NOT SHOWN)

Whole plant

Erigeron philadelphicus L. Aster Family (Asteraceae or Compositae)

⚠ Slender, hairy annual or biennial; 1–3 ft. Basal leaves oblong; stem leaves smaller, *clasping at base*. Flowers less than 1 in. across., usually white, sometimes pinkish, with numerous slender rays and a yellow disk; Apr.–July. **WHERE FOUND:** Thickets. Most of our area. **USES:** Plant tea is used as a diuretic, astringent; a folk remedy for diarrhea, "gravel" (kidney stones), diabetes, painful urination; also used to stop hemorrhages of stomach, bowels, bladder, kidneys, and nose. Once used for fevers, bronchitis, tumors, hemorrhoids, and coughs. This and other *Erigeron* species have been found to have anti-inflammatory and antioxidant activity. **RELATED SPECIES:** Other common and widespread Fleabanes including *E. pulchellus* Michx., easily identified with its densely hairy stems, *E. strigosus* Muhl., and *E. annuus* (L.) Pers. All are easily confused with and used interchangeably with Daisy Fleabane. They occur throughout much of our range. All were generally used as a mild diuretic and mild astringent. Approximately 170 variable species of *Erigeron* are found in N. America. **WARNING:** May cause contact dermatitis.

Also called Daisy Fleabane, Erigeron pulchellus *is confused with and used similarly to* E. philadelphicus.

Ox-eye Daisy, the familiar roadside daisy, is a widely naturalized weed across the U.S.

OX-EYE DAISY

Whole plant, roots

Leucanthemum vulgare (Vaill.) Lam.
(*Chrysanthemum leucanthemum* L.)

Aster Family
(Asteraceae or Compositae)

⚠ Familiar white daisy of roadsides; perennial; 1–3 ft. tall. Leaves more or less *cleft, lobed, or toothed*; reduced above, lance- to spatula-shaped at base. Flowers have white petals and yellow centers, 2 in. across; May–Sept. **WHERE FOUND:** Fields, roadsides, waste places. Throughout our range. Alien (Europe). Very common weed. **USES:** In European traditions, used as a tonic, antispasmodic (in whooping cough), diuretic, astringent; for asthma, to regulate menses, and induce vomiting. Root formerly used to reduce night sweats; flowers used in tea as an antispasmodic. Native Americans adopted tea for fever and as a spring tonic; externally, a wash for chapped hands; eyewash. Effects considered similar to those of Chamomile. Contains anti-inflammatory and spasm-relieving components. Known to concentrate polychlorinated biphenyls (PCBs) from contaminated soil and suggested as a plant to grow for phytoremediation of contaminated sites. **WARNING:** May cause contact dermatitis or cross-allergic reactions with other Aster family members.

GERMAN, HUNGARIAN, OR WILD CHAMOMILE

Flowers, leaves

Matricaria recutita L.
Aster Family (Asteraceae or Compositae)
[*Chamomilla recutita* (L.) Rauschert; *Matricaria chamomilla* L.]

Smooth, apple or musk-scented annual; 6–24 in. Leaves finely divided, with 3 fine veins. Flowers daisylike, ¾ in. across, disk flowers mostly 5-lobed.; *receptacle hollow within*; May–Oct. **WHERE FOUND:** Locally abundant. Much of our area. Alien. **USES:** In Germany, Chamomile's reputation as a medicinal herb is reflected in its name, which translates

German Chamomile has hollow flower receptacles.

into "capable of anything." Dried flowers make a famous beverage tea, traditionally used for colic, diarrhea, insomnia, indigestion, gout, sciatica, headaches, colds, fevers, flu, cramps, and arthritis. Flowers also a folk remedy for cancer. Experimentally, essential oil is antifungal, antibacterial, anodyne, antispasmodic, anti-inflammatory, and antiallergenic. One component in the leaves, apigenin, has been shown to have anti-anxiety and sedative activity. Essential oil contains at least two dozen different compounds with anti-inflammatory action, which may work in concert with one another to reduce inflammation. Suggested by some researchers as an alternative treatment for phlebitis. Approved in Germany and elsewhere for internal use to treat inflammation or spasms of the gastrointestinal tract and inflammation of the respiratory tract. Externally, approved for inflammatory conditions of the mouth and gums or bacterial-induced skin diseases. Recent research suggests that consumption of Chamomile tea may help to prevent progression of diabetic complications. Among the most widely used medicinal plants and herbal tea ingredients in the world. **WARNING:** Ragweed allergy sufferers may react to Chamomile, too.

FEVERFEW

Leaves, whole plant

Tanacetum parthenium (L.) Shultz-Bip. — Aster Family
[*Chrysanthemum parthenium* (L.) Bernh.] — (Asteraceae or Compositae)

⚠ Bushy, tap-rooted perennial; 1–3 ft. Leaves pinnately divided into *ovate rounded divisions*, with upper leaves more deeply lobed; coarsely toothed. Flowers daisylike (but smaller), with a *large disk* and *stubby white rays*; June–Sept. **WHERE FOUND:** Roadsides, waste places. Alien native to the Balkan Peninsula; escaped from cultivation in N. and S. America, most of Europe. Widely grown as a garden ornamental and in herb gardens. Highly variable both in looks and chemistry. Some chemical types contain a purported active constituent, parthenolide, one of several closely related compounds with anti-migraine potential. **USES:** Tea of whole plant a folk remedy for arthritis, colds, fevers, cramps, worms; regulates menses; sedative. Proven effective (1–4 fresh leaves chewed per day) as a preventive (reducing number, duration, and increasing time between) of migraine attacks. British studies suggest that Feverfew can prevent 70 percent of migraines. Backed by five clini-

LEFT: *Feverfew is commonly grown as an ornamental.* ABOVE: *Feverfew has daisylike flowers with relatively wide ray petals.*

cal studies (four positive, one negative). One study revealed that Feverfew can profoundly reduce pain intensity and typical symptoms associated with migraine attacks. Most research was conducted in the 1980s and 1990s, without additional recent studies. Approved in Canada and England for use against migraines. Also shown to be antiseptic, anti-inflammatory. The essential oil is reportedly anti-inflammatory, analgesic, and a potential insect-repellent. **WARNING:** May cause dermatitis or allergic reactions. Mouth sores common; a systemic reaction resulting from chewing the leaves.

AMERICAN SWEETFLAG, AMERICAN CALAMUS

Rhizome

Acorus americanus (Raf.) Raf. Sweet Flag Family (Acoraceae)
Acorus calamus var. *americanus* (Raf.) H. Wolf. [Formerly in Arum Family (Araceae)]

Now recognized as a separate species, *A. americanus* differs from *A. calamus* by leaves with *prominent single midrib with 1–5 additional midribs* equally raised above the leaf surface. This is a fertile diploid species, producing viable seeds. *A. calamus* in N. America is a sterile triploid not producing viable seeds. **WHERE FOUND:** Wet soils, marshes, banks of slow-moving water. PE, NL to s. New England, NY, n. PA, west to IA, e. NE, SD, ND and adjacent Canadian provinces. **USES:** German studies show that for maximum efficacy and safety against spasms, the diploid American species, devoid of beta-asarone, should be used. Oils devoid of beta-asarone showed spasmolytic properties comparable to those of standard antihistaminic drugs. More studies are needed to discern chemical differences between *A. americanus* and *A. calamus* with properly authenticated specimens.

SWEETFLAG, CALAMUS

Rootstock

Acorus calamus L. Sweet Flag Family (Acoraceae)
[Formerly in Arum Family (Araceae)]

⚠ Strongly aromatic, colony-forming perennial; 1–4 ft. Root jointed. Cattail-like leaves, with *a single vertical midrib*. Flowers tightly packed on a *fingerlike spadix*, jutting at an angle from leaflike stalk, flowers sterile; May–Aug. **WHERE FOUND:** Pond edges, wet fields. Most of our area. Often found in depressions of pastures, pond edges, ditches, and moist meadows, with grasslike leaves; produces fleshy rhizomes that form large clumps. It is the aromatic rhizomes that are used in traditional medicine. The fingerlike flowerheads, which grow about one-third up the stem, are usually found on a low percentage of plants in a given population. **USES:** Dried-root tea (or fresh or dried rhizome chewed) as an aromatic bitter for gas, stomachaches, indigestion, heartburn, fevers, colds, and coughs; antispasmodic, anticonvulsant, and CNS-depressant. In India, used as an aphrodisiac. Native Americans nibbled root for stomach ailments, to assuage thirst, and as a stimulant on long journeys, and were probably key to spreading the Old World strains of *A. calamus* once introduced by European settlers. *A. calamus* is frequently found in disjunct populations near sites of historic Indian villages and trails. The roots were also an item of trade among Native American groups. Science confirms a wide range of activities, including analgesic, antibacterial, antioxidant, anti-inflammatory, and neuroprotective effects. Oils devoid of beta-asarone showed spasmolytic properties comparable to those of standard antihistaminic drugs. Controlled dosage of root helped lower serum cholesterol levels in rabbit studies. Essential oil from roots repels some insects. **WARNING:** Some strains said to contain the carcinogen beta-asarone.

American Sweetflag, American Calamus has aromatic, grasslike leaves with 2–5 raised midribs.

Sweetflag. Note the single, prominent, raised midrib.

ASPARAGUS

Root, shoots, seeds

Asparagus officinalis L.

Asparagus Family (Asparagaceae)
[Formerly in Lily Family (Liliaceae)]

⚠ Perennial; 6 ft. or more. Leaves *finely fernlike* (actually branches functioning as leaves). Flowers seldom noticed. Fruits reddish; June–July. **WHERE FOUND:** Garden escape. Throughout our area. Alien. **USES:** Spring shoots a popular vegetable. Asian Indians report asparagine (in shoots) is a good diuretic in dropsy and gout. Japanese report green asparagus aids protein conversion into amino acids. Has been suggested as a food to treat gout, as asparagus contains at least ten anti-inflammatory minerals or compounds, but also rumored to trigger gout. Roots considered diuretic, laxative (because of fiber content), induce sweat-

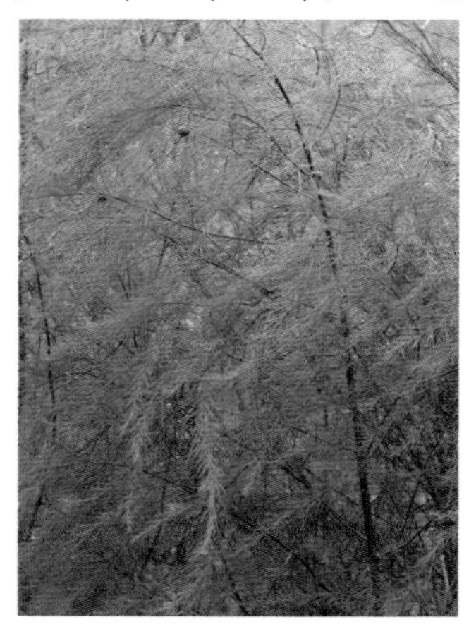

Asparagus is fernlike as it matures, producing tiny yellowish flowers.

ing. Chinese report roots can lower blood pressure and blood glucose parameters. Seeds possess antibiotic activity. Use of root is approved in European phytomedicine as a diuretic for irrigation therapy in the treatment of urinary tract inflammation; also to prevent kidney stones. **WARNING:** May cause rare contact dermatitis or allergic skin reactions; berries potentially toxic.

CUPLIKE FLOWERS (AT LEAST WHEN YOUNG)

MARSH-MARIGOLD, COWSLIP — Root, leaves

Caltha palustris L. — Buttercup Family (Ranunculaceae)

⚠ Aquatic perennial with a *succulent, hollow stem*. Leaves glossy, *heart- or kidney-shaped*. Flowers like a large buttercup (to 1½ in. wide); deep yellow-orange, with 5–9 "petals" (actually sepals); Apr.–July. **WHERE FOUND:** Swamps, wet ditches; restricted to slow-moving water in mountains toward south. PE, south to NC, TN mtns.; west to IA, Canadian Provinces, AK. **USES:** Root tea induces sweating; emetic, expectorant. Leaf tea diuretic, laxative. Ojibwa mixed tea with maple sugar to make a cough syrup that was popular with colonists. Syrup used as a folk antidote to snake venom. Contains anemonin and protoanemonin, both with marginal antitumor activity. Flowerbuds once pickled as a caper substitute. **WARNING:** All parts may irritate and blister skin or mucous membranes. Sniffing bruised stems induces sneezing. Intoxication has

Marsh-marigold, Cowslip has yellow buttercup-like flowers.

Marsh-marigold occurs in wet habitats. Note the round, toothed leaves and deeply notched base.

resulted from the use of the raw leaves in salads and from using the raw flower buds as substitutes for capers. Do not confuse with American White or False Hellebore, which can be fatally toxic (see p. 278).

AMERICAN LOTUS, YELLOW LOTUS — Roots, leaves, seeds

Nelumbo lutea Willd. — Lotus Family (Nelumbonaceae)
[Formerly Water-Lily Family (Nymphaeaceae)]

Large aquatic perennial, with *leaf stalks up to 6 ft.* tall. Leaves *large, cupped in center, umbrella-like*, up to 2 ft. across. Flowers whitish to pale yellow; June–Oct. See p. 19.

SPATTERDOCK, YELLOW POND LILY — Root

Nuphar lutea (L.) Sm. — Water-lily Family (Nymphaeaceae)
[*Nuphar advena* R. Br.; *Nuphar lutea* var. *advena* (Aiton) Kartesz & Gandhi]

⚠ Aquatic perennial. Leaves round-oval; *base V-notched*; leaves submersed or erect above water. Flowers yellow, cuplike; *stigma disklike*; May–Sept. **WHERE FOUND:** Ponds, slow-moving water. Canada south to SC; west to IL, IA. **USES:** Native Americans used root tea for "sexual irritability," blood diseases, chills with fever, heart trouble; poulticed on swellings, inflammations, wounds, contusions, boils. Elsewhere, roots used for gum, skin, and stomach inflammations. A folk remedy for

Flowers of American Lotus are whitish to pale yellow.

Spatterdock, Yellow Pond Lily. Yellow Pond Lilies have cup-shaped, yellow flowers. Historically, species were used interchangeably.

impotence; rhizome contains steroids. Alkaloids reportedly hypotensive, antispasmodic, cardioactive, tonic; vasoconstrictor, antimicrobial; experimental anticancer activity. Like many other species, this plant contains antagonistic alkaloids, one hypotensive, one hypertensive. Can the human body select the one it needs? **REMARKS:** Historically treated as a single highly variable species, but molecular studies indicate that the once polymorphic single species probably represents several species in N. America, treated in herbal traditions as a single species. Probably most eastern N. American material can be assigned to *Nuphar advena* R. Br. **WARNING:** Large doses of root potentially toxic.

PRICKLY-PEAR CACTUS

Pads, fruits

Opuntia humifusa (Raf.) Raf. — Cactus Family (Cactaceae)

Cactus; to 1 ft. Jointed pads have tufts of bristles, usually sharp-spined. Large, showy yellow flowers; May–Aug. **WHERE FOUND:** Dry soils. MA to FL; TX to MN, west to NE, SD. Our most common eastern cactus. **USES:** The Pawnee and other native groups poulticed peeled pads on wounds, applied juice of fruits to warts, and drank pad tea for lung ailments. In folk medicine, peeled pads poulticed for rheumatism, juice used for "gravel" (kidney stones), baked pads used for gout, chronic ulcers, and wounds. The edible red fruit pulp was considered diuretic and cooling. African slaves in the South ate the peeled, split pads; also poulticed for rheumatism, gout, wounds, and chronic ulcers. Laboratory research suggests potential for increasing bone density. A yogurt/Prickly-pear extract facial mask, the subject of a Korean study, suggests the combination improved moisture, brightness, and elasticity of treated skin. Experimentally shown to reduce blood glucose and cholesterol levels. Antioxidant and anti-inflammatory activity also reported. Introduced to East Asia; interestingly, most commercial products containing this cactus or its extracts are cosmetic and dietary supplements (drink products) made for the Korean market. **RELATED SPECIES:** *Opuntia ficus-indica* (L.) P. Mill. has been the subject of a number of Mexican clinical trials exploring its potential to help non-insulin-dependent diabetics. The results have been mixed. An Israeli research group found the dried flowers useful in reducing the urgency to urinate in cases of benign prostatic hyperplasia. The fruit and leaf pads of various *Opuntia* species are food items available wherever Mexican foods are sold in the U.S.; now widely available.

Prickly-pear Cactus has beautiful yellow-orange flowers and produces edible fruits.

4-PETALED FLOWERS IN TERMINAL CLUSTERS; MUSTARDS

WINTER CRESS

Leaves

Barbarea vulgaris R. Br. Mustard Family (Brassicaceae or Cruciferae)

Highly variable, smooth-stemmed mustard; 1–2 ft. Lower leaves with 4–8 *lateral, earlike lobes*; uppermost leaves clasping, cut, toothed. Flowers deep yellow; Mar.–Aug. Seedpods (silique) mostly erect, short-beaked; fruit stalks more slender than the pods. **WHERE FOUND:** Wet fields, ditches, roadsides. Throughout our range. Alien (Europe); widely naturalized throughout temperate climates. **USES:** Cherokee ate greens as a blood purifier. The leaf tea was taken once every half hour to suppress coughs. Tea thought to stimulate appetite; diuretic, used against scurvy. Europeans poulticed leaves on wounds. Saponins with insect-resistant and feeding-deterrent activity have recently been isolated from Winter Cress. **WARNING:** Although this plant has been described as an edible wild food, studies suggest that it may cause kidney malfunctions.

Winter Cress. Note the earlike lobes on lower leaves.

BLACK MUSTARD

Seed, leaves, oil

Brassica nigra (L.) Koch Mustard Family (Brassicaceae or Cruciferae)

Annual; 2–3 ft. Lower leaves bristly, coarsely lobed; upper leaves *lance-shaped, with no hairs*. Flowers yellow; June–Oct. Pods closely hug stem. **WHERE FOUND:** Waste places. Throughout our area; absent from AR, SC, GA. Alien, sw. Asia, Europe; widely cultivated elsewhere. **USES:** Leaves and seeds irritant, emetic. Leaf poultice used for rheumatism, chilblains, toothaches, headaches. Seeds eaten as a tonic and appetite stimulant, for fevers, croup, asthma, bronchial conditions. Ground seeds used as a snuff for headaches. Leafy *Brassica* species have emerged as important subjects in the study of cancer prevention. Contains compounds shown to have strong anticancer activity and may prevent (even at relatively small levels) breast and colon cancers; polyphenols in leaf extracts strongly antioxidant. Works in part by encouraging the self-destruction of cancer-damaged cells, and also by increasing intracellular glutathione. **WARNING:** Allyl isothiocyanate (responsible for mustard

LEFT: *Black Mustard, with hairless, lance-shaped upper leaves.*
ABOVE: *The seeds of Black Mustard.*

flavor) is a strong irritant. May blister skin. Eating large quantities may cause red, burnlike skin blotches that occasionally develop into ulcers.

FIELD OR WILD MUSTARD (NOT SHOWN) — Seeds

Brassica rapa L. — Mustard Family (Brassicaceae or Cruciferae)

Succulent, *gray-green* annual herb; 24–32 in. Lower leaves sparsely toothed or divided. Differs from other *Brassica* species in that the upper leaves *clasp the stem,* with *earlike lobes*. Flowers pale yellow; June–Oct. Pods erect, slender-beaked. **WHERE FOUND:** Fields. Throughout our area. Alien (Eurasia). **USES:** Crushed ripe seeds poulticed on burns. Like some other Mustard family members, it contains factors that the National Cancer Institute has suggested may prevent certain cancers. Leafy vegetables of many wild and cultivated *Brassicas* (cabbage, cauliflower, broccoli, collards, kale, kohlrabi, mustard, rape, turnips) are rich in vitamins A and C, fiber, and isothiocyanates, all cited by the National Cancer Institute as having some cancer-preventing activity.

FLOWERS WITH 4 PETALS

CELANDINE — Stem juice, roots, leaves

Chelidonium majus L. — Poppy Family (Papaveraceae)

⚠ Smooth-stemmed biennial; 1–2 ft. Stems brittle; yellow juice within. Leaves divided, round-toothed or lobed, also with *yellow juice in leaf veins*. Flowers yellow, 4-petaled; to ¾ in. across; Apr.–Aug. Seedpods *smooth, linear, 2-valved*. **WHERE FOUND:** Waste places. Much of our area. Alien weed (Europe). **USES:** Fresh stem juice a folk remedy (used ex-

Celandine flowers are 4-parted.

Celandine produces yellow sap in its leaves and stems.

ternally) for warts, eczema, ringworm, corns. The root juice is bright orange. A folk medicine of N. America, Europe, and China. Root tincture once used by physicians for inflammations, hemorrhoids; taken internally for jaundice, lung ailments, diuresis. Fresh leaves once used for amenorrhea, poulticed for wounds. Folk remedy for cancer in China. Mentioned as a potential cancer remedy in the *Journal of the American Medical Association* as early as 1897, Celandine contains at least 4 antitumor compounds, including chelerythrine, citric acid, coptisine, and sanguinarine. Ukrain, a semisynthetic version of chelidonine, causes regression of tumors. Approved in European phytomedicine for treatment of discomfort caused by spasms of bile ducts and the gastrointestinal tract in controlled dosages. Chelidonic acid is a COX-2 inhibitor; shown to have experimental potential in ulcerative colitis. In experiments has shown blood pressure–lowering activity. Experimentally, Celandine and its compounds have been shown to have anti-inflammatory, antimicrobial, analgesic, and liver-protectant activity. **WARNING:** Can be toxic; although thought to help protect the liver, Celandine is associated with several cases of liver toxicity, heightening debate about its safety. Stem juice highly irritating, allergenic; may cause paralysis.

ST. ANDREW'S CROSS (NOT SHOWN) — Root, leaves

Hypericum hypericoides (L.) Crantz. St. John's Wort Family (Hypericaceae)

Similar to St. John's Wort (which see p. 157), except St. Andrew's Cross has 4 petals. Variable, smooth subshrub (somewhat woody); 1–2½ ft. Leaves linear-oblong; in pairs. Flowers terminal, solitary, with *4 sepals*;

1 pair of sepals large and leaflike, the other pair tiny or lacking; narrow yellow petals form a cross; July–Aug. **WHERE FOUND:** Sandy soil. MA to FL; e. TX, NE, to IL. **USES:** Cherokee poulticed for snakebite, made tea for fevers, bowel complaints; crushed leaves used to stop nosebleed. Other native groups, including the Alabama, Choctaw, and Houma, used root tea for colic, fevers, pain, toothaches; externally, as a wash for ulcerated breasts. Leaf tea used by herbalists for bladder and kidney ailments, skin problems, and children's diarrhea. Like St. John's Wort, in laboratory experiments St. Andrew's Cross has shown some potential against HIV-infected cells. **WARNING:** May cause photodermatitis (see *H. perforatum,* p. 157).

COMMON EVENING-PRIMROSE — Seeds, root

Oenothera biennis L. — Evening-primrose Family (Onagraceae)

Biennial; 1–8 ft. Leaves numerous, lance-shaped. Flowers yellow, with 4 broad petals; June–Sept. *Sepals drooping, stigma X-shaped.* Flowers bloom after sunset, unfolding before the eyes of those who watch them open, hence the common name. **WHERE FOUND:** Roadsides, fields. Native, sometimes weedy, throughout our area; naturalized and weedy in China, Europe, S. America, and New Zealand. **USES:** Cherokee used root tea for obesity, bowel pains; poulticed root for hemorrhoids. Iroquois used root for bruises; rubbed root on muscles to give athletes strength. Recent research suggests seed oil may be useful for atopic eczema, allergy-induced eczema, asthma, migraines, inflammations, premenstrual syndrome, breast problems, metabolic disorders, diabetes, arthritis, and alcoholism. Research has demonstrated that seed oil preparations can alleviate imbalances and abnormalities of essential fatty acids in prostaglandin production. Evening-primrose oil is a natural source of gamma-linolenic acid (GLA). Cold-pressed seed oil is produced commercially. Components of seed oils are anti-inflammatory; one of the top 10 ingredients used in anti-aging creams. Approved in Britain for treatment of atopic eczema, premenstrual syndrome, and prostatitis.

CELANDINE POPPY — Leaves

Stylophorum diphyllum (Michx.) Nutt. — Poppy Family (Papaveraceae)

⚠ Perennial herb; to 20 in. tall, with *saffron-colored juice in stems and leaves.* Leaves mostly basal, deeply lobed into 5–7 segments, with obtuse lobes or toothed margins; one pair of smaller leaves on stem. Similar to Celandine (above) when flowers are absent. Flowers yellow, to 2 in. across; late Mar.–May. **WHERE FOUND:** Rich, moist woods. PA south to TN, west to AR.; north to MI. Increasingly available in the native perennial plant trade. **USES:** Suggested as an antispasmodic by nineteenth-century physicians. Tincture used externally for skin eruptions and hemorrhoids. Internally it was used for liver and spleen afflictions. Historically, the root has also been found as an adulterant to Goldenseal. Contains the alkaloids chelidonine and chelerythrine, also found in Celandine. **RELATED SPECIES:** The only other species, *Stylophorum lasiocarpum*

ABOVE: *Common Evening-primrose generally blooms in the evening and on overcast days.* RIGHT: *Celandine Poppy occurs in rich woods, blooming in early spring; it has thick yellow juice within stems.*

(Oliver) Fedde and *S. sutchuenense* (Franch.) Fedde, both endemic to China, where they are considered obscure medicinal plants, of interest because of their alkaloids. **WARNING:** Considered potentially poisonous.

YELLOW OR YELLOWISH ORCHIDS

ADAM-AND-EVE ROOT, PUTTY ROOT

Root

Aplectrum hyemale (Muhl. ex Willd.) Nutt. Orchid Family (Orchidaceae)

Perennial; 7–16 in. Large single leaf with *distinct pleats (folds)* or often *white* lines; leaf lasts through winter, shrivels before plant flowers. Flowers yellowish to greenish white; *lips purple, crinkle-edged*; May–June. **WHERE FOUND:** Rich woods. QC, VT to GA mtns.; west to AR, e. KS, s. MN. Leave it alone! Too rare to harvest. **USES:** Cherokee proffered root to children as a ceremonial "gift of eloquence" to make them fleshy and fat. Roots poulticed on boils. Root tea formerly used for bronchial troubles. The root is gummy and adhesive, hence the name Putty Root.

SPOTTED CORALROOT

Root

Corallorhiza maculata (Raf.) Raf. Orchid Family (Orchidaceae)
(Often misspelled "Corallorrhiza")

Brownish—no or little chlorophyll; 8–20 in.; stalk *sheathed,* leafless. Flowers grayish yellow to dull purple, with purple-red spots; July–Aug. **WHERE FOUND:** On leaf mold, in woods. NL to NC, GA mtns.; OH, SD, and westward. Six species (apparently used interchangeably) occur in our area. **USES:** Folk remedy for colds, "breaking fevers"; induces profuse

Adam-and-Eve Root. Note the red-tinged petals.

Spotted Coralroot lacks chlorophyll.

sweating, which reduces temperature. Root tea also used for bronchial irritation, coughs. In the American West, the Paiute and Shoshone used root tea as a blood "strengthener." Iroquois ritually used root as a wash for hunting implements and as a love charm and to thwart witchcraft. All tuberous-bulbed orchids at one time or another have been believed to be aphrodisiacs and uterine tonics, based on long-standing historical use in European and Asian traditions of various tubers of terrestrial orchids traditionally ground into a flour known as "salep."

YELLOW LADY'S-SLIPPER, AMERICAN VALERIAN — Root

Cypripedium parviflorum Salisb. — Orchid Family (Orchidaceae)
[*Cypripedium calceolus* var. *pubescens* (Willd.) Correll; *C. calceolus* var. *parviflorum* (Salisb) Fern., and other designations]

Highly variable, with much morphological plasticity; also hybridizes with other species in the genus, resulting in confused taxonomy. Here all N. American, yellow Lady's-slippers are treated collectively as varieties of *C. parviflorum.* Individuals growing in different habitats also highly variable. Plants in the Southeast generally are smaller in size, whereas plants from more northerly areas are larger (historically called *Cypripedium pubescens* Willd. Large Yellow Lady's-slipper). Perennial; 8–36 in. Leaves broadly lance-shaped; *alternate*, hugging stem. Flowers yellow, often purple-streaked; May–July. **WHERE FOUND:** Rich woods, bogs. All of the lower 48 states and adjacent Canada except for the extreme Southeast; northern Europe and Asia. Too rare to harvest. Yellow Lady's-slippers were probably once more common, but heavy harvesting for medicinal use in the nineteenth century likely greatly decreased populations. **USES:** Lady's-slippers, called American Valerian, were widely used in nineteenth century America and promoted as a sedative for nervous headaches, hysteria, insomnia, nervous irritability, mental depression from sexual abuse, and menstrual irregularities accompanied by despondency (PMS?). Active compounds not water-soluble. Once considered a useful substitute for opium, but without narcotic effects. **RELATED SPECIES:** The Pink Lady's-slipper (*C. acaule,* p. 32) was considered a substitute for the more commonly used Yellow Lady's-slipper; its properties were considered analogous. **WARNING:** All Lady's-slippers may cause dermatitis. All Lady's-slippers are too rare for harvest and some are protected as federally listed or state-listed endangered species. All terrestrial orchids are protected in international trade.

Large Yellow Lady's-slipper is one of our most striking orchids.

FLOWERS WITH 5 PETALS; LOW-GROWING PLANTS

INDIAN STRAWBERRY

Whole plant, flowers

Duchesnea indica (Andr.) Focke

Rose Family (Rosaceae)

Small, creeping perennial; to 6 in. Leaves strawberry-like. Flowers *yellow*; 3-toothed bracts longer than petals and sepals; Mar.–July. Strawberry-like shiny red fruit; *spongy* within; flavor insipid. **WHERE FOUND:** Yards, waste places. Most of our area. Asian alien. **USES:** In Asia, whole plant a folk medicine used as a poultice or wash (astringent) for abscesses, boils, burns, insect stings, eczema, ringworm, rheumatism, traumatic injuries. Whole-plant tea

Indian Strawberry, a 5-petaled, yellow-flowered alien, has distinct, strawberry-like, 3-parted leaves.

Indian Strawberry fruit (note highly textured surface) does not live up to its namesake in flavor.

All of our yellow-flowered Wood-sorrels, with shamrocklike leaves, are used similarly.

used for laryngitis, coughs, lung ailments. Flower tea traditionally used to stimulate blood circulation. Ethanolic extract of plant found to reduce herpes simplex encephalitis inflammatory injury in brain neurons; antioxidant protecting against skin damage. Inhibits an enzyme on the surface of influenza A virus, producing an antiviral effect.

CREEPING WOOD-SORREL — Leaves

Oxalis corniculata L. and related species — Wood-sorrel Family (Oxalidaceae)

Creeping perennial; 6–10 in. Leaves *cloverlike*, with large brownish stipules. Flowers yellow; Apr. to frost. Seedpods deflexed. **WHERE FOUND:** Waste places. Throughout our range. Alien. **USES:** Sour, acidic leaves were once chewed for nausea, mouth sores, sore throats. Fresh leaves were poulticed on cancers, old sores, ulcers. Leaf tea used for fevers, urinary infections, and scurvy. Phenolic compounds in plant associated with antioxidant activity and possible cardioprotective effects. **RELATED SPECIES:** Other yellow-flowered Wood-sorrels such as Common Yellow Wood-sorrell *Oxalis stricta* L. used similarly. **WARNING:** Large doses may cause oxalate poisoning.

PURSLANE

Leaves, whole plant

Portulaca oleracea L. Purslane Family (Portulacaceae)

Prostrate, smooth, fleshy annual; to 1 ft. Stems often reddish and forking. Leaves *spatula-shaped, fleshy*. Flowers yellowish to about ½ in. across; in leaf rosettes; May–Nov. **WHERE FOUND:** Waste ground. Throughout our area. Alien. A widespread weed in many parts of the world. **USES:** Native Americans adopted the plant as a poultice for burns, juice for earaches, tea for headaches, stomachaches. Plant juice said to alleviate caterpillar stings; used in Europe (as poultice) for inflammation, sores, eczema, abscesses, pruritus, and painful urination (strangury). Reportedly hypotensive and diuretic. Leaves best known as a wild edible; very nutritious. This palatable powerhouse is a very good source of omega-3 fatty acids, especially alpha-linolenic acid, and beta-carotene, vitamin C, alpha-tocopherol, magnesium, and potassium. Also reportedly contains dopa and noradrenalin (norepinephrine). This neurohormone reduces hemorrhage at the tissue level, perhaps accounting for use in Traditional Chinese Medicine to stop postpartum bleeding. Considered a good antioxidant and has antibiotic activity; suggested potential in kidney fibrosis and inflammation due to diabetes. Used in traditional medicine systems throughout the world, though few commercial products available. An important medicinal plant worldwide in need of further commercial development.

Purslane, with red stems and succulent leaves, produces small yellow flowers.

DWARF CINQUEFOIL

Whole plant

Potentilla canadensis L. Rose Family (Rosaceae)

Prostrate perennial, on cylindrical rhizomes; rooting at nodes, nodes producing leaves and flower stalks. 2–10 in. Leaves palmate; leaflets rounded, *sharply toothed above middle*, strongly wedge-shaped at base. Flowers yellow with 5 rounded petals; first flowers found on the season's first well-developed leaf; Mar.–June. **WHERE FOUND:** Fields, woods. PE to GA; TX to ON. **USES:** The Iroquois used tea of pounded roots to treat diarrhea; considered astringent. In nineteenth-century medical practice, the plant was touted as a remedy for colds. Other

Dwarf Cinquefoil. Most cinquefoils are used interchangeably for astringent properties.

cinquefoils are considered astringent as well. As with many astringent Rose family members, tannins may explain many of the activities.

FLOWERS IN AXILS, WITH 5 PETALS OR PARTS; LEAVES ALTERNATE

VELVET LEAF (NOT SHOWN)

Leaves, roots, seeds

Abutilon theophrasti Medik. — Mallow Family (Malvaceae)

Annual; 3–6 ft. Entire plant velvety. Leaves heart-shaped, large (4–10 in. long), irregularly toothed. *Single* yellow flowers, each 1–1½ in. across, in leaf axils; June–Nov. Fruit sections beaked. **WHERE FOUND:** Waste places. Throughout our area. Alien (India). **USES:** Chinese use 1 ounce dried leaf in tea for dysentery, fevers; poultice for ulcers. Dried root used in tea for dysentery and urinary incontinence. Seed powder diuretic; eaten for dysentery, stomachaches. CNS-depressant in mice experiments.

FROSTWEED (NOT SHOWN)

Whole plant

Helianthemum canadense (L.) Michx. — Rockrose Family (Cistaceae)

Like that of the American Dittany, the roots of this perennial sometimes form "volcanoes" of ice, hence the name Frostweed. Perennial; 6–20 in. Leaves alternate, lance-shaped, toothless, green on upper surface; basal leaves absent. First flower solitary, yellow, 5-petaled, with many stamens; later flowers without petals, few stamens; May–June. **WHERE FOUND:** Dry, sandy soil, rocky woods. QC, s. ME to NC; TN to IL, MN.

USES: Native Americans used leaf tea for kidney ailments, sore throats; "strengthening" medicine. Patients were covered by a blanket "tent" to hold steam; feet soaked in hot tea for arthritis, muscular swellings, and rheumatism. Historically, physicians once used a strong tea for scrofula (tuberculous swelling of lymph nodes), for which it was reported to produce astonishing cures; also diarrhea, dysentery, and syphilis. Externally, used as a wash for skin diseases such as prurigo, and eye infections; gargled for throat infections. Leaves poulticed on "scrofulous tumors and ulcers."

CLAMMY GROUND-CHERRY

Leaves, root, seeds

Physalis heterophylla Nees.

Nightshade Family (Solanaceae)

⚠ Perennial; 1–3 ft. Stem sticky-hairy; upper part of stem with slender, soft, wide-spreading hairs. Leaves oval, coarsely toothed; *base rounded,* with *few teeth*. Flowers bell-like, greenish yellow with a brownish center; June–Sept. Fruit enclosed in a papery bladder. **WHERE FOUND:** Dry clearings. NS to FL; TX to MN. The most abundant of about 12 highly variable species in our range. **USES:** Native Americans used tea of leaves and roots for headaches; wash for burns, scalds; in herbal compounds to induce vomiting for bad stomachaches; root and leaves poulticed for wounds. Seed of this and other *Physalis* species are considered useful for difficult urination, fevers, inflammation, various urinary disorders. Fruit was eaten in small quantities, mostly as a condiment. Plant compounds researched for antitumor and antimicrobial activity. **RELATED SPECIES:** Other species of *Physalis* used similarly. **WARNING:** Potentially toxic.

The 12 species of Ground-cherry in our range are used similarly. Missouri Ground-Cherry shown here.

*The inflated pod of Chinese-lantern, (*Physalis alkekengi*) a commonly cultivated perennial.*

NODDING, BELL-LIKE, 6-PARTED FLOWERS; LEAVES BASAL OR IN WHORLS

CLINTONIA, BLUEBEARD LILY
Leaf, root

Clintonia borealis (Ait.) Raf.
Lily Family (Liliaceae)

Perennial; 4–12 in. Leaves basal; *leathery, shiny*, entire (not toothed). Flowers bell-like, on a leafless stalk; yellow (or greenish); May–July. Berries blue. **WHERE FOUND:** Cool woods. NL to GA mtns.; west to WI, MN. **USES:** Native Americans poulticed fresh leaves on burns, old sores, bruises, infections, rabid-dog bites; drank tea of plant for heart medicine, diabetes; root used to aid labor in childbirth. Root contains anti-inflammatory and estrogenic diosgenin, from which progesterone, testosterone, and other hormones can be made in the laboratory.

Clintonia produces blue-black berries.

YELLOW TROUT-LILY
Leaves, root

Erythronium americanum Ker-Gawl.
Lily Family (Liliaceae)

Perennial; to 1 ft., with 1–2 mottled, lance-shaped leaves. Flowers lily-like, yellow; Mar.–May. *Petals strongly curved back*. **WHERE FOUND:** Moist woods, often in colonies. NS to GA; AR, OK to MN. **USES:** Cherokee used root tea for fevers; leaf poultice for hard-to-heal ulcers and scrofula. Iroquois women ate raw leaves to prevent conception. Root poultice was used to draw out splinters, reduce swelling. Fresh and recently dried leaves and roots were considered emetic, expectorant. Externally,

White Trout-lily, a species with white flowers, is used interchangeably with E. americanum.

Trout-lily produces yellow flowers and mottled leaves.

Yellow Trout-lily (E. rostratum) *is similar to* E. americanum; *however, the petals are not as strongly recurved; more westerly distribution.*

Indian Cucumber has stem leaves in whorls beneath flowers.

leaves are considered softening to skin tissue, though they may cause immediate allergic reactions. Water extracts are active against gram-positive and gram-negative bacteria. Trout-lilies take up and concentrate phosphorus in the leaves. **RELATED SPECIES:** In a historical context, most *Erythronium* species were used for similar purposes. White Trout-lily (*E. albidum* Nutt.) flowers are white, leaves often mottled. It is found from ON to GA; KY, AR, OK to MN. Used similarly; historically, fresh leaves and roots stewed in milk were applied to sores. Beaked Trout-lily, also called Yellow Trout-lily (*E. rostratum* W. Wolf), is similar to *E. americanum*; however, petals are not as strongly recurved; more westerly distribution from OH to AL; TX to MO. Components of *E. grandiflorum* Pursh (not shown), a plant that grows in w. N. America, have been shown to be slightly antimutagenic. An Asian species used as a wild spring green in S. Korea has been shown to have antioxidant activity due to phenolic compounds. **WARNING:** *Erythronium* spp. are implicated in causing birth defects in livestock.

INDIAN CUCUMBER

Root, leaves, fruit

Medeola virginiana L.

Lily Family (Liliaceae)

Perennial; 1–3 ft. Leaves oval; 6–10, in 1 or 2 whorls (plants with only 1 whorl usually do not flower). Flowers yellow, *drooping*; stamens *red to purplish*, petals and sepals curved back; Apr.–June. **WHERE FOUND:** Rich, moist wooded slopes. NS to FL; AL, LA north to MN. **USES:** Cucumber-flavored root is crisp, edible; Iroquois chewed root and spat it on hook to make fish bite. Leaf and berry tea administered to babies with convulsions. Root tea once used as a diuretic for dropsy. Little studied.

NODDING, BELL-LIKE, 6-PARTED FLOWERS; LEAVES PERFOLIATE OR SESSILE; BELLWORTS

LARGE BELLWORT

Root, plant

Uvularia grandiflora Sm.

Asparagus Family (Asparagaceae)
[Formerly in Lily Family (Liliaceae)]

Perennial; 6–20 in. Solitary or growing in small stands. Leaves oval to lance-shaped, not glaucous; *clasping or perfoliate*; white-downy beneath. Leaves have a wilted appearance. Flower a yellow-orange drooping bell, petals smooth within, twisted; Apr.–June. **WHERE FOUND:** Rich woods. S. QC to GA mtns.; AR to e. KS, e. ND. **USES:** Ojibwa and Potawatomi used the root tea as a wash for rheumatic pains; in fat as an ointment for sore muscles and tendons, rheumatism, and backaches. Plant poulticed to relieve toothaches and swellings. Root tea once used for stomach and lung ailments, also for sore mouth, inflamed larynx and gums. Shoots said to be edible like Asparagus. Another of the many rattlesnake bite remedies in colonial America.

Large Bellwort leaves have a wilted appearance.

PERFOLIATE BELLWORT

Root

Uvularia perfoliata L.

Asparagus Family (Asparagaceae)
[Formerly in Lily Family (Liliaceae)]

Perennial, forming colonies; 6–18 in. Leaves long-oval; smooth beneath. Stem *perfoliate—appears to pass through leaf*. Flowers yellow-orange bells; petals rough-granular on inner surface; Apr.–June. **WHERE FOUND:** Thin woods. S. VT to FL; LA to ON. **USES:** Iroquois used root tea as a cough medicine and for sore mouth, sore throat, inflamed gums, and snakebites. Formerly used in tea or ointment for herpes, sore ears, mouth sores, mild cases of erysipelas (acute local skin inflammation and swelling).

Note how Perfoliate Bellwort stems appear to pass through edge of leaf.

Wild Oats. Leaves do not surround stem. Should not be confused with the grasses Avena sativa *or* A. fatua, *also called Wild Oats.*

WILD OATS

Roots

Uvularia sessilifolia L.

Asparagus Family (Asparagaceae)
[Formerly in Lily Family (Liliaceae)

Perennial; 6–12 in. Leaves sessile, *not surrounding stems* as in above species; glaucous beneath. Stem forked about ⅔ of way up from the ground. Flowers pale, straw-colored bells; May–June. **WHERE FOUND:** Alluvial woods, thickets. NB to GA; AL to AR, north to ND. **USES:** Native Americans used the root tea to treat diarrhea and as a blood purifier; taken internally to aid in healing broken bones. Poulticed for boils and broken bones. Root a folk medicine for sore throats and mouth sores; said to be mucilaginous (slimy) and somewhat acrid-tasting when fresh.

MANY 6-PARTED FLOWERS, IN A SPIKE OR PANICLE

DEVIL'S-BIT

Root

Chamaelirium luteum (L.) Gray

Melanthium Family (Melanthiaceae)
[Formerly placed in Lily Family (Liliaceae)]

 Flowers whitish at first, then turning yellow. See p. 100.

FALSE ALOE, RATTLESNAKE-MASTER

Root

Manfreda virginica (L.) Salisb. ex Rose — Asparagus Family (Asparagaceae)
(*Agave virginica* L.) — [Formerly in Amaryllis Family (Amaryllidaceae)]

⚠ Leaves radiating from root; lance-shaped, smooth, *fleshy* (mottled purple in form *tigrina*); to 16 in. long. Flowers greenish white to yellow, tubular, fragrant at night. Flowers 6-parted, scattered in a *loose spike on a 3- to 6-ft. stalk*; June–July. **WHERE FOUND:** Dry soils. OH, NC, WV to FL; west to TX; north to s. IL. **USES:** The Catawba used diuretic root tea for dropsy; externally a wash used for snakebites. Cherokee nibbled for severe diarrhea, worms; laxative; externally a snakebite remedy. Like species of Agave, this plant might be used as a source for steroid synthesis, but is little researched. **WARNING:** May produce a strongly irritating latex.

False Aloe, Rattlesnake-master produces succulent leaves.

Devil's-bit flowers tend to be more whitish than yellow, despite the species name luteum, *meaning yellow.*

False Aloe's blue-green flower stalks grow to 6 ft. or higher, with narrow tubular flowers.

AMERICAN WHITE OR FALSE HELLEBORE — Root

Veratrum viride Ait. — Melanthium Family (Melanthiaceae)
[Formerly placed in Lily Family (Liliaceae)]

Perennial; 2–8 ft. Leaves large, broadly oval; *strongly ribbed* and pleated. Flowers yellowish, turning dull green; small, *star-shaped*, in a many-flowered panicle; Apr.–July. **WHERE FOUND:** Wet wood edges, swamps. New England to GA mtns.; TN to WI. **USES:** Historically valued as an analgesic for pain, epilepsy, convulsions, pneumonia; heart sedative; weak tea used for sore throat, tonsillitis. Used in pharmaceutical drugs to slow heart rate, lower blood pressure; also for arteriosclerosis and forms of nephritis. Components of the plant (alkaloids) are known to slow heart rate, reduce systolic and diastolic pressure, and stimulate peripheral blood flow to the kidneys, liver, and extremities. Powdered root used in insecticides. The Shakers, famous for herb products in the nineteenth century, continued to make a *Veratrum viride* product well into the twentieth century. Widely used as a treatment for hypertension (in minute, controlled dosages) in the 1960s. **WARNING:** All parts, especially the root, are highly or fatally toxic. Leaves have been mistaken for Pokeweed (see p. 8), Marsh-marigold (see p. 126), Ramps (see p. 40) and eaten with dangerous results. Most recent publications on the plant report on cases of poisoning from mistakenly ingesting the plant when confused with wild edibles. Too dangerous for use. Even handling the plant is dangerous, as the alkaloids can be absorbed through the skin.

Highly toxic American White or False Hellebore produces strongly pleated leaves. In early spring, the large, broad, pleated leaves are seen in wet woods.

FLOWERS IRREGULAR, LIPPED, TUBULAR

YELLOW CORYDALIS — Whole plant, root

Corydalis flavula (Raf.) DC. — Poppy Family (Papaveraceae)
[*Capnoides flavula* (Raf.) Kuntze]

Perennial; 6–16 in. Leaves finely dissected. Flowers yellow, ½ in. long, with a *toothed projecting wing* on top of petal; Mar.–May. Other yellow

Corydalis species in our area have a projecting wing on the top petal, except Golden Corydalis (*C. aurea* Willd.). **WHERE FOUND:** Sandy soils, open woods. ON, NY to FL; LA, AR, e. OK, KS, and NE, to IA. **USES:** Native Americans used tea of this and other yellow-flowered *Corydalis* sp. for painful menstruation, backaches, diarrhea, bronchitis, heart diseases, sore throats, stomachaches; inhaled fumes of burning roots for headaches. Historically, physicians used tea for menstrual irregularities, dysentery, diarrhea, recent syphilitic nodes, and related afflictions. **RELATED SPECIES:** Other yellow-flowered *Corydalis* species in our range, including *C. micrantha* (Englm.) A. Gray and *C. aurea,* separated on technical characters such as length of spurred petal and a flat median wing along flower edge, used similarly, probably without regard to species identity. Roots of several Chinese *Corydalis* species are used for menstrual irregularities, pain, and hemorrhage. Chinese studies show that alkaloids from the genus work as muscle relaxants, painkillers; inhibit gastric secretions, suggesting usefulness against ulcers. **WARNING:** *Corydalis* species are potentially toxic in moderate doses.

YELLOW JEWELWEED, PALE TOUCH-ME-NOT — Leaves, stem juice, flowers

Impatiens pallida Nutt. — Touch-me-not Family (Balsaminaceae)

⚠ Annual; 3–5 ft. Similar to Spotted Touch-me-not (*I. capensis,* p. 184) but flowers are *yellow*, spurs are shorter. Difficult to tell the two apart with-

Yellow Corydalis is among several yellow-flowered Corydalis *species that bloom before foliage of trees.*

Yellow Jewelweed. Its leaves are used to treat Poison Ivy rash. In most areas, it is much less common than Jewelweed.

out flowers present. Seedpods explode when touched. **WHERE FOUND:** Wet, shady, limey soils. NL to GA mtns.; AR to KS. **USES:** Crushed leaves are poulticed on recent Poison Ivy rash. Cherokee rubbed on child's abdomen for sour stomach. Leaf tea taken to treat measles. The juice of 7 flowers rubbed on Poison Ivy rash. **WARNING:** Rare case of severe dermatitis with hives have been reported from application of leaves in alcohol tincture.

BUTTER-AND-EGGS, TOADFLAX

Leaves, flowers

Linaria vulgaris Mill.

Plantain Family (Plantaginaceae)
[Formerly in Figwort Family (Scrophulariaceae)]

Perennial; 1–3 ft. Many lance-shaped leaves. Flowers are yellow, *orange-marked; snapdragon-like, with drooping spurs*; June–Oct. The snapdragon-like flowers appear in two shades of yellow, hence the name Butter-and-eggs. **WHERE FOUND:** Waste places. Throughout our area. Alien (Eurasia). **USES:** In folk medicine, leaf tea used as a laxative, strong diuretic; for dropsy, jaundice, enteritis with drowsiness, skin diseases, hemorrhoids. Ointment made from flowers used for hemorrhoids, skin eruptions. A "tea" made in milk has been used as an insecticide. Science confirms antioxidant, diuretic, and fever-reducing activities. Contains various alkaloids and an iridoid glucoside called antirrhinoside, a compound especially concentrated in young leaves believed to deter insects from feeding on the plant.

Butter-and-eggs has bicolored flowers.

LOUSEWORT, WOOD BETONY

Root, plant

Pedicularis canadensis L.

Broomrape Family (Orobanchaceae)
[Formerly in Figwort Family (Scrophulariaceae)]

Perennial; 5–10 in. Leaves mostly basal; lance-shaped, deeply incised; fernlike. Flowers *hooded*, like miniature snapdragons; yellow, reddish (or both), in tight terminal clusters; Apr.–June. **WHERE FOUND:** Open woods. QC, ME to FL; TX to MB and westward. **USES:** The Cherokee used root tea for diarrhea, stomachaches, ingredient in cough remedies, and as an insecticide dip to rid sheep and dogs of lice. Iroquois used tea of whole plant for heart troubles; poulticed for swellings, tumors, sore muscles. Ojibwa grated roots, which were secretly added to food as an alleged aphrodisiac. In Menominee tradition, root was carried by person wishing to make advances to another; also added to food of estranged lovers to bring them back together. **REMARKS:** Hemiparasitic, with roots attaching to nearby plants at seedling stage to derive some of their nutrition from host plants. The relationship is mutually beneficial. For example, Lousewort produces much nectar that attracts bumblebees. Conversely, nearby Mayapple plants, blooming at the same time, produce little nectar but depend on bumblebees for pollination. If Lousewort and Mayapple are in the same plant community, Mayapple sets more fruit.

The leaves of Lousewort, Wood Betony are fernlike and persist through summer (flowers sometimes red as well as yellow).

5 OR MORE PETALS; LEAVES COMPOUND OR STRONGLY DIVIDED

AGRIMONY, CHURCHSTEEPLES — Whole plant

Agrimonia eupatoria L. — Rose Family (Rosaceae)

Erect perennial; to 3 ft. tall. Leaves alternate, pinnate; leaflets in 3–6 pairs, bluntly toothed; densely or softly hairy, with long stiff hairs on the veins. Flowers yellow, 5-parted, tightly clustered in long terminal racemes; May–July. **WHERE FOUND:** Naturalized, escaped from gardens, MA, NY, PA, OH, NC; IA to MN. Increasingly commonly grown in herb gardens, with range expanding. Alien (Europe). **USES:** *Agrimony* species are used similarly, valued as astringents. In France it is drunk as much for its flavor as a beverage tea as for its medicinal virtues. Tea traditionally used in diarrhea, blood disorders, fevers, gout, hepatitis, pimples, sore throats, and worms. Water extract shown to have antioxidant and anti-inflammatory activity; flavonoids found to protect nerve cells in ethanolic extract; antibacterial and diuretic activity also reported. In European phytomedicine, Agrimony is used in the treatment of mild diarrhea and inflammations of the throat and mouth; used externally for mild skin inflammations.

Close-up of Agrimony flowers.

Agrimony has alternate, pinnate leaves with 3–6 pairs of leaflets.

SMALL-FLOWERED AGRIMONY

Whole plant

Agrimonia parviflora Ait.

Rose Family (Rosaceae)

Perennial; 3–6 ft. Stem hairy. Leaves divided; main stem leaves with *11–19 unequal leaflets.* Leaflets smooth above, hairy below; strongly serrated, 1–3 in. long. Tiny yellow flowers widely spaced, in slender, branched wands; June–Sept. **WHERE FOUND:** Damp thickets, in clumps. MA, NY to FL, e. TX north to NE, WI, s. ON. **USES:** Cherokee and other native groups used seed heads for diarrhea and fevers; root tea to build blood. Herbal tea (made from whole plant) astringent, stops bleeding; used for wounds, diarrhea, inflammation of gallbladder, urinary incontinence, jaundice, and gout. Thought to "strengthen" blood and aid food assimilation. Gargled for mouth ulcers, throat inflammation.

Native Small-flowered Agrimony has small flowers, and its main stem leaves produce 11–19 uneven leaflets.

GOAT'S BEARD

Root

Aruncus dioicus (Walter) Fern.
(*Aruncus vulgaris* Raf.)

Rose Family (Rosaceae)

⚠ Shrublike; 4–6 ft. Leaves mainly basal; divided into large, serrated, oval leaflets. Tiny yellowish to white flowers, crowded in spikes on *a pyramidal plume*; Mar.–May. **WHERE FOUND:** Rich woods, streambanks. KY to GA; west to OK; north to IA, WI; West Coast states; range extends to Asia and Europe. **USES:** Cherokee poulticed pounded root on bee stings. Root tea used to allay bleeding after childbirth and to reduce profuse urination. Tea also used externally, to bathe swollen feet. Korean research suggests that extracts may reduce ultraviolet light-induced skin damage; of possible use in cosmetics; antioxidant, and potential for anticancer research. **WARNING:** One report of mistaking this plant for a dissimilar *Aconite* (Monkshood) caused a case of poisoning, which highlights the importance of plant identitification.

A widespread species of rich woods, Goat's Beard has yellow to white flowers.

COMMON OR TALL BUTTERCUP

Root, leaves

Ranunculus acris L.

Buttercup Family (Ranunculaceae)

⚠ Erect annual or perennial; 2–3 ft. Leaves palmately divided into 5–7 stalkless, *lance-shaped, toothed segments*. Flowers shiny; golden yellow within, lighter outside; May–Sept. Fruits flat, smooth, with distinct margins. **WHERE FOUND:** Fields. Throughout our area, but mostly absent from Midsouth (AR, LA, TX, OK, FL). Alien (Europe). **USES:** Fresh leaves historically used as external rubefacient in rheumatism, arthritis, neuralgia. Native Americans poulticed root for boils, abscesses. Action based on irritating affected part. **WARNING:** Can be extremely acrid, causing intense pain and burning of mouth, mucous membranes; blisters skin. Similar warning applies to other buttercups and many other plants in the Buttercup family.

Close-up of the familiar waxy yellow-petaled Common Buttercup.

Buttercups are considered potentially toxic.

FLOWERS IN UMBRELLA-LIKE CLUSTERS (UMBELS)

DILL

Leaves, seeds

Anethum graveolens L. Parsley Family (Apiaceae or Umbelliferae)

⚠ Smooth, branched annual; often glaucous; 2–4 ft. Leaves finely dissected into linear segments with *characteristic fragrance of dill* (the flavoring for dill pickles). Flowers in umbels, with 30–40 spreading rays, to 6 in. Individual flowers yellowish; June–Sept. Fruits half as wide as long, to 3/8 in. long, also with characteristic dill fragrance; Aug.–Oct. **WHERE FOUND:** Garden soils. Occasional throughout the U.S.; absent in Southeast. Alien (s. Europe). **USES:** Dill leaves are typically used for flavoring; considered digestive, carminative; a folk medicine for conditions of the gastrointestinal and urinary tract; reduces

Dill produces leaves with the distinct fragrance of dill.

spasms, diuretic. Contains numerous bioactive compounds. Seed (fruit) widely used as a carminative and stomachic. Considered antispasmodic and antibacterial. Used in European phytomedicine as a tea for dyspepsia. Widely used in Ayurvedic medicine in India (one of the world's largest suppliers of dill seed) as a carminative, stomachic, and diuretic. **WARNING:** Proper identification essential. Do not confuse with fatally toxic Parsley family members.

FENNEL

Seeds, leaves

Foeniculum vulgare P. Mill. Parsley Family (Apiaceae or Umbelliferae)

Smooth herb; 4–7 ft. *Strongly anise-* or *licorice-scented*. Leaves *threadlike*. Yellowish flowers in flat umbels; June–Sept. **WHERE FOUND:** Roadsides. ME, MA to FL; TX, NE to MI. Very common weed in CA. Alien (Europe). Several different types of fennel, including annual, biennial, and perennial varieties varying in form and leaf color, are grown in gardens for food, flavoring, and ornamental and medicinal use. **USES:** Seeds (actually fruits) or tea taken to relieve gas, infant colic, stimulate milk flow. Reportedly diuretic, expectorant, carminative, laxative; soothing to stomach; used to improve the flavor of other medicines. Seeds have been shown to increase gastrointestinal motility, and to increase expectorant action from the lungs by 12 percent. Boiled water extracts of leaves were shown to produce a reduction in arterial blood pressure without reducing heart rate or respiratory rate. Seeds eaten in Middle East to increase milk secretion, promote menstruation, and increase libido. Powdered seeds poulticed in China for snakebites. Experimentally, seed oil relieves spasms of smooth muscles, kills bacteria; antioxidant, antimutagenic, anti-inflammatory, among other effects. Fennel seed is our best source of anethole, used commercially as "licorice" (actually anise) flavor. Seeds widely used in European phytomedicine and elsewhere for treatment of gastrointestinal fullness and spasms, catarrh of the upper respiratory tract. Seed extracts stimulate gastrointestinal motility. Widely used in world herbal traditions as a spice, vegetable, and medicinal plant. **WARNING:** Fennel or its seed oil may cause rare contact dermatitis. Ingestion of oil may cause vomiting, seizures, and pulmonary edema.

Fennel produces anise-scented leaves.

Parsley is a garden biennial with yellow flowers in the second year.

Sometimes escapes from cultivation. Typical, curly-leaved Parsley shown here.

PARSLEY

Root, leaves, seeds

Petroselinum crispum (Mill.) Nyman ex A.W. Hill — Parsley Family (Apiaceae or Umbelliferae)

The familiar parsley of commerce; biennial, often grown as an annual, to 2½ ft. in flower; first-year plants remain green after frost. Leaves pinnately divided, sharp-toothed; often with crisped (wavy) margins or flat. Flowers yellow, in loose umbels; May–Aug. **WHERE FOUND:** Cultivated, escaped from gardens; naturalized in scattered populations throughout our range. **USES:** Not limited to culinary use. Fresh leaves chewed as a breath freshener. Traditionally, leaves valued as a diuretic for dropsy (water retention); leaf tea for painful menstruation. Externally, bruised fresh leaves poulticed for enlarged glands, insect bites, and contusions; roots also diuretic. Leaves are used in European phytomedicine for lower urinary tract infections and "gravel" (kidney stones). Topical preparations used for cracked, chapped skin. Seed (fruit) considered stronger; however, the seed is not used in commercial preparations generally because of possible toxicity or confusion with other Parsley family members. Science confirms potential diuretic, smooth muscle relaxant (antispasmodic), liver-protectant, anti-inflammatory, antioxidant activity. **WARNING:** Seed (fruit) and essential oils derived from plant potentially toxic.

WILD PARSNIP

Root

Pastinaca sativa L. — Parsley Family (Apiaceae or Umbelliferae)

Biennial; 2–5 ft. Stalk *deeply grooved* (ribbed). Leaves in stout rosette in first year, divided into 5–15 stalkless, toothed, oval leaflets. Tiny golden flowers with 5 petals, in umbels; May–Oct. **WHERE FOUND:** Roadsides, waste places. Much of our area. Alien (Eurasia). **USES:** Native Americans used roots to treat sharp pains, tea in small amounts for "female" disorders; poulticed roots on inflammations, sores. **WARNING:** May cause contact dermatitis or photodermatitis due to xanthotoxin, which

Wild Parsnip. Note its deeply grooved stalk.

Early-spring-blooming Golden Alexanders occurs in dry habitats.

is used to treat psoriasis and vitiligo. Avoid contact and exposure to sunlight. Contains furanocoumarins, which cause photodermatitis.

GOLDEN ALEXANDERS Root

Zizia aurea (L.) W. D. J. Koch Parsley Family (Apiaceae or Umbelliferae)

Smooth perennial; 1–3 ft. Leaves thin; all *twice-compound* with 35 leaflets or divisions, divided again; leaflets *finely sharp-toothed*. Tiny yellow flowers in an umbel with 6–20 rays; Apr.–June. **WHERE FOUND:** Dry woods, rocky outcrops. New England to GA; TX to SK. **USES:** The Meskwaki used root tea for fevers. Historically, the plant has been referred to as a vulnerary (agent used to heal wounds) and a sleep inducer; it was also used for syphilis. **WARNING:** Possibly toxic—eating a whole root has caused violent vomiting, which itself was believed to mitigate further adverse reaction. Amateurs fooling with plants in the Parsley family are playing herbal roulette.

IRREGULAR 2-LIPPED FLOWERS; MINT FAMILY

YELLOW GIANT HYSSOP Leaves

Agastache nepetoides (L.) Kuntze Mint Family (Lamiaceae or Labiatae)

Perennial; 3–5 ft. Stem *square*, mostly smooth; branching above. Leaves opposite; narrowly ovate to lance-shaped, saw-toothed. Flowers yellow to whitish, in dense, usually continuous spikes; July–Sept. **WHERE FOUND:** Rocky, wooded hillsides, wood edges. S. QC to GA, AL; ne. TX, e. OK to SD, MN. **USES:** The Iroquois used the leaves in a compound mixture to apply to Poison Ivy rash. **REMARKS:** The name *nepetoides* refers to the plant's resemblance to *Nepeta* (Catnip).

STONEROOT, HORSE-BALM — Roots, leaves

Collinsonia canadensis L. — Mint Family (Lamiaceae or Labiatae)

Branching, square-stemmed perennial herb; 2–4 ft. Leaves large, oval, coarsely toothed. Root hard, broader than long. Flowers greenish yellow, *lower lip fringed, stamen strongly protruding*; lemonlike scent; July–Sept. **WHERE FOUND:** Rich woods. ON, VT to FL; MO to WI. **USES:** Cherokee used fresh crushed flowers and leaves as an ingredient in a poultice for swollen breasts. The Iroquois used a handful of the root in warm water to bathe babies to make them strong; powdered leaves a poultice for headaches; also diuretic and to strengthen blood. The warm, pungent fragrance of the leaves attracted interest of early settlers, as native groups of Virginia and elsewhere used the plant for sores and wounds and considered it a useful tonic or panacea

ABOVE: *Yellow Giant Hyssop produces small yellow flowers in tightly packed, whorled flowerheads.*
TOP RIGHT: *Stoneroot has larger, mostly unscented leaves.*
BOTTOM RIGHT: *The flower petals of Stoneroot, Horse-balm are fringed.*

for many disorders. Folk uses include leaf poultice for burns, bruises, wounds, sores, sprains; root tea for hemorrhoids, laryngitis, indigestion, diarrhea, dysentery, dropsy, kidney and bladder ailments. Contains alkaloids; strongly diuretic, useful in cystitis. Roots contain more than 13,000 parts per million of rosmarinic acid, the same antioxidant found in Rosemary. Studies also report diuretic, stomachic, and tonic effects. The root is extremely hard, especially when dried, hence the name "Stoneroot." **WARNING:** Minute doses of fresh leaves may cause vomiting.

HORSEMINT — Leaves

Monarda punctata L. — Mint Family (Lamiaceae or Labiatae)

Strongly aromatic biennial or short-lived perennial; 1–4 ft. Leaves lance-shaped. Flowers like gaping mouths; *yellowish, purple-dotted; in tiered whorls*, with yellowish to lilac bracts beneath; July–Oct. **WHERE FOUND:** Dry soils. QC, VT, MA, NY to FL; LA, TX, AR, KS to MN. **USES:** The Delaware Indians used a leaf tea wash to bathe the face of those with fevers; tea also drunk for fevers. The Ojibwa used the leaf decoction for treating nausea, stomach problems, bowel ailments, and as a mild laxative. Various native groups used leaf tea for colds, fever, flu, and cough. Historically, physicians used leaf infusion as a carminative, stimulant, digestive, diuretic, and to regulate menses. Essential oil high in thymol; antiseptic, expels worms. Thymol, now manufactured synthetically, was once commercially derived from Thyme (*Thymus* species). During World War I, commercial Thyme fields were destroyed in Europe and Horsemint was grown in the U.S. as a substitute source of thymol. Also contains carvacrol-rich extracts experimentally shown to suppress elevated levels of blood lipids. At least 22 compounds, especially glycosides, identified from the whole plant.

Horsemint produces a thymelike scent and yellow flowers with violet spots.

COMMON ST. JOHN'S WORT

Leaves, flowers

Hypericum perforatum L. St. John's Wort Family (Hypericaceae)

⚠ Perennial; 1–3 ft. Leaves oblong, dotted with translucent glands. Flowers yellow, stamens in a bushy cluster, 5 petals with *black dots on margins*; June–Sept. **WHERE FOUND:** Fields, roadsides. Throughout our range. Alien (Europe). **USES:** St. John's Wort has emerged as the best-known herbal treatment for mild to moderate forms of depression. Reportedly outselling the conventional antidepressant Prozac by as much as 20 to 1 in Germany, where used clinically for treatment of mild depression. Various controlled clinical trials have both confirmed and questioned its safety and effectiveness. St. John's Wort brings us a great example of synergy of different chemical compounds; several compounds are believed to contribute different mechanisms to help relieve depression. Its compounds regulate brain levels of such important compounds as dopamine, interleukins, melatonin, monoamine-oxidases, and serotonin. It has been shown to be a selective serotonin reuptake inhibitor. Fresh flowers in tea, tincture, or olive oil were once a popular domestic medicine for treatment of external ulcers, wounds (especially those with severed nerve tissue), sores, cuts, and bruises. Tea is a folk remedy for bladder ailments, depression, dysentery, diarrhea, and worms. Experimentally, antidepressant, sedative, anti-inflammatory, and antibacterial. Contains many biologically active compounds, including rutin, sitosterol, hypericin, pseudohypericin, and

Common St. John's Wort. Note the black dots on flower margins.

Common St. John's Wort, a frequent roadside herb.

hyperforin. Studies in the 1980s found that hypericin and pseudohypericin have potent anti-retroviral activity, without serious side effects. Being researched for AIDS treatment. **WARNING:** Taken internally or externally, hypericin may cause photodermatitis (skin burns) on sensitive skin that is exposed to light. As an antidepressant, it is used only for mild to moderate forms of depression, not severe forms. Various studies confirm that, like grapefruit juice, St. John's Wort extracts produce an herb-drug interaction by inducing a reaction in the intestine and liver that can reduce the oral bioavailability of many drugs, making them less effective.

FALSE SPOTTED ST. JOHN'S WORT — Tops

Hypericum pseudomaculatum Bush. St. John's Wort Family (Hypericaceae)
[*Hypericum punctatum* var. *pseudomaculatum* (Bush) Fern.]

Erect perennial herb, stems sometimes woody at base; 1–3 ft. tall. Leaves oblong to elliptical, about 2 in. long, ½ in. wide, firm, sharp-tipped, with numerous black glandular dots. Flowers yellow, *strongly black-dotted* (representing glands with hypericin—red when crushed between fingers), about half the size of Common St. John's Wort (above); May–Aug. **WHERE FOUND:** Upland prairies, glades, wood edges. SC to FL; TX to OK, IL. **USES:** Likely used similar to Common St. John's Wort for wounds and nervous conditions by early settlers, given its resemblance to its European relative. Yellow-flowered *Hypericum* were often used generically for the treatment of wounds. Science confirms antibacterial activity. **RELATED SPECIES:** Spotted St. John's Wort (*Hypericum punctatum* Lam.) is a smaller, more sprawling plant, with rounded-tipped leaves and smaller flowers. It has a broader range from QC to FL; TX to NE, ON. Often confused with *H. pseudomaculatum.*

COMMON MULLEIN — Leaves, flowering tops

Verbascum thapsus L. Figwort Family (Scrophulariaceae)

Biennial; 1–8 ft. in flower; produces a rosette of large, fuzzy, gray-green leaves the first year, and an attractive spike of light yellow flowers the second year. Leaves are large, broadly oval, *very hairy (flannel-like); hairs branching.* Flowers yellow, in tight, long spikes; July–Sept. **WHERE FOUND:** Poor soils. Common throughout our area. Widespread alien. A roadside weed naturalized from Europe, it also grows in sandpits and gravel pits; it seems to thrive in the poorest of soils. **USES:** Traditionally, leaf and flower tea expectorant, demulcent, antispasmodic, diuretic, for chest colds, asthma, bronchitis, coughs, kidney infections, tuberculosis; leaves poulticed for ulcers, tumors, hemorrhoids. In tribal areas of Pakistan the leaf is used to expel worms; also used as an expectorant. Flowers soaked in olive or mineral oil are used as earache drops. Historically, and in modern Europe, flowers are preferred over leaves; both are used in cough remedies. Leaves high in mucilage, soothing to inflamed mucous membranes; experimentally, strongly anti-inflammatory. Science confirms mild expectorant and antiviral activity against herpes simplex and influenza viruses. Contains verbascoside, which

False Spotted St. John's Wort has black-spotted yellow flowers, about half the size of those of Common St. John's Wort.

Common Mullein. Its yellow flowers appear in its second year.

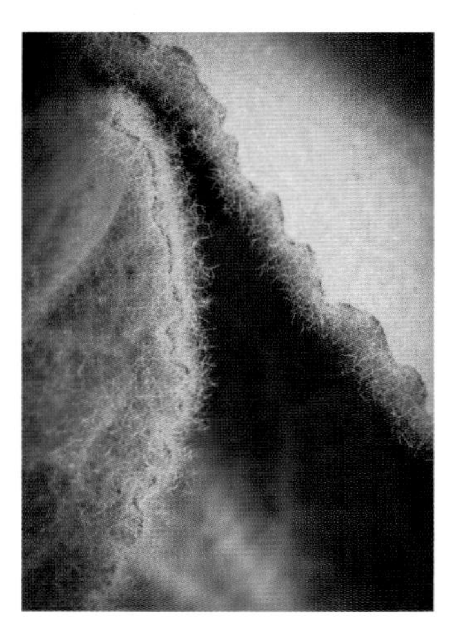

ABOVE: *Close-up of Mullein's yellow flowers.* RIGHT: *Close-up of the thick, branched hairs on Mullein's fuzzy gray leaves.*

has antiseptic, antitumor, antibacterial, and immunosuppressant activity. In European phytomedicine the flowers are approved as an expectorant in inflammations of the upper respiratory tract. Asian Indians used the stalk for cramps, fevers, and migraine. The seed is a narcotic fish poison. **WARNING:** The leaves contain potentially unsafe rotenone and coumarin. Hairs may irritate skin.

CREAM WILD INDIGO

Seeds, roots, leaves

Baptisia bracteata var. *leucophaea* (Nutt.) Kartesz & Gandhi — Pea Family
(*Baptisia leucophaea* Nutt.) — (Fabaceae or Leguminosae)

Hairy, bushy perennial; 10–30 in. Leaves with 3 spatula-shaped leaflets. Flowers *cream-yellow*, on showy lateral drooping racemes, with *large leaflike bracts beneath*; Apr.–June. Seedpods inflated to 3 in. long, with pubescent tapered beak. **WHERE FOUND:** Dry soils. AR, TX, NE to MN. **USES:** Pawnees used ointment of seed powder mixed in buffalo fat applied to stomach for colic. Root tea formerly used for typhoid and scarlet fever; leaf tea in "mercurial salivation"; used externally on cuts and wounds (astringent). Recent research suggests immune-system stimulant activity.

WILD INDIGO

Whole plant, root

Baptisia tinctoria (L.) Vent. — Pea Family (Fabaceae or Leguminosae)

⚠ *Smooth, blue-glaucous* perennial; 1–3 ft. Leaves narrowly cloverlike, nearly stalkless. Flowers yellow, few, on numerous racemes on upper branchlets; May–Sept. **WHERE FOUND:** Dry open woods, clearings. VA to FL; less common from s. ME to IN, se. MN. **USES:** The Cherokee used a cold infusion of the leaves to induce vomiting (emetic); also as a pur-

Cream Wild Indigo's flowers spread laterally, drooping toward the ground.

Wild Indigo has small yellow pealike flowers.

gative. A poultice of the root was used for toothaches, to allay inflammation; wash used for cuts, wounds, bruises, and sprains. Historically, fresh-root tea was considered strongly laxative, astringent, antiseptic. Used for typhus and scarlet fever; gargled for sore throats. Tea used as a wash for leg, arm, and stomach cramps, wounds. Said to stimulate bile secretion. Root poultice for gangrenous ulcers. In early nineteenth-century medical practice the whole plant, particularly the root, was considered valuable for gangrenous and septic conditions and every ulcerous affliction, suggesting antiseptic or immunostimulant activity long before antibiotics were in use. German studies have shown the extract stimulates the immune system. Recent research found that an extract produces an antibody in the bloodstream similar to the reaction induced by salmonella bacteria exposure, suggesting that the plant could have research potential for typhoid, still one of the most serious bacterial infections in third-world countries that have grown resistant to antibiotics. Compounds called arabinogalactan-proteins are among the compounds responsible for immunostimulating activity. **WARNING:** Large or frequent doses are potentially harmful, given historical use to induce vomiting or purging.

YELLOW SWEET-CLOVER — Flowering plant

Melilotus officinalis (L.) Pall. — Pea Family (Fabaceae or Leguminosae)

⚠ Straggly biennial; 2–6 ft. Leaves in cloverlike arrangement; leaflets narrow, elongate, slightly toothed. Flowers small, yellow, pealike, in long, tapering spikes; *fragrant when crushed*; Apr.–Oct. **WHERE FOUND:** Roadsides. Throughout our area. Alien (Europe). **USES:** Dried flowering plant traditionally used in tea for neuralgic headaches, painful urination, nervous stomach, colic, diarrhea with flatulence, painful menstruation with lameness and cold sensation, aching muscles; poulticed for inflammation, ulcers, wounds, rheumatism; smoked for asthma. Recent clinical trials support use in cyclic mastalgia, also as an ingredient in an external cream for chronic venous insufficiency with anti-inflammatory and anti-edema effects. Tea of leaves used to treat varicose veins. **WARNING:** Moldy hay causes uncontrollable bleeding in cattle due to coumarins. Science developed compounds such as warfarin

Yellow Sweet-clover commonly grows along roadsides.

from such coumarins to prevent blood clotting in rodents. See White Sweet-clover, p. 104.

LEGUMES (PEA FAMILY); MANY LEAFLETS

PARTRIDGE-PEA

Roots

Chamaecrista fasciculata (Michx.) Greene
(*Cassia fasiculata* Michx.)

Pea Family
(Fabaceae or Leguminosae)

Annual; 6–15 in. Leaves fold when touched. *Leaflets in 6–15 pairs.* Flowers yellow, *reddish at base,* ½ in. across, with 10 stamens; July–Sept. **WHERE FOUND:** Sandy soil. MA, NY to FL; TX to SD, MN; MO to OH. **USES:** Cherokee used root tea with other plants to relieve fatigue and keep ball players from tiring. **RELATED SPECIES:** Wild Sensitive Plant [*C. nictitans* (L.) Moench], which has 5 stamens (rather than 10), and petals significantly unequal in length, used similarly.

WILD SENNA

Leaves, seedpods

Senna marilandica (L.) Link
(*Cassia marilandica* L.)

Pea Family (Fabaceae or Leguminosae)

⚠ Erect perennial; 3–6 ft. Leaves compound, with 4–8 pairs of elliptical leaflets. Note *rounded gland at base of leafstalk.* Yellow flowers in loose clusters at leaf axils; July–Aug. Seedpods with joints; seeds twice as wide as they are long. **WHERE FOUND:** Dry thickets. NY to FL; TX to NE, WI. **USES:** Cherokee used dried powered leaf tea to treat cramps, and as a strong laxative; also for fevers. Given its close resemblance to the

Partridge-pea has 10 anthers per flower, with petals of equal length.

Wild Senna produces a rounded gland at base of leafstalk.

Close-up of Wild Senna flowers.

commercial source of Alexandrian Senna (*Senna alexandrina* Mill., *Cassia senna* L.), the leaves and pods of which have been used as stimulant laxatives since ancient times, early settlers immediately made use of our native Wild Senna. Leaves and pods a strong laxative. One teaspoon of ground Coriander seeds was added to leaf tea to prevent griping (cramps); tea of pods milder, slower-acting. Laxatives made from Alexandrian Senna are found in every pharmacy. The compounds responsible for laxative action develop upon drying and are not found in the fresh herb. **RELATED SPECIES:** The native *Senna hebecarpa* (Fern.) Irwin & Barneby, separated by technical details, used similarly. **WARNING:** Contraindicated in cases of diarrhea, abdominal pain (of unknown origin), and inflammatory conditions of the bowels. Consult a licensed health care practitioner.

SICKLEPOD, COFFEEWEED — Leaves, seeds

Senna obtusifolia (L.) Irwin & Barneby (*Cassia obtusifolia* L.) — Pea Family (Fabaceae or Leguminosae)

Annual; to 3 ft. Leaflets in 3 pairs; obovate; tip rounded, often with an abrupt point. Note *cylindrical gland between 2 lowermost leaflets*. Flowers July–Sept. Pods to 9 in. *long, curved, sickle-shaped*. **WHERE FOUND:** Waste places. NY to FL; TX to NE, WI. **USES:** Pesticides have been developed to eradicate this weed from Midwestern corn and soybean fields. Seeds have 5–7 percent seed oil, which even in small quantities can taint the quality of soybean oil. Both a food and medicinal plant.

ABOVE: *Sicklepod, Coffeeweed has sickle-shaped seedpods.*
RIGHT: *Goat's Rue has pale yellow and pink-red blooms.*

Seed tea used for headaches, fatigue, and stomachaches. In North Africa, nomadic tribes use the leaves as food in the form of a fermented protein paste. Seeds of *S. obtusifolia* and the closely related *S. tora* (L.) Roxb. (*Cassia tora* L.) are roasted as a coffee substitute, eaten during famine. Chinese use seeds of both species as source plants for seeds called *jue ming zi* in Traditional Chinese Medicine; used for boils (internal and external), eye diseases. Seed tea used for headaches, hepatitis, herpes, and arthritis. A novel protein isolated from the seed has been shown to decrease serum cholesterol concentrations. **RELATED SPECIES:** *S. obtusifolia* is considered both distinct from and synonymous with *S. tora.* Taxonomy is confused. Both have foul-smelling leaves that have been used for leprosy, psoriasis, and ringworm.

GOAT'S RUE, VIRGINIA TEPHROSIA — Root, leaves

Tephrosia virginiana (L.) Pers. — Pea Family (Fabaceae or Leguminosae)

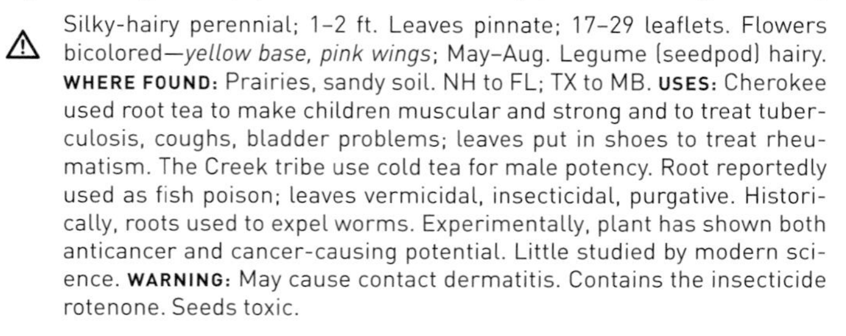

⚠ Silky-hairy perennial; 1–2 ft. Leaves pinnate; 17–29 leaflets. Flowers bicolored—*yellow base, pink wings*; May–Aug. Legume (seedpod) hairy. **WHERE FOUND:** Prairies, sandy soil. NH to FL; TX to MB. **USES:** Cherokee used root tea to make children muscular and strong and to treat tuberculosis, coughs, bladder problems; leaves put in shoes to treat rheumatism. The Creek tribe use cold tea for male potency. Root reportedly used as fish poison; leaves vermicidal, insecticidal, purgative. Historically, roots used to expel worms. Experimentally, plant has shown both anticancer and cancer-causing potential. Little studied by modern science. **WARNING:** May cause contact dermatitis. Contains the insecticide rotenone. Seeds toxic.

THISTLES AND OTHER COMPOSITES WITH FERNLIKE LEAVES OR LACERATED UPPER LEAVES

SPANISH NEEDLES, SOAPBUSH NEEDLES

Leaves

Bidens bipinnata L. Aster Family (Asteraceae or Compositae)

Square-stemmed annual; 1–3 ft. Leaves *strongly dissected, fernlike*. Flowers yellow, rays absent; Aug.–Oct. Elongate seeds ("needles") *in spreading clusters*; each seed topped with 2–4 barbs. **WHERE FOUND:** Waste places, sandy soil, disturbed forests and fields. MA, NY to FL; TX to NE, IA. Alien (China); widely naturalized worldwide. **USES:** Cherokee used leaf tea as a worm expellent; leaves chewed for sore throat. Plant juice once used for eardrops and as a styptic. In Traditional Chinese Medicine, the herb is well known as an ingredient in prescriptions for fevers and for anti-inflammatory activity in rheumatic complaints. Science confirms antidiarrheal, liver-protective, and anti-inflammatory activity. Various flavonoids are linked to biological effects. **RELATED SPECIES:** A related species has CNS-depressant and blood sugar–lowering activity. **WARNING:** May be an irritant.

Spanish Needles, Soapbush Needles has small flowers with elongated seed clusters.

BLESSED THISTLE

Whole plant

Cnicus benedictus L. Aster Family (Asteraceae or Compositae)

Hairy annual herb; 10–30 in. Both leaves and stems hairy. Stems 5-sided. Leaves *broadest at base*; lacerated, spiny-toothed. Flowers yellow with a *large leafy bract beneath*; Apr.–Sept. *Reddish, spinelike projections surround yellow tufts* of flowers. **WHERE FOUND:** Roadsides, waste places. Escaped from cultivation; naturalized scattered from NS, MA, NY to FL; TX to WI. Alien (Europe). **USES:** Weak tea (2 teaspoons to 1 cup of water) of dried flowering plant traditionally used in Europe to stimulate sweating, appetite, milk production; diuretic. Folk reputation as remedy for boils, indigestion, colds, deafness, gout, headaches, migraines, suppressed menses, chilblains, jaundice, and ringworm. Experimentally, it increases gastric and bile secretions; antibacterial, anti-inflammatory. Contains two lignans, arctigenin and trache-

Blessed Thistle produces spider web–like hairs in the flowers.

Golden Ragwort, Squaw-weed produces heart-shaped leaves at the base and divided leaves on stem. Blooms very early.

ologenin, that have been experimentally useful against retroviruses. These compounds are also found in Burdock. Highly bitter principles in the leaf, such as cnicin, stimulate secretion of gastric juices and saliva, and are responsible for Blessed Thistle's use as an appetite stimulant. Used in European phytomedicine as a bitter digestive stimulant for treatment of loss of appetite and dyspeptic discomfort. Seeds have served as emergency oil seeds.

GOLDEN RAGWORT, SQUAW-WEED — Leaves, roots

Senecio aureus L. — Aster Family (Asteraceae or Compositae)

Perennial; 2–4 ft. Has two leaf types, unlike other *Senecio* species in our range—*basal leaves heart-shaped, rounded*; upper leaves lance-shaped, incised. Highly variable. This early spring wildflower, often growing in clumps among exposed rocks in streams, is easily distinguished from other *Senecio* species by the heart-shaped leaves at the base of the plant. Flowers yellow, in flat-topped clusters; late Mar.–July. **WHERE FOUND:** Streambanks, moist soil, swamps. Most of our area. **USES:** Root and leaf tea traditionally used by Native Americans, settlers, and herbalists to treat delayed and irregular menses, leukorrhea, and childbirth complications; also used for lung ailments, dysentery, difficult urination. Its traditional use in treating a variety of female diseases led to its common name, Squaw-weed. **WARNING:** Many ragworts (*Senecio* species) contain highly toxic pyrrolizidine alkaloids and are no longer used.

COMPOSITES WITH PROMINENT OVERLAPPING BRACTS

GUMWEED, ROSINWEED

Leaves, flowers

Grindelia squarrosa (Pursh) Dunal Aster Family (Asteraceae or Compositae)

Highly variable biennial, perennial or woody at base; 1–3 ft. Leaves *strongly aromatic*, mostly serrate, linear-oblong; mostly clasping. Flowers yellow, with *very gummy bracts*; bract tips recurved; July–Sept. Plant tissue has glands just below surface that exude sticky, resinous terpene mixture, with turpentine-like fragrance. **WHERE FOUND:** Prairies, roadsides. Prairie states; spreading locally eastward and westward. **USES:** Plant tea traditionally used for asthma, coughs, kidney ailments, bronchitis. Tea of flowering tops used for colic, stomachaches. Externally, a leaf poultice or wash was used for sores, skin eruptions, wounds. This and other *Grindelia* species have been used in folk remedies for cancers of the spleen and stomach, burns, colds, fever, gonorrhea, pneumonia, rashes, rheumatism, smallpox, and tuberculosis. Said to be sedative, spasmolytic, and antibacterial. Herb used in European phytomedicine for treatment of catarrhs of the upper respiratory tract. **RELATED SPECIES:** *G. lanceolata* Nutt. differs in that the individual bracts surrounding the flowerhead are spreading rather than strongly curved back, among other technical details. Its general range is more eastern than that of *G. squarrosa*. For all practical historical medicinal purposes, both plants were used interchangeably.

Gumweed, Rosinweed produces gummy flowerheads.

ELECAMPANE

Roots

Inula helenium L. Aster Family (Asteraceae or Compositae)

⚠ Perennial; 4–8 ft. Leaves large, burdocklike, but narrower and woolly beneath; upper leaves reduced. Flowers yellow, *large—to 4 in. across; rays slender*; July–Sept. Broad bracts beneath flowerhead. **WHERE FOUND:** Fields, roadsides. Locally established and escaped from gardens. ME to SC, TN; MO to MN. Alien (Europe). **USES:** Root tea (½ ounce to 1 pint water) a folk remedy for pneumonia, whooping cough, asthma, bronchitis, upset stomach, diarrhea, worms; used in China for certain cancers. Wash used for facial neuralgia, sciatica. Experimentally, tea strongly sedative to mice; antispasmodic, expectorant, worm expellent, anti-inflammatory, antibacterial, and fungicidal. In studies of mice, the root infusion (tea) had pronounced sedative effects. Contains alantolactone, which is a better wormer than santonin and less toxic. In small doses it lowers blood-sugar levels, but in larger doses it raises blood sugar, at least in experimental animals. Roots contain 20 to 44 percent inulin, a dietary fiber that is recently recommended for leaky gut syndrome. Components of the essential oil promising in the search for new antibacterial agents as they damage cell membranes of *Staphylococcus aureus.* **WARNING:** Contains toxic sesquiterpene lactones, notably alantolactone, which can irritate mucous membranes. This compound attaches to skin proteins, causing sensitization and contact allergic dermatitis. Because potential risks outweigh benefits, use is not recommended.

Elecampane produces large yellow flowers with narrow rays.

GOLDENRODS AND BUTTONLIKE COMPOSITES

PINEAPPLE-WEED

Leaves, flowers

Matricaria discoidea DC. Aster Family (Asteraceae or Compositae)
[*Chamomilla suaveolens* (Pursh) Rydb.; *Matricaria matricarioides* (Lessing) Porter has often been misapplied to the native *M. discoidea*]

Pineapple-scented annual; 4–16 in. Leaves *finely dissected; segments linear.* Flowers tiny, *without rays*; yellow "button" is a rayless composite flower; May–Oct. **WHERE FOUND:** Waste places, roadsides. Often listed as an alien species (debated), some botanists consider it native with occurrence throughout N. America except the extreme Southeast. Introduced to Eurasia and Australia. Confusion on origin stems in part from the plant's weedy nature in rural and urban environments. **USES:** Many western and northern Native American groups used the leaves or plant in flower as a tea for indigestion, stomach pains from gas, as a mild laxative, and for colds and fevers. Sometimes considered a general tonic and panacea, much like Chamomile is revered in Europe. Traditionally, plant tea used for stomachaches, flatulence, colds, menstrual cramps;

wash used externally, for sores, itching. **WARNING:** Some individuals may be allergic to this plant.

CANADA GOLDENROD

Roots, leaves, seeds

Solidago canadensis L. Aster Family (Asteraceae or Compositae)

⚠ Our most common and widespread goldenrod; variable, 1–5 ft. Stem smooth at base, hairy below lower flower branches. Leaves many; lance-shaped, 3-veined, sharp-toothed. Flowers in a *broad, triangular panicle*; July–Sept. **WHERE FOUND:** Fields, roadsides. Most of our area, especially in the Northeast; PE to NC; NE to SK. Widely cultivated as an ornamental outside its natural range and established elsewhere, including w. N. America, Europe, and China. **USES:** Native Americans used root for burns; flower tea for fevers, snakebites; crushed flowers were chewed for sore throats. Contains quercetin, a compound reportedly useful in treating hemorrhagic nephritis. Seeds eaten as survival food. Leaf extracts are diuretic and mildly antispasmodic. Like European Goldenrod *Solidago virgaurea* L., the leaves of Canada Goldenrod are used in European phytomedicine as a diuretic in treatment of inflammatory diseases of the lower urinary tract. Also used in irrigation therapy to both prevent and treat urinary and kidney gravel. Contains various bioactive flavonoids; various diterpenes responsible for characteristic fragrance. Of interest in research for antibacterial activity against antibiotic-resistant bacterial strains. **WARNING:** Causes allergies, though most allergies attributed to goldenrods are due to Ragweed pollen. Some may experience dermatitis from handling the plant.

Pineapple-weed often grows in poor dry soils, even in cracks of pavement.

Canada Goldenrod. Our most common goldenrod.

Sweet Goldenrod. Leaves, when crushed, emit a pleasant, tarragon-like fragrance.

Common Tansy. Leaves are fernlike; flowers are yellow buttons.

SWEET GOLDENROD

Leaves, flowers

Solidago odora Ait. Aster Family (Asteraceae or Compositae)

Anise- or tarragon-scented perennial; 2–5 ft. Leaves lance-shaped, *not toothed* (but tiny prickles catch skin when rubbed backward along edge of leaf). Flowers *on one side of branch*; July–Oct. **WHERE FOUND:** Dry, open woods. S. NH to FL; TX to OK, MO. **USES:** Cherokee used plant tea to treat diarrhea, colds, coughs, fevers, and for tuberculosis, measles, and neuralgia. Leaf tea pleasant tasting. Formerly used as a digestive stimulant, diaphoretic, diuretic, mild astringent; to regulate menses, for colic, stomach cramps, colds, coughs, fevers, dysentery, diarrhea, measles; externally, a wash for rheumatism, neuralgia, headaches. Tea formerly used to mask the taste of other medicines. The flowering tops were claimed to be exported to China as a beverage tea in the early nineteenth century (disputed by some because of lack of evidence). Contains significant amounts of methyl chavicol in the essential oil, the same compound that gives French Tarragon its characteristic flavor and fragrance. **WARNING:** May cause allergic reactions.

COMMON TANSY

Whole plant

Tanacetum vulgare L. Aster Family (Asteraceae or Compositae)

⚠ Strong-scented perennial; 1–4 ft. Leaves *fernlike*. Flowers to ½ in., in flat terminal clusters; June–Sept. **WHERE FOUND:** Roadsides, fields. Scattered throughout our range, common in the Northeast. Alien (Europe). **USES:** Traditionally, weak cold leaf tea was used for dyspepsia, flatulence, jaundice, worms, suppressed menses, weak kidneys; externally, as a wash for swelling, tumors, inflammations; spray or inhalant of tea for sore throats. Experiments have confirmed antioxidant, anti-inflammatory, antispasmodic, antiseptic, and diuretic activity. A recent study found that components of the aerial parts, particularly parthenolide, have potential activity against herpes simplex virus by interfering with virus replication. Leaves insecticidal and may repel ticks. **WARNING:** Oil is lethal—½ ounce can kill in 2–4 hours. May cause dermatitis in some individuals.

DAISYLIKE FLOWERS, BLACK-EYED SUSANS

GARDEN COREOPSIS, TICKSEED

Whole plant

Coreopsis tinctoria Nutt. Aster Family (Asteraceae or Compositae)

Smooth-stemmed annual; 2–4 ft. Leaves divided into slender segments. Flowers yellow; *base of rays brown*; flowering year-round, depending upon location; mostly July–Sept. **WHERE FOUND:** Moist sandy or clay soils, ditches; widely cultivated and naturalized. Throughout our range and beyond. Naturalized in Europe. **USES:** The Cherokee used the root tea for diarrhea and as an emetic. In Navajo tradition the root is used as a general tonic. In Portugal a tea of the flowering tops is used to treat symptoms of diabetes. An infusion of flowering tops, rich in bioactive flavonoids, when taken for three weeks was found to increase glucose tolerance in animal experiments.

Garden Coreopsis produces bicolored flowers.

SNEEZEWEED

Disk florets, flowers

Helenium autumnale L. Aster Family (Asteraceae or Compositae)

⚠ Perennial; 2–5 ft. Winged, angled stems. Leaves lance-shaped to ovate-oblong, coarsely toothed. Flowers yellow; *disk globular; rays wedge-shaped, 3-toothed*; July–Nov. **WHERE FOUND:** Rich thickets, wet fields. QC, MA to FL and westward. **USES:** The Cherokee used powdered dried disk florets as a snuff for head colds and catarrh by inducing sneezing. Meskwaki used flowers as a cold remedy and drank tea for "catarrh of

Sneezeweed produces winged, angled stems.

Black-eyed Susan. Note its prominent dark center.

stomach." Flower tea used to treat intestinal worms. Folk remedy for fevers. Helenalin, a lactone found in this and other species of *Helenium,* has shown significant antitumor activity in the cancer-screening program of the National Cancer Institute. **WARNING:** Poisonous to cattle. May cause contact dermatitis. Helenalin is poisonous to fish and worms as well as to insects.

BLACK-EYED SUSAN — Root

Rudbeckia hirta L. — Aster Family (Asteraceae or Compositae)

Annual, biennial, or short-lived variable perennial; 1–3 ft. Leaves lance-shaped to oblong, *bristly-hairy*. Flowers yellow, daisylike, with 8–21 rays around a deep brown center; June–Oct. **WHERE FOUND:** Fields, roadsides, waste places. Most of our area and beyond. Four distinct varieties are recognized in N. America. **USES:** The Cherokee used root tea externally as a wash for sores, snakebites, swelling; root juice for earaches. The Iroquois used root tea for worms in children. The Potawatomi used a root infusion to treat colds. Contains chlorogenic acids with antioxidant, anti-inflammatory, antimutagenic, and antiviral activity. Like *Echinacea*, Black-eyed Susan has been found to have immunostimulant activity. **WARNING:** Contact sensitivity to the plant has been reported.

GREEN-HEADED CONEFLOWER

Root, flowers, leaves

Rudbeckia laciniata L.

Aster Family (Asteraceae or Compositae)

Large, branched perennial; 3–12 ft. Leaves deeply divided into 3–5 sharp-toothed lobes. Flowers yellow; rays drooping, disk *greenish yellow*; June–Sept. Four varieties are recognized. **WHERE FOUND:** Moist, rich thickets. NS to FL; AZ to MT and beyond. **USES:** Cherokee ate cooked leaves in spring as a spring tonic for "good health"; dried leaves stored for winter food; young shoots cooked in fat as food. Chippewa used root tea (with Blue Cohosh, p. 275) for indigestion. Flower poultice (with Blue Giant Hyssop or Anise-Hyssop, p. 253, and a Goldenrod species) applied to burns. Was once sold as a balsamic diuretic and tonic, recommended for urinary catarrh. Little researched.

Green-headed Coneflower has sharply incised leaves.

TALL PLANTS WITH DANDELION-LIKE FLOWERS

WILD LETTUCE

Leaves, sap

Lactuca canadensis L.

Aster Family (Asteraceae or Compositae)

⚠ Biennial; usually over 30 in. tall. Stem smooth, branched, with whitish film. Highly variable. Leaves lance-shaped, *wavy-margined to deeply lobed*. Flowers in panicles of small yellow dandelion-like heads; June–Oct. **WHERE FOUND:** Thickets, roadsides, wet soils. Much of N. America. **USES:** Cherokee and other Native American groups used plant tea as a nerve tonic, sedative, pain reliever. Milky latex from stem used for warts, pimples, Poison Ivy rash, and other skin irritations. Used similarly by settlers. The milky juices of this and other lettuces, wild and "tame," have been used to make the so-called let-

Wild lettuce leaves are highly variable from lance-shaped to deeply lobed.

tuce opium, which has been sold as a sedative in the U.S. We suspect it might make a better substitute for rubber than for opium or chicle. **WARNING:** This species and other *Lactuca* species may cause dermatitis or internal poisoning.

PRICKLY LETTUCE

Leaves, sap

Lactuca serriola L.
(*Lactuca scariola* L.)

Aster Family (Asteraceae or Compositae)

Annual or biennial; 2–7 ft. Leaves oblong to lance-shaped or dandelion-like but *prickly*; margins with weak spines. Flowers yellow, dandelion-like; June–Oct. **WHERE FOUND:** Waste places. Much of our area; indeed much of the world; a noxious weed in se. China. Alien (Europe). **USES:** Leaf tea to stimulate milk flow; diuretic. Also used like other wild lettuces. **WARNING:** May cause dermatitis or internal poisoning.

SPINY-LEAVED SOW-THISTLE

Root, sap, leaves

Sonchus asper (L.) Hill.

Aster Family (Asteraceae or Compositae)

Annual or biennial; 1½–4 ft. Leaves divided, dandelion-like; leaf base with rounded earlike lobes, or *curled lobed, clasping*; prickly toothed. Flowers dandelion-like; Mar.–Nov. Bracts often glandular-hairy. **WHERE FOUND:** Fields, waste places. Throughout our range and beyond. Alien (Europe). **USES:** Leaves used as a famine food. Asian Indians use the root of this species and other sow-thistles in tea for asthma, bronchitis, cough, and whooping cough; leaves (poultice or wash) for swellings; latex (juice) for severe eye inflammations. In Europe, leaves are poulticed

LEFT: *Prickly Lettuce produces small dandelion-like flowers.* ABOVE: *The leaves of Prickly Lettuce have prickly margins.*

as an anti-inflammatory. Science confirms antibacterial, antioxidant, liver-protective, and anti-inflammatory activity. Folk remedy for cancer. Young shoots of this and other species were eaten as a salad or potherb. **RELATED SPECIES:** *Sonchus arvensis* L. (Field Sow-Thistle) and *S. oleraceus* (L.) L. are both alien species from Europe common in N. America. The Cherokee used a tea of *S. arvensis* to calm nerves. The leaves of *S. oleraceus* were used to treat diarrhea and stimulate menstruation. Once thought useful as a cure for opium addiction.

Spiny-leaved Sow-thistle is a common garden weed.

SMALL PLANTS WITH DANDELION-LIKE FLOWERS

DANDELION

Roots, leaves

Taraxacum officinale F. H. Wigg Aster Family (Asteraceae or Compositae)

Highly variable, familiar weed; 2–18 in. Flowering stalk *hollow, with milky juice*. Leaves *jagged-cut*. Flowers yellow; Mar.–Sept. and sporadically throughout the year. Bracts reflexed. **WHERE FOUND:** Lawns, fields, waste places. Throughout our range and beyond. **USES:** Fresh-root tea traditionally used for liver, gallbladder, kidney, and bladder ailments, diuretic (not indicated when inflammation is present). Also used as a

The Dandelion's seed head is a familiar sight on lawns (even the day after mowing).

The Dandelion's flowers are familiar to all. Its leaves are eaten as a green and used in medicine.

tonic for weak or impaired digestion, constipation. Dried root thought to be weaker, often roasted as coffee substitute. Dried-leaf tea is a folk laxative. Leaf tea a folk remedy for anemia and a blood purifier. A recent study confirmed that an extract increased red blood cells in laboratory experiments and at high doses (200 mg/kg) normalized white blood cell counts, providing a scientific basis for blood purifying claims and immunostimulant activity. Experimentally the root is hypoglycemic, a weak antibiotic against yeast infections (*Candida albicans*), and stimulates flow of bile and weight loss, plus has antioxidant, liver-protectant, and diuretic activity. Folkloric use in cystitis and other inflammations is explained by a large number of anti-inflammatory compounds in the leaves and root, particularly taraxasterol, a triterpene. All plant parts have served as food. Leaves and flowers are rich in vitamins A and C. Dandelion leaf used in European phytomedicine for treatment of loss of appetite and dyspepsia with a feeling of fullness and flatulence. The root is used for treatment of bile flow disturbances, as a diuretic, to stimulate appetite, and to treat dyspepsia. **WARNING:** Contact dermatitis reported from handling the plant, probably caused by latex in stems and leaves.

COLT'S FOOT

Leaves, flowers

Tussilago farfara L. — Aster Family (Asteraceae or Compositae)

⚠ Perennial; 4–8 in. Leaves rounded, slightly lobed, toothed; base strongly heart-shaped. Flowers *appear before* leaves; Mar.–Apr. Flowers yellow with many slender rays, on a *reddish-scaled stalk*. **WHERE FOUND:** Fields. NS to NC, TN; OH to IL to ON, QC. Alien (Europe). **USES:** Leaf and flower tea traditionally used as a demulcent and expectorant for sore throats, coughs, asthma, bronchitis, lung congestion. One of Eu-

LEFT: *Colt's Foot leaves appear after the flowers.* ABOVE: *Colt's Foot flowers appear before leaves in early spring, sometimes sprouting through snow.*

rope's mostly popular cough remedies historically; dried leaves smoked for coughs and asthma. Smoke is believed to impede impulse of fibers of parasympathetic nerves and to act as an antihistamine. Research indicates that leaf mucilage soothes inflamed mucous membranes and that leaves have spasmolytic activity. Leaf is used in European phytomedicine for treatment of cough and hoarseness, inflammation of the respiratory tract, and mild inflammation of the mouth or throat. Naturalized in China, there the flower buds are used for similar purposes and preferred over the leaves. **WARNING:** Contains traces of liver-affecting pyrrolizidine alkaloids; potentially toxic in large doses. In Germany, use is limited to 4 to 6 weeks per year, except under advice of a physician.

PLANTS WITH SUNFLOWER-LIKE FLOWERS

SUNFLOWER

Whole plant

Helianthus annuus L. — Aster Family (Asteraceae or Compositae)

Familiar, well-known annual; only North American native plant to become a major agronomic crop throughout the world; 6–10 ft. Leaves mostly alternate, rough-hairy, broadly heart- or spade-shaped. Flowers orange-yellow; disk flat, to 1 ft. or more in diameter; July–Oct. **WHERE FOUND:** Prairies, roadsides. Throughout N. America. Wild parent of our domesticated sunflower. Cultivated worldwide. **USES:** Widely used by Native American groups, particularly in the West; flower tea for lung ailments, malaria. Leaf tea taken for high fevers; astringent; poultice on snakebites and spider bites. Seeds and leaves said to be diuretic, expectorant; tea of the stem pith once considered diuretic. Seed oil traditionally used for coughs, inflamed bladder and kidneys. Leaves astringent, formerly used to treat diarrhea. Seed sprouts are rich in cymarin; considered antioxidant and hypoglycemic; suggested as a beneficial food choice for diabetics. Components in flower petals (triterpene glycosides) shown to have marked anti-inflammatory activity. Seeds eaten since ancient times in the Americas. **WARNING:** Pollen or plant extracts may cause allergic reactions.

Sunflower is an annual with familiar flowers.

LEFT: *Jerusalem Artichoke flowers.* ABOVE: *Jerusalem Artichoke produces edible tubers.*

JERUSALEM ARTICHOKE Tubers, stalks, flowers
Helianthus tuberosus L. Aster Family (Asteraceae or Compositae)

Hairy, tuber-bearing perennial; 5–10 ft. Leaves oval, thick, hard; sandpapery above, 3-nerved; leafstalk winged. Yellow flowers; Aug.–Oct. **WHERE FOUND:** Thickets, fields. Throughout our area. Dispersed by humans since ancient times in the Americas, obscuring original native range. Widely cultivated and naturalized in Europe. **USES:** Native Americans drank leaf and stalk tea or ate flowers to treat rheumatism. Folk use has suggested that the edible tubers, which contain inulin, may aid in treating diabetes. Inulin also used for leaky gut syndrome. Like those of beans, the roots are known to induce intestinal gas. Plant was cultivated by early settlers for the edible tubers and for making beer. Introduced to European horticulture at an early date; a French patent of 1848 enumerates numerous food uses for not only the tubers, but also the plant stalks as fodder for animals, particularly to keep horses in good condition to sustain hard labor. Experimentally antioxidant and antidiabetic.

COMPASS-PLANT Root, leaves, resin
Silphium laciniatum L. Aster Family (Asteraceae or Compositae)

Perennial; 3–10 ft. Leaves large, stiff, upright, *deeply divided, pinnately lobed*, with blades lance-shaped, ovate, or triangular; with uneven-toothed margins or entire, *rough-hairy; aromatic*. Leaves and stems exude gummy, translucent tears of a turpentine-scented oleoresin. Flowers yellow, with few rays on tall stiff stalks; July–Sept. **WHERE FOUND:** Prairies, glades. NY to AL; TX to SD, MN, ON. **USES:** Native Americans

Compass-plant produces showy yellow flowers on long stalks.

Compass-plant leaf edges follow the sun, hence the name Compass-plant.

such as the Dakota, Meskwaki, Omaha, and Pawnee used root tea as a general tonic for debility; worm expellent. Leaf tea emetic, once used for coughs, lung ailments, asthma. Resin said to be diuretic and expectorant. Root formerly used for dry cough and various lung afflictions, such as asthma. Surprisingly little researched. **WARNING:** Of unknown safety.

CUP-PLANT, INDIAN CUP-PLANT — Root

Silphium perfoliatum L. — Aster Family (Asteraceae or Compositae)

Tall, bold, square-stemmed perennial; 3–8 ft. Upper leaves *united at base, forming a cup*. Flowers like small sunflowers; July–Sept. **WHERE FOUND:** Rich, moist thickets. ON, VT to NC, AL, LA; TX, OK to ND. **USES:** Widely used by Native American groups; root tea for lung bleeding, back or chest pain, profuse menstruation, and to induce vomiting; inhaled smoke for head colds, neuralgia, and rheumatism. Historically, root tea was used for enlarged spleen, fevers, internal bruises, debility, liver ailments, and ulcers. Considered tonic, diaphoretic, and useful as a restorative for recovery from fevers and liver afflictions. Gum from plant once considered a useful stimulant and antispasmodic for dry cough. Aromatic volatile components in plant mimic sex pheromones of female wasps helping males find them hidden in dried leaves. Leaves contain various bioactive flavonoids. **WARNING:** Of unknown toxicity.

Cup-plant flower subtended by a cup-shaped leaf.

Cup-plant produces strongly perfoliate leaves that are cup-shaped at the stem.

LEFT: *Prairie-dock flowers are on a thick round stem.* ABOVE: *The leaves of Prairie-dock emit a turpentine odor when crushed.*

PRAIRIE-DOCK Root

Silphium terebinthinaceum Jacq. Aster Family (Asteraceae or Compositae)

Perennial; 2–9 ft. Leaves *huge, to 10 in. wide and 18 in. long*, heart-shaped, odor of turpentine when crushed. Flowers like small sunflowers; July–Oct. Frequently in populations with many leaf rosettes with few flowering plants. **WHERE FOUND:** Prairies, glades. ON to GA, MS; AR to IA; WI. **USES:** Same as for *S. laciniatum* (above). **WARNING:** Of unknown safety.

ORANGE LILIES OR LILYLIKE FLOWERS

DAYLILY

Root, flower buds

Hemerocallis fulva (L.) L.

Daylily Family (Hemerocallidaceae)
[Formerly in the Lily Family (Liliaceae)]

⚠ Perennial; 3–6 ft. Leaves in clumps, swordlike. Flowers face *upward or outward*, not downward; striped in middle, petals curved back. Large, showy flowers; May–July. **WHERE FOUND:** Escaped from gardens. Grows near abandoned houses and gardens throughout our range. Alien (Asia). A familiar garden perennial from e. Asia, Daylily has become widely naturalized in the U.S. and is now thought to be more common in the wild in N. America than in its native China. **USES:** The roots and young shoots are an ancient medicinal of Traditional Chinese Medicine, used for more than 2,000 years for mastitis, breast cancer, and a variety of other ailments. In China, the root tea is used as a diuretic in turbid urine, edema; to treat poor or difficult urination, jaundice, nosebleeds, leukorrhea, uterine bleeding; poultice for mastitis. A folk remedy for breast cancer. Experimentally, Chinese studies indicate that root extracts are antibacterial, useful against blood flukes (parasites), and diuretic. The edible flower buds are used for diuretic and astringent properties in jaundice and to "relieve oppression and heat in the chest"; poulticed for hemorrhoids; flower extracts are strongly antioxidant and have recently been shown to improve mobility of fat tissue, possibly leading to suppression of body fat accumulation. **WARNING:** The roots and young leaf shoots are considered potentially toxic. Chinese reports indicate that the toxin accumulates in the system and can adversely affect the eyes, even causing blindness in some cases. Chinese studies hint that the roots may also contain the carcinogen and teratogen colchicine, which, though poisonous, has long been used in the treatment of acute gout crises. Foragers beware.

Daylily is a commonly naturalized lily from e. Asia.

BLACKBERRY LILY

Root

Iris domestica (L.) Goldblatt & Mabb.
[*Belamcanda chinensis* (L.) DC.]

Iris Family (Iridaceae)

⚠ Though called a lily, the Blackberry Lily is actually an Iris family member. Perennial; 1–2 ft. Leaves swordlike. Flowers in cymes, *orange, mottled with purple or brownish marks*; each flower lasts only one day; May–August. Fruits are 3-lobed capsules, splitting to reveal black fleshy seeds (hence the name Blackberry Lily). **WHERE FOUND:** Pastures, road-

Blackberry Lily. Note the overlapping, irislike leaves. Its fruits superficially resemble blackberries.

Canada Lily is a tall plant with bell-like, hanging flowers.

sides, ditches. Alien from e. Asia, introduced into Europe in the 1730s. NH south to FL; west to TX, SD, MN. Widely grown as a perennial garden flower elsewhere and escaped from cultivation. **USES:** Roots (*she gan*) used in Traditional Chinese Medicine in prescriptions for sore throats, cough, asthma, wheezing, bronchitis. Externally, poulticed for sprains, boils, contusions, rheumatism; folk remedy for breast cancer. Science confirms antioxidant, blood pressure–lowering, antifungal, antibacterial, antiviral, and potential anticancer activity. Contains phytoestrogenic and antiandrogenic compounds, plus the isoflavonoids tectorigenin and tectoridin as major components with COX-2-inhibitory activity. **WARNING:** Usually used under medical supervision in China. Avoid during pregnancy. Contains toxic iridoid components, including belamcandin and iridin.

CANADA LILY

Root

Lilium canadense L.

Lily Family (Liliaceae)

Perennial; 2–5 ft. Leaves lance-shaped, usually in whorls. *Nodding* yellow, orange, or reddish flowers; spotted, bell-shaped; July–Aug. **WHERE FOUND:** Moist meadows, openings. Se. Canada. to e. MD, VA mtns., NC, GA, FL; AL to KY. **USES:** The Algonquin used root tea for stomach ailments. The Cherokee used root preparations for dysentery and rheumatism. Micmac use root for irregular menses; Chippewa herbalists used root poultice for snakebites.

WOOD LILY

Root, flowers

Lilium philadelphicum L.

Lily Family (Liliaceae)

Perennial; 1–3 ft. Leaves in whorls. Flowers bright orange, *upturned*, spotted; June–July. **WHERE FOUND:** Acid woods, openings, clearings. ME to WV, GA mtns., north to KY, ON. **USES:** The Algonquin used root tea for

stomach disorders. Used in colonial America for coughs, consumption, fevers; to expel placenta; externally, for swelling, bruises, wounds, sores. Flowers poulticed for spider bites.

Wood Lily produces upturned, spotted flowers.

MISCELLANEOUS ORANGE-RED FLOWERS

COLUMBINE
Roots, seeds, whole plant

Aquilegia canadensis L.
Buttercup Family (Ranunculaceae)

Perennial; 1–2 ft. Leaves divided in 3s. Flowers drooping, bell-like, with *5 spurlike appendages at top*; Mar.–July. **WHERE FOUND:** Moist, rich woods. S. Canada southward. Throughout our area. **USES:** Astringent, diuretic, anodyne. Meskwaki chewed root of plant for stomach and bowel complaints. Mixed with tobacco, the dried seed capsules were smoked to make young people smell good and appear refined. Seeds used as a love charm. Boiled root and leaves made into decoction to treat diarrhea; also to induce power of persuasion in trade or council. The Omaha, Pawnee, and Ponca used minute amounts of crushed seeds for headaches, "love charm" (uses related?), fevers. Ojibwa used seed for stomach troubles. Seeds rubbed into hair to control lice. Various Native American groups used the roots as incense or as a sachet to make clothes smell pleasant. Root chewed or weak tea for diarrhea, stomach troubles, diuretic. Root tea for uterine bleeding. Seldom used by European settlers, though compared with *A. vulgaris* of Europe as a potential diuretic, emmenagogue, and stomachic; seeds in wine infusion for jaundice. **WARNING:** Of unknown safety.

Columbine. Note the 5 spurlike appendages atop the flower.

BUTTERFLYWEED, PLEURISY-ROOT

Roots

Asclepias tuberosa L. Milkweed Family (Asclepiadaceae)
[Modern works place in Dogbane Family (Apocynaceae)]

Easily recognized at first sight. Densely hairy perennial; 1–3 ft., with large deep-set root (rhizomes). Stem *without milky juice* (but with clear latex). Leaves alternate, lance-shaped, entire, mostly without stalks. Flowers showy bright orange or reddish orange (rarely yellow) in one to many terminal or axillary flowerheads; May–Sept. **WHERE FOUND:** Dry roadsides, prairies. S. NH to FL; TX, KS to MN. **USES:** Most Native American groups in the plant's range used root tea for pleurisy; also for influenza. Cherokee used the large root for lung inflammation, pleurisy; expectorant; root infusion for heart troubles; mild laxative. Delaware used root for rheumatism. Iroquois used a wash of the roots applied to limbs to gain strength; poulticed on legs to increase strength of runners. Menominee used a poultice of roots for bruises, swelling, cuts, and wounds. Tea or tincture of large, tuberous root widely used historically for lung inflammations (pleurisy), asthma, bronchitis, colds, and pneumonia; anodyne, laxative, diuretic, expectorant. Soaked in brandy, the root was a popular folk remedy for rheumatism in the South. Helps to ease breathing and increase expectoration. Root poultice used for bruises, swellings, rheumatism, and lameness. Contains various glycosides and steroidal compounds responsible for biological activity. Seed oil contains compounds that have been shown to have excellent viscosity, oxidation stability, and cold-flow properties. A glycoside fraction of the root increased proliferation of skin cells. **WARNING:** Larger doses have a laxative effect.

Butterflyweed, Pleurisy-root produces striking orange blossoms.

JEWELWEED, SPOTTED TOUCH-ME-NOT

Leaves, juice

Impatiens capensis Meerb. Touch-me-not Family (Balsaminaceae)

Smooth annual; 3–5 ft. Leaves oval, toothed; lower ones opposite, upper ones alternate. Flowers *pendantlike*, with *red spots*; June–Sept. **WHERE FOUND:** Wet, shady soil. Most of our area. **USES:** Crushed leaves are poulticed on recent Poison Ivy rash—a well-known folk remedy. Mucilaginous stem juice, harvested before flowering, also applied to rash. A 1957 study by a physician found it effective (in 2–3 days) in treating 108 of 115 patients. Some people swear by the leaf tea as a Poison Ivy rash preventive; others rub on frozen tea (in the form of ice cubes)

ABOVE: *Jewelweed is commonly used to prevent and treat Poison Ivy rash.* RIGHT: *Hoary Puccoon produces orange-yellow flowers in midspring.*

as a remedy. Poultice a folk remedy for bruises, burns, cuts, eczema, insect bites, sores, sprains, warts, and ringworm. A component in the leaves, lawsone, may explain reported antihistamine and anti-inflammatory activities. A poultice of the fresh crushed leaves can help prevent development of dermatitis after coming into contact with Poison Ivy. However, a recent study that compared Jewelweed with soap as a preventive for Poison Ivy dermatitis found that washing with soap was more effective, and the purported active compound lawsone did not correlate with Poison Ivy rash prevention.

HOARY PUCCOON

Whole flowering plant

Lithospermum canescens (Michx.) Lehm. Borage Family (Boraginaceae)

⚠ Perennial, with *very fine, soft white hairs*; 4–18 in. Leaves alternate, lance-shaped. Flowers orange to yellow; Apr.–June. Flowers 5-petaled, in curled or flat clusters; stamens concealed in tube. **WHERE FOUND:** Dry soils, prairies. S. ON to GA, MS; TX to SK. **USES:** Menominee used leaf tea (as a wash) for fevers accompanied by spasms. Wash rubbed on persons thought to be near convulsions. Ponca children used reddish root bark as a colorant, chewed with gum (from *Silphium laciniatum*) to impart a red color. Similarly, flowers used as a yellow colorant. A European species, *Lithospermum arvense* L., Bastard Alkanet or Field Gromwell, was used to color paper and linen; during festivals, eighteenth-century Swedish women used root as a cosmetic rouge. Some species of *Lithospermum* have been used for thyroid problems and as a contraceptive. At least 7 pyrrolizidine alkaloids have been isolated from the plant. These compounds, common in the Borage family, are associated with causing veno-occlusive liver disease. Other compounds isolated from the roots have been found to increase immunostimulant activity and exhibit potential anticancer activity. **WARNING:** Pyrrolizidine alkaloids in the plant may cause veno-occlusive disease of the liver.

WILD GINGER Rhizome, leaves

Asarum canadense L. Birthwort Family (Aristolochiaceae)

⚠ Creeping perennial, with spicy aromatic rhizomes. Leaves *strongly heart-shaped*. Flowers maroon, *urn-shaped*, with 3 "petals" (actually sepals); between crotch of leaves; Apr.–May. Root strongly aromatic. **WHERE FOUND:** Rich woods. Canada to SC, AL; OK north to ND. **USES:** Native American groups throughout the plant's range highly valued root tea for indigestion, coughs, colds, heart conditions, "female ailments," throat ailments, nervous conditions, and cramps. Relieves gas, promotes sweating; expectorant; used for fevers, colds, sore throats. The strongly aromatic, spicy-fragranced root attracted the attention of early settlers; a cordial or syrup was used for fevers, whooping cough, stimulant to digestion, and flavoring to mask the flavor of other medicines. Powdered dried leaves were used as a snuff for head congestion; induce sneezing, among other uses. Contains the antitumor compound aristolochic acid. However, like so many compounds, it is reported to both cause and "cure" cancer. Historically, a frontier ginger substitute. **WARNING:** Aristolochic acid is considered an insidious toxin, inducing mutations.

PINK LADY'S-SLIPPER Root

Cypripedium acaule Ait. Orchid Family (Orchidaceae)

Perennial orchid; 6–15 in. Leaves 2; basal. Flower pink (rarely white), strongly veined; pouch with a deep furrow; May–June. **WHERE FOUND:** Acid woods. NL to GA; AL, TN to MN. Too rare to harvest. **USES:** Called

ABOVE: *Wild Ginger, with heart-shaped leaves, spreads by runners.* RIGHT: *Wild Ginger's urn-shaped flowers are found at ground level.*

American Valerian. Widely used in the nineteenth century as a sedative for nervous headaches, hysteria, insomnia, and "female" diseases. See Yellow Lady's-slipper (p. 134). **WARNING:** May cause dermatitis.

INDIAN PAINTBRUSH, PAINTED CUP — Flower

Castilleja coccinea (L.) Spreng. — Broomrape Family (Orobanchaceae)
[Formerly in Figwort Family (Scrophulariaceae)]

Parasitic annual or biennial; sends out slender shoots from roots that attach to roots of surrounding plants; to 2 ft. tall. Leaves in basal rosettes. Flowers with 3-*lobed, scarlet-tipped bracts* (rarely yellow); Apr.–July. **WHERE FOUND:** Meadows, prairies. S. NH to FL; TX, OK to s. MB. **USES:** The Chippewa used weak flower tea for rheumatism, "female diseases." Menominee used plant as a love charm, applying the herb to the person to whom advances were intended. Conversely, tea used by Cherokee to "destroy your enemies." **WARNING:** Of unknown safety.

RED TRILLIUM, WAKEROBIN, BETHROOT — Root, whole plant

Trillium erectum L. — Melanthium Family (Melanthiaceae)
[Formerly placed in the Trillium Family (Trilliaceae) or the Lily Family (Liliaceae)]

Perennial; 6–16 in. Leaves triangular-oval; 3, *in a single whorl.* Flowers dull red (to white), with 3 triangular petals and sepals; Apr.–June.

Pink Lady's-slipper is common in the Northeast.

Indian Paintbrush flowers are typically orange-red and occasionally pale yellow.

WHERE FOUND: Rich woods. NS to GA mtns., FL; TN to MI, ON. **USES:** The root was traditionally used as an aid in childbirth, hence the name "Bethroot" (a corruption of "birth root"). Cherokee used root tea for menstrual disorders, to induce childbirth, to aid in labor; for "change of life" (menopause), uterine astringent, aphrodisiac (root contains steroids). Used for coughs and bowel troubles. Whole plant poulticed for tumors, inflammation, and ulcers. Used as a folk medicine in colonial America. C. S. Rafinesque claims to have introduced the genus into medical practice in 1830, treating the use of most *Trillium* species generically. He observed that Native Americans called those species with red flowers "male," while those species with white blossoms were considered "female" and were considered best for female conditions. Historically, physicians used root tea as above, and for hemorrhages, asthma, difficulty in breathing, chronic lung disorders; externally, for snakebites, stings, skin irritations. At least a dozen steroidal saponins have been isolated from the roots, which are likely responsible for reported traditional uses.

Red Trillium produces dull red flowers.

MISCELLANEOUS PLANTS WITH PINK OR RED FLOWERS

WINDFLOWER, RUE ANEMONE — Root

Anemonella thalictroides (L.) Spach [*Thalictrum thalictroides* (L.) Boivin.] — Buttercup Family (Ranunculaceae)

⚠ Delicate perennial; 4–8 in. Leaves in *whorls*; small, 3-lobed. Flowers mostly white, sometimes pink, with 5–11 "petals" (sepals); Mar.–May. See p. 65.

CYPRESS-VINE — Leaves

Ipomoea quamoclit L. — Morning-glory Family (Convolvulaceae)

⚠ Slender, smooth, weak-stemmed, twining annual. Vine to 15 ft. long. Leaves broadly oval overall, pinnately divided into many narrow, linear segments. Flowers 1 to several, *scarlet, funnel-shaped*, to 1¾ in. long; July–Sept. **WHERE FOUND:** Fields and waste places. Alien (tropical America). Cultivated as an ornamental and occasionally naturalized, especially in South. **USES:** A South American folk medicine for pain relief and purgative effects. Externally applied to carbuncles, snakebites, sores, and hemorrhoids. In India the leaves are eaten as a potherb; also for treatment of ulcer and breast pain; externally applied to carbuncles,

Windflower, Rue Anemone is often pink, with 5–11 "petals" (sepals).

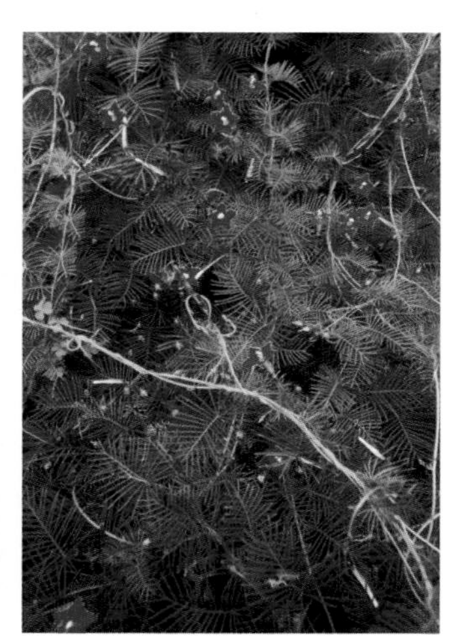

Cypress-vine produces scarlet flowers.

bleeding hemorrhoids, snakebite, and as a purgative. In Australia it has been used as a purgative and snakebite remedy. Reported to contain various alkaloids, cyanogenic glycosides, and jalapin. **WARNING:** Leaves contain potentially toxic hydrocyanic acid.

OPIUM POPPY

Latex from flower capsule

Papaver somniferum L.

Poppy Family (Papaveraceae)

Stout annual; 2–3 ft., *blue-green*, smooth stem. Leaves lettuce leaf–like, *stalkless, coarsely toothed, margins wavy or shallowly lobed*, heart-shaped and clasping at base. Flowers highly variable, usually 4-petaled, white, pink, red, to purple; Apr.–June. Large *radiating disk within*, producing an iconic dried oval to broadly rounded dried seed capsule. **WHERE FOUND:** Escaped from gardens throughout much of our range. Surprisingly common. Alien originating from Central Asia. **USES:** One of the world's most famous medicinal plants, used since prehistoric times; today derivatives of Opium Poppy are staples of pain management in virtually every medical clinic and hospital in the world. The white latex extracted from the flowerheads is crude opium. Poppy heads were infused in wine, made into pills or syrup, and used to induce sleep and relieve pain. Contains more than 25 alkaloids, notably morphine, codeine, narcotine, and others. Morphine, first isolated in 1803, became widely used to relieve pain in the 1850s; used for rapid pain relief. In the Civil War, morphine addiction was called the "soldier's disease." Also widely added to nineteenth-century patent medicines; by 1900, 1 in 400 Americans was addicted to opiates. In 1898, a refined

ABOVE: *Opium Poppy, with red flowers. Petal has a dark base.* TOP RIGHT: *Opium Poppy's immature seed capsule, from which opium is derived.* BOTTOM RIGHT: *Opium Poppy, with crinkled pink petals.*

form of morphine—heroin—was developed; available as an over-the-counter cough medicine well into the twentieth century. The physical addictive effects of poppy's crude opium and extractive alkaloids are well known. Still, no widely available synthetic drug has replaced morphine for relief of severe pain. **WARNING:** Poppy extractives, including crude opium and alkaloids such as morphine, codeine, and the morphine-derivative heroin, are dangerously addictive and, of course, illegal narcotics except when used in a controlled medical setting.

The flowers of Wood Betony are usually yellow, but they can be red like those pictured here.

Feverwort, Coffee Plant produces small red flowers in leaf axils.

LOUSEWORT, WOOD BETONY

Root, leaves

Pedicularis canadensis L. — Broomrape Family (Orobanchaceae)
[Formerly in Figwort Family (Scrophulariaceae)]

Flowers hooded, like miniature snapdragons; reddish, yellow (or both). See p. 147.

FEVERWORT, COFFEE PLANT

Root, leaves

Triosteum perfoliatum L. — Honeysuckle Family (Caprifoliaceae)

Perennial; 2–4½ ft. Leaves opposite; *connected around stem* (perfoliate). Flowers greenish to dull purple, in leaf axils; *5 prominent sepals*; May–July. Fruit bright red to orange. **WHERE FOUND:** Moist woods. QC, MA to GA, AL; TX to MN, ON. **USES:** Cherokee used the root tea for fevers, as an emetic; externally a wash for sore feet and leg swelling. The Iroquois used root tea for irregular to profuse menses, urinary disorders; cold tea for bad colds and sore throats. Root poulticed for snakebites, sores, and abcess at the end of fingers. Leaf tea taken to induce sweating. Historically, root was used by physicians for headaches, colic, vomiting, diarrhea, and indigestion. Diuretic for chronic rheumatism. In larger doses it is cathartic and emetic. Seeds used as coffee substitute. Little researched but found to have mild antibacterial activity.

VALERIAN

Root

Valeriana officinalis L.

Honeysuckle Family (Caprifoliaceae)
[Formerly in Valerian Family (Valerianaceae)]

Perennial; 4–5 ft. Leaves strongly divided, *pinnate*; lower ones toothed. Tiny, pale pink to whitish flowers, in tight clusters; June–July. **WHERE FOUND:** Escaped, along roadsides, especially in ne. U.S. PE, QC, ME to MD, WV; IA to SK; Pacific Northwest. Alien (from Europe). **USES:** A well-known herbal calmative, antispasmodic, nerve tonic; used for hypochondria, nervous headaches, irritability, mild spasmodic afflictions, depression, despondency, insomnia. Research has confirmed that teas, tinctures, and/or extracts are CNS-depressant, antispasmodic, and sedative when agitation is present, but also a stimulant in fatigue; antibacterial, antidiuretic, liver-protective. Valerian is a leading over-the-counter tranquilizer and sleep aid in Europe. Several controlled clinical studies on Valerian preparations or single components from Valerian show both positive and negative results as a short-term sleep aid. One study found it worked best as a sleep aid over a period of a month rather than on a single-dose basis. Recently evaluated with positive results as a possible treatment for obsessive-compulsive disorder and for generalized anxiety disorder. Commonly used in European phytomedicine as a mild sedative, in sleep-inducing preparations for nervous restlessness, and as an aid in falling asleep. Cats are said to be attracted to the scent of the root, as they are to Catnip. In eighteenth-century apothecaries, the quality of Valerian root was determined by the way in which cats reacted to it. According to folklore, the root repels rats. **RELATED SPECIES:** *Valeriana sitchensis* Bong. (not shown), native to the w. U.S., is thought to have higher levels of valepotriates, thus stronger medicinal activity. *Valeriana uliginosa* (Torr. & Gray) Rydb., a subarctic species found in the ne. U.S., was used as a sedative by the Menominee. A number of the more than 300 species of *Valeriana* worldwide have been used in traditional medicine systems for equivalent antispasmodic and sedative effects. Still, science is confused on effectiveness, best types of preparations, and difference in chemistry between fresh roots and dried roots.

Valerian flowers are usually white, but they can be pale pink.

MISCELLANEOUS PLANTS OF BOGS, SWAMPS, LAKES

WATER OR PURPLE AVENS

Root

Geum rivale L.

Rose Family (Rosaceae)

Perennial; 1–2 ft. Basal leaves much divided, leaflets toothed, *outermost one largest*; stem leaves divided into *3 parts*. Nodding, dull reddish (rarely yellow) globular flowers, mostly in 3s; May–Aug. Fruits hooked. In spring, the roots have a pleasant, clovelike fragrance. **WHERE FOUND:** Bogs, moist ground. NL to WV; MN, IL, west to Rocky Mtns., BC; widespread in Arctic and cooler regions of the Northern Hemisphere. **USES:** The Algonquin, Malecite, and Micmac used a root decoction for diarrhea and dysentery; also for fevers and to suppress coughs. Powdered root was once used as astringent and styptic for hemorrhage, fevers, diarrhea, dysentery, indigestion, leukorrhea; also for fevers and stomach problems. In colonial America, the roots were prepared with sugar for a drink called "Indian chocolate" used for dysentery, diarrhea, colic, sore throat; ingredient in ale as a stomach tonic. Various compounds associated with astringent effects have been isolated from the root; antimicrobial and anti-inflammatory. **RELATED SPECIES:** In China and Japan, a tea of the whole plant of *Geum japonicum* Thunb. is used as a diuretic and as an astringent to treat coughs and the spitting up of blood. The root and leaves of *G. japonicum* are used as a poultice or wash for skin diseases and boils. Possesses proven antiviral activity, probably due to tannins. Other *Geums* are used similarly.

Water or Purple Avens is often found in wooded bog habitats.

LOTUS, SACRED LOTUS

Entire plant

Nelumbo nucifera Gaertn.

Lotus Family (Nelumbonaceae)
[Formerly Water-Lily Family (Nymphaeaceae)]

Large aquatic perennial; to *3 ft. above water surface* in flower. Leaves large, umbrella-shaped to 2 ft. across. Flowers white to pink, large to 8 in. across, sweetly fragrant; June–Aug. See p. 20.

The flowers of Sacred Lotus, an Asian alien, range from white to pink-red.

Pitcher-plant flowers range from green to reddish in color.

PITCHER-PLANT Leaves, root

Sarracenia purpurea L. Pitcher-plant Family (Sarraceniaceae)

The tubular, pitcher-shaped leaves have evolved to capture and digest insects. The "pitchers," often containing water (atypical of most *Sarracenia* species), have smooth surfaces, making it difficult for insects to crawl out. Downward-pointing hairs at the top of the pitcher further prevent insects from escaping. Once the insect drowns, the leaves secrete enzymes that digest soft body parts, obtaining nutrients that are otherwise unavailable in its habitat. Unique perennial; 8–24 in. Leaves *red-veined, pitcherlike*. Flowers dull red or greenish, nodding, with a *large flat pistil*; May–July. **WHERE FOUND:** Peat or sphagnum bogs, savannas, and wet meadows. NL to GA mtns., OH to MN, throughout Canada; scattered elsewhere. May be a threatened species; best left undisturbed in the wild. **USES:** The Cree of Northern Quebec traditionally used the plant for diabetes symptoms. A Canadian research team recently isolated 11 compounds from the plant, 3 of which were shown to potentiate glucose uptake, while decreasing liver glucose, providing an experimental basis for the traditional use. Various northern Native American groups used root for lung

Pitcher-plant has distinctive pitcher-like leaves. Flowers range from green to reddish in color.

and liver ailments, spitting up of blood; childbirth aid; diuretic. Dried-leaf tea used for fevers, chills, and shakiness. Historically, physicians considered the herb to be a stimulating tonic, diuretic, and laxative. The root famously emerged as a smallpox treatment in 1861, based on a report published in the *London Pharmaceutical Journal.* Some reported success with use; others declared the cure to be mere humbug. European physicians researched the plant as a possible smallpox cure in the nineteenth century, but without success. Medicinal merit neither proved nor disproved.

SWAMP ROSE-MALLOW
Leaves, root

Hibiscus moscheutos L.
Mallow Family (Malvaceae)

Musky-scented perennial; 5–7 ft. Lower leaves often 3-lobed; median leaves lance-shaped. Flowers to 8 in. across; white, with a *purple-red center*; June–Sept. **WHERE FOUND:** Marshes. MD to FL; AL to IN. **USES:** Traditionally used for the treatment of inflamed bladder, both internally and externally. Abounds in mucilage. Leaves and roots of this plant, like those of related species and genera, used as demulcent and emollient in dysentery and lung and urinary ailments.

CARDINAL FLOWER
Root, leaves

Lobelia cardinalis L.
Bellflower Family (Campanulaceae)

⚠ Many consider this one of our showiest wildflowers. The vibrant scarlet flower spikes flag its presence along stream and pond edges in late summer. Perennial; 2–3 ft. Leaves oval to lance-shaped, toothed. Flowers scarlet (rarely white), in brilliant spikes; July–Sept. **WHERE FOUND:** Moist soil, streambanks. NB to FL; TX to MN. **USES:** Cherokee used root tea for stomach troubles and to expel worms; leaf tea for fevers, rheu-

ABOVE: *Swamp Rose-mallow produces pink flowers with a darker center.* RIGHT: *Cardinal Flower is one of the showiest wildflowers.*

matism, colds; externally, root tea for hard-to-heal sores; leaf poultice for headaches. The Meskwaki used as a ceremonial herb, tossing dried leaves toward a storm to reduce winds; also strewn on grave for last rites. Iroquois used tea to relieve breast abscess; also poulticed. As a wash, the crushed roots were used as a face wash to attract the opposite sex. Pulverized roots used for cramps. Pawnee also used root and flowers in love charms. Historically this plant was considered a substitute for Lobelia or Indian-tobacco (*L. inflata,* p. 246), but with weaker effects. It was seldom used by the mid-nineteenth century. Contains alkaloids and anthocyanins; poorly researched. **WARNING:** Potentially toxic; degree of toxicity unknown.

4–5 PETALS; FRUITS UPTURNED; "STORK'S BILLS"

FIREWEED, WILLOW-HERB

Leaves, root

Epilobium angustifolium L. — Evening-primrose Family (Onagraceae)
Chamerion angustifolium (L.) Holub

Perennial; 1–7 ft. Leaves lance-shaped. Flowers rose pink; July–Sept. Fireweed has *4 rounded petals; drooping buds.* **WHERE FOUND:** This common weed is invasive in burned areas and land that has been recently cleared. It occurs throughout northern temperate regions and is used as a traditional medicine by native peoples of N. America, Europe, and Asia. Subarctic to GA mtns.; IN to IA. **USES:** Native Americans poulticed peeled root for burns, skin sores, swelling, boils, carbuncles. Leaf and root tea a folk remedy for dysentery, abdominal cramps, "summer bowel troubles." Leaf poultice used for mouth ulcers. Leaves used in Russia as *kaporie* tea (10 percent tannin). Leaf extract is antibacterial, antioxidant; shown to reduce inflammation. Leaf and root used in Baltics to treat benign prostatic hyperplasia, with benefits attributed to polyphenolic content. Oenothein B is shown to have immunostimulating activity.

STORK'S BILL, ALFILERIA

Leaves

Erodium cicutarium (L.) L'Her. ex Ait. — Geranium Family (Geraniaceae)

Winter annual or biennial; 3–12 in. Leaves fernlike, twice pinnate, often in a basal rosette. Flowers pinkish, 5-petaled; less than ½ in. long; Apr.–Oct.; blooms most of the year in the South. Seeds smooth, elongate, sharp—like a stork's bill. **WHERE FOUND:** Waste places. Much of our area. Alien. **USES:** Leaf tea a folk medicine used to induce sweating, allay uterine hemorrhage; diuretic. Seed poultice used for gouty tophus. Source of vitamin K. Leaves soaked in bath water for rheumatic patients.

WILD OR SPOTTED GERANIUM

Root

Geranium maculatum L. — Geranium Family (Geraniaceae)

Perennial; 1–2 ft. Leaves *broad, deeply 5-parted*; segments toothed. Flowers pink to lavender (rarely white), 5-petaled; Apr.–June. Distinct

Fireweed is often invasive after burning.

Stork's Bill has fernlike leaves and petals with tridentlike marking at base.

"crane's bill" in center of flower enlarges into a seedpod. **WHERE FOUND:** Woods. QC, ME to GA; AR, KS to MB. **USES:** Cherokee used root to stop bleeding on open sores. Various Native American groups used infusion of plant or root as an astringent for diarrhea. Tannin-rich (10–20 percent) root is highly astringent, styptic; once widely used to stop bleeding, diarrhea, dysentery, relieve hemorrhoids, gum diseases, kidney and stomach ailments; diuretic. Root tea used as a gargle for sore throats. Powdered root applied to canker sores. Externally used as a folk remedy for cancer.

Wild or Spotted Geranium is an early spring woodland wildflower.

HERB ROBERT

Leaves

Geranium robertianum L. — Geranium Family (Geraniaceae)

Annual or biennial; 6–18 in. Stems often *reddish*, scent bitter-aromatic. Leaves pinnate, with 3–5 toothed segments; *end segment long-stalked*. Flowers pinkish, usually in pairs; *petals not notched*; May–Oct. **WHERE FOUND:** Rocky woods. NL to NC, TN, MO; OH, IL to MB. Alien; widespread in Eurasia. **USES:** Leaf tea once used for nosebleed, malaria, tuberculosis, stomach and intestinal ailments, jaundice, kidney infections; to stop bleeding; gargled for sore throats. Fresh-herb extract active against vesicular stomatitis virus. Ethanolic extracts are antibacterial. Externally, wash or poultice used to relieve pain of swollen breasts; folk remedy for cancer; applied externally to fistulas, tumors, and ulcers. Fresh leaves chewed for stomatitis. Linnaeus suggested that the fresh bruised herb is a useful insect repellent. Recently found to be a potential research target for lowering blood sugar and improving antioxidant efficiency in the liver. Observed anti-inflammatory activity is linked to antioxidant effects.

Herb Robert has smaller flowers than does Spotted Geranium.

5-PARTED FLOWERS WITH A TUBULAR CALYX

MADAGASCAR PERIWINKLE — Leaves

Catharanthus roseus (L.) G. Don — Dogbane Family (Apocynaceae)

⚠ Flowers often light to deep pink or, if white, with pink centers. See p. 56.

BOUNCING BET, SOAPWORT — Leaves, roots

Saponaria officinalis L. — Pink Family (Caryophyllaceae)

⚠ Stem *thick-jointed*; smoothish perennial; 1–2 ft. Leaves opposite, oval to lance-shaped. Flowers white or rose, 1 in. across; July–Sept. Petals *reflexed, notched*. **WHERE FOUND:** Throughout our area. Alien (Europe). **USES:** Crushed leaves and roots make lather when mixed with water. Cherokee poulticed leaves for spleen pain, boils; used squeezed juice from roots as a hair tonic; leaves also used for soap. In European tradition, plant tea is used as a diuretic, laxative, expectorant; poulticed on acne, boils, eczema, psoriasis, Poison Ivy rash. Root tea used as above; also for lung disease, asthma, gall disease, and jaundice. A ribosome-inactivating protein called saporin 6 inhibits the growth of breast cancer and other cancer cell types in laboratory experiments. The saporin protein is a research target for "suicide gene" therapy, in which the protein is used to irreversibly block protein synthesis in tumor cell pathways, thwarting the ability of cancer cells to reproduce.

ABOVE: *Madagascar Periwinkle, with white, pink, or violet flowers, is grown as a garden annual in the U.S. and is a widespread pantropic weed.* TOP RIGHT: *Bouncing Bet, Soapwort ranges from white to rose.* RIGHT: *Fire Pink has notched petals.*

Experimentally an immunostimulant. Another component, quillaic acid, provides diuretic activity. Considered expectorant (by irritating gastric mucosa). Root used in European phytomedicine to treat catarrhs of the upper respiratory tract. **WARNING:** Contains saponins. Large doses may cause poisoning (toxic to cells). Causes rare stomach irritation in small doses (1.5 g per day).

FIRE PINK — Root

Silene virginica L. — Pink Family (Caryophyllaceae)

Perennial; 8–20 in. Leaves *opposite*, in 2–6 pairs on stems. Flowers brilliant scarlet; 5 petals, *notched at tips*; Apr.–June. **WHERE FOUND:** Rocky woods. S. ON to GA; AR, OK to MN. **USES:** Historical reports speak of worm-expellent properties for this and others of the 70 N. American *Silene* species. **WARNING:** Reports state Native Americans considered the plant poisonous. It might have been confused with Pink-root (see below).

PINK-ROOT

Root

Spigelia marilandica (L.) L.

Logania Family (Loganiaceae)

⚠ Perennial; 12–24 in. Leaves opposite, united by stipules; ovate to lance-shaped. Flowers *scarlet*, 5-lobed flaring trumpets, with a *cream-yellow interior*; May–June. **WHERE FOUND:** Rich woods, openings. MD to FL; TX, e. OK to MO, IN. **USES:** Southern tribes such as the Cherokee and the Creek used root tea as a worm expellent. This plant was also once used by physicians for worms, especially in children. In the nineteenth century, it was heavily harvested and became threatened. By 1830 C. S. Rafinesque reported it had already become rare in MD and VA because of overharvesting. Surprisingly little researched, given its historical importance. Combined with Senna, the root tea was popularly known as "worm tea." **WARNING:** Side effects include increased heart palpitations, vertigo, and convulsions. Early medical works described it as a narcotic and purgative in addition to well-known worm expellant effects.

Pink-root is easily recognized by its bicolored flowers, which are cream-yellow-colored within the throat.

FLOWERS WITH 5 SHOWY PETALS; MALLOW FAMILY

MARSHMALLOW

Leaves, root

Althaea officinalis L.

Mallow Family (Malvaceae)

⚠ Erect, branched, perennial; 2–5 ft. tall. Leaves *velvety*, mostly *3-lobed*, margins serrated. Flowers in clusters from upper leaf axils, typically pale pink, to 1½ in. across; June–Sept. **WHERE FOUND:** Moist soils, naturalized in salt marshes along Mid-Atlantic States (MA to VA); west to KY, AR, WI; escaped from cultivation. Alien (Europe). **USES:** Marshmallow roots and leaves traditionally used in tea for sore throat and expectorant in bronchitis; also for gastritis, gastric or peptic ulcers, enteritis, and mild inflammation of the gastric mucosa. Used since Roman times. Externally poulticed for bruises, sprains, aching muscles, and inflammations. Root (up to 30 percent) and leaves (up to 16 percent) high in mucilage, responsible for demulcent or soothing effect to irritated mucous membranes and skin. Has immunostimulating activity. A poly-

Marshmallow has white to pink flowers.

Hollyhock is used similarly to the Marshmallow, though its roots are considered an adulterant to Marshmallow root.

saccharide from the root experimentally a cough suppressive agent in allergic inflammation of the throat. Leaf and root preparations widely used in European phytomedicine to relieve local irritation (such as digestive-tract inflammatory conditions) and to soothe mucous membrane irritation, such as a sore throat accompanied by dry cough. **RELATED SPECIES:** The common Hollyhock, *Alcea rosea* L., used similarly, though roots are higher in tannins. In Europe, considered an adulterant to Marshmallow root (see below). **WARNING:** Root is high in sugar compounds; diabetics should avoid it. Like other mucilage-containing plants, when taken internally, Marshmallow may hamper absorption of other drugs.

HOLLYHOCK

Flowers, leaves, roots

Alcea rosea L.
[*Althaea rosea* (L.) Cav.]

Mallow Family (Malvaceae)

Erect biennial or perennial herb; 6–9 ft. tall; stems densely hairy, especially on youngest growth. Leaves alternate, *papery*, heart-shaped to rounded, *with 3–5 irregular lobes* or coarsely crenate-toothed, 2½–6½ in. long and wide, stellate tomentose above, sparsely hairy beneath. Flowers solitary, or grouped in *leaf axils* forming a tall spikelike display; 2½–4 in. in diameter; normally pink, but ranging from white to red, purple, yellowish, or black-purple. **WHERE FOUND:** Alien, cultivated since ancient times, origin obscure, though most of the 60 species of *Alcea* are native to cen. to se. Europe, or cen. Asia. Garden escape and sporadically naturalized throughout. **USES:** Dried flowers (with or without the calyces) formerly used in teas or decoction as antispasmodic, expectorant, and demulcent for coughs, hoarseness, and gastrointestinal irritation; gargled for sore throat. Externally, decoction used as a wash to soothe sores. Water extracts of flowers show immunomodulatory activity, stimulating B-lymphocytes. The dried flowers have the highest concentration of mucilage (polysaccharides) compared with leaves. Roots formerly used as substitute (or adulterant) of marshmallow (*Althaea officinalis*).

PURPLE POPPY-MALLOW — Root

Callirhoe involucrata (Torr. & Gray) Gray — Mallow Family (Malvaceae)

Creeping herb. Leaves palmate, divided into 5–7 parts with pointed toothed lobes (especially lower leaves). Flowers *poppylike*, reddish purple; May–Aug. **WHERE FOUND:** Prairies. S. MO to TX; ND west to UT, WY. **USES:** The Teton Dakotas crushed the dried root, burned it, and inhaled the smoke to treat head colds. Aching limbs were exposed to the smoke to reduce pain. Root boiled, then tea drunk for pains. One Dakota name for the plant means "smoke-treatment medicine."

COMMON MALLOW, CHEESES — Leaves, root

Malva neglecta Wallr. — Mallow Family (Malvaceae)

Deep-rooted herb; to 1 ft. Stems trailing. Leaves rounded, toothed; *slightly 5- to 7-lobed.* Flowers pale rose-lavender to whitish, in axils; Apr.–Oct. Petals notched (heart-shaped) on ends; pistils smoothish,

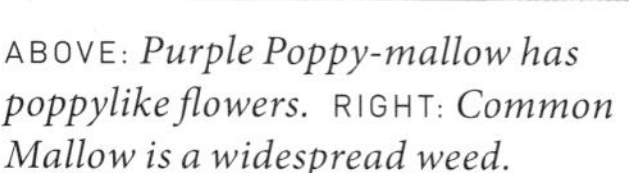

ABOVE: *Purple Poppy-mallow has poppylike flowers.* RIGHT: *Common Mallow is a widespread weed.*

not veined. The common name "Cheeses" is derived from the similarity of the shape of the flat, rounded fruits to a "round" of cheese. **WHERE FOUND:** Yards. Throughout our area. Alien (Eurasia). **USES:** Leaves edible, highly nutritious. Used as a soup base in China. As with Okra, also of the Mallow family, the mucilaginous properties tend to thicken soup. Leaf or root tea of this mallow soothing to irritated membranes, especially of digestive system. Tea also used for angina, coughs, bronchitis, stomachaches; anti-inflammatory, mildly astringent. Poulticed on wounds and tumors. Root extracts show activity against tuberculosis. The leaves of Common Mallow and High Mallow are used in European phytomedicine in the treatment of irritations of the mouth and throat associated with irritating, dry cough.

HIGH MALLOW

Leaves, flowers, root

Malva sylvestris L. — Mallow Family (Malvaceae)

Erect, hairy biennial; 8–36 in. Leaves long-stalked; rounded, with *5–7 distinct lobes*. Flowers rose-purple, with darker veins; *pistils wrinkle-veined*; May–Aug. **WHERE FOUND:** Waste places. Frequently cultivated, then escaped, but not considered naturalized. Scattered over much of our area. Alien (Europe, N. Africa and Asia). **USES:** Leaves edible. Leaf or root tea soothing to irritated membranes, especially of digestive system; also used for coughs, bronchitis, stomachaches; anti-inflammatory, mildly astringent. In China the leaves and flowers have been used as an expectorant, a gargle for sore throats, and a mouthwash. Diuretic properties are also attributed to the plant, and it is said to be good for the stomach and spleen. Flowers (and leaves) widely used in European phytomedicine to treat irritation of the mouth and throat, with associated dry cough. Science confirms anti-inflammatory, antioxidant, potential anticancer, wound-healing, and antiulcer activity. The dried flowers are the primary form found in commerce. Other *Malva* species are used similarly.

High Mallow has pink to bluish flowers with dark stripes.

SPREADING DOGBANE, BITTER DOGBANE — Root

Apocynum androsaemifolium L. — Dogbane Family (Apocynaceae)

⚠ Shrublike; 1–4 ft. Milky latex within. Leaves oval, opposite; smooth above. Flowers are drooping pink bells, *rose-striped within*; in leaf axils *and* terminal; June–July. **WHERE FOUND:** Scattered throughout our area. Fields, roadsides. Absent from KS, south of NC highlands; westward. **USES:** Native American groups widely used root of this plant for many ailments. Induces sweating and vomiting; laxative. Used in headaches with sluggish bowels, liver disease, indigestion, rheumatism, and syphilis. **RELATED SPECIES:** Indian Hemp (*A. cannabinum*), a close relative, is also used medicinally but is also considered poisonous. **WARNING:** Poisonous. Cymarin, a cardioactive glycoside, poisons cattle. Nonetheless, the plant has shown antitumor activity.

TWINFLOWER — Whole plant

Linnaea borealis L. — Honeysuckle Family (Caprifoliaceae)

Delicate creeper; 3–5 in. Leaves paired. Flowers fragrant, *nodding bells, in pairs on a slender stalk*; June–July. **WHERE FOUND:** Cold, moist woods. Canada to Long Island; WV, TN mtns., OH to n. IN; all of Canada; mtns. of w. N. America, n. Europe, n. Asia. Locally too rare for harvest. **USES:** Algonquins used plant tea as a tonic for pregnancy and in difficult or painful menstruation; also for children's cramps, fevers. In Europe, a

The flower of Spreading Dogbane has rose stripes within.

Twinflower, with "twin" drooping flowers.

folk medicine for rheumatism and scabies; considered diuretic and a mild astringent. The favorite wildflower of Linnaeus, which he named after himself; he is seen in portraits holding a sprig of the herb.

HEART-LEAVED FOUR O'CLOCK

Root, leaves

Mirabilis nyctaginea (Michx.) MacMill. [*Oxybaphus nyctagineus* (Michx.) Sweet]

Four O'Clock Family (Nyctaginaceae)

Perennial; 1–5 ft. Leaves opposite, heart-shaped. Pink to purple flowers *atop a 5-lobed, green, veiny cup or bracts*; June–Oct. **WHERE FOUND:** Prairies, rich soil. WI to AL; TX to MT; escaped and weedy eastward. **USES:** Cherokee used the mashed roots for boils and as a fly poison. Chippewa used root tea or poultice for sprains or muscle strains. The Meskwaki used root as burn poultice; whole herb for bladder problems. Dakota used roots to treat worms, also for fevers. Other Native groups also used as poultice or wash for sprains and swelling. **WARNING:** Considered poisonous.

TOBACCO

Leaves

Nicotiana tabacum L.

Nightshade Family (Solanaceae)

⚠ Acrid, rank, clammy-hairy, large annual; 3–9 ft. Leaves lance-shaped to oval. *Funnel-shaped flowers, to 3 in. long*; greenish to pink, 5-parted; Aug.–Sept. **WHERE FOUND:** Escaped from recent cultivation. Alien (from tropical America). **USES:** Well-known addictive narcotic. Native Americans employed it in rituals; leaf tea diuretic, emetic, strongly laxative, worm expellent, anodyne; used for cramps, sharp pains, toothaches, dizziness, dropsy, colic. Poulticed for boils, snakebites, insect stings.

LEFT: *Heart-leaved Four O'Clock. Note its heart-shaped leaves.*
ABOVE: *Heart-leaved Four O'Clock flower close-up.*

Returning to our Puritan roots, modern laws read like old Massachusetts colony laws penalizing anyone "who shall smoke tobacco within twenty poles of any house; or who shall take tobacco in any inn or common victualating house, except in a private room, so as that neither the master of the said house nor any other guest shall take offence thereat" (quoted from Bigelow's *American Medical Botany,* 1818, vol. 2, p. 176). **WARNING:** Hazards of Tobacco use are well known. Still, the toxic insecticidal alkaloid, nicotine, is offered in pills and in skin patches to help curb the nicotine habit.

Tobacco is cultivated in the Southeast.

FLOWERS IN DOMED CLUSTERS; MILKWEEDS

SWAMP MILKWEED — Root

Asclepias incarnata L. — Milkweed Family (Asclepiadaceae)
[Modern works place in Dogbane Family (Apocynaceae)]

⚠ Strongly branched, smooth perennial; 2–4 ft. Leaves numerous, *opposite; narrowly lance-shaped* or oblanceolate (wider at base), veins ascending; soft-hairy, especially beneath; when broken, leaves exude small amounts of milky latex. Flower clusters in branched, flat-topped groupings (corymb). Flowers reddish to deep rose, in small umbels; June–Sept. **WHERE FOUND:** Wet areas, marshes, streambanks, and moist meadows. Scattered throughout our area. **USES:** The Chippewa used a root tea as a bath to strengthen weak children and adults. In Iroquois tradition, fibers from the stem were used to extract loose baby teeth, and as cordage to sew moccasins. Among the Meskwaki, the root was used to remove tapeworms, also as a cathartic, emetic, and a diuretic. American colonists used it for asthma, rheumatism, syphilis, worms, and as a heart tonic. In larger doses used as a laxative and emetic. **WARNING:** Potentially toxic.

COMMON MILKWEED — Root, latex

Asclepias syriaca L. — Milkweed Family (Asclepiadaceae)
[Modern works place in Dogbane Family (Apocynaceae)]

⚠ Milky-juiced, downy perennial; 2–4 ft. Stems usually solitary. Leaves opposite, large, widely elliptical; to 8 in. long. Flowers pink-purple (variable), in globe-shaped (often drooping) clusters from leaf axils; June–Aug. Pods *warty.* **WHERE FOUND:** Fields, roadsides. S. Canada to GA, AL; OK, KS to ND. The most common milkweed in the Northeast.

Swamp Milkweed is found near wet habitats.

Common Milkweed is our most often seen milkweed.

USES: Cherokee applied milky latex to warts. Root infusion taken as a laxative, to treat venereal disease, and to treat mastitis. Fresh latex from stems also applied to moles and ringworm. Various Native American groups used the flowerbuds, shoots, unripe seedpods, and flowers as food, a common practice in the Upper Midwest. Latex is chewed as gum, a potentially dangerous practice—see warning below. Silky seed tassels used in pillows, feather beds; developed commercially in recent years. The silk was used in the nineteenth century for thread, netting, tapes, socks, and mixed with cotton for other uses. Paper made from plant fibers was developed in Germany in 1785. Folk remedy for cancer. One Mohawk antifertility concoction contained Milkweed and Jack-in-the-Pulpit, both considered dangerous and contraceptive. Formerly used as a diuretic for gonorrhea; also as a laxative. Monarch butterfly larvae feeding on the plant sequester potentially toxic cardenolides as an antifeedant, bitter agent to thwart predators. Various flavonoid compounds have been isolated from the flowers. **WARNING:** Potentially toxic—contains cardioactive compounds.

SLENDER SPIKES OF MANY FLOWERS

PURPLE OR SPIKE LOOSESTRIFE — Flowering plant

Lythrum salicaria L. — Loosestrife Family (Lythraceae)

Downy, highly variable perennial; 2–4 ft. Leaves whorled or opposite; *rounded or heart-shaped at base*. Purple-pink, *6-petaled* flowers, in

Purple or Spike Loosestrife. Despite its beauty, it is a rampant invasive alien.

Lopseed, with small snapdragon-like flowers.

spikes; June–Sept. **WHERE FOUND:** Invasive in swampy meadows, often forming large stands and blanketing moist meadows in a sea of color. New England to NC; MO to MN. Alien (Europe). **USES:** Tea made from whole flowering plant, fresh or dried, is a European folk remedy (demulcent, astringent) for diarrhea, dysentery; used as a gargle for sore throats, a douche for leukorrhea, and a cleansing wash for wounds. In Europe also a folk remedy for hemorrhoids, eczema, and to treat varicose veins, bleeding gums, and venous insufficiency. Experimentally, plant extracts stop bleeding and kill some bacteria. Recent research suggests a polysaccharide complex from the plant has antitussive and bronchodilator activity; also antioxidant, anti-inflammatory. Contains tannins and other components, with bactericidal activity in the gastrointestinal tract. The vast majority of research focuses on how to control this highly invasive alien.

LOPSEED — Roots

Phryma leptostachya L. — Lopseed Family (Phrymaceae)

Slender-branched perennial; 1–3 ft. Stems swollen for a short distance (1 in. or less) above each pair of leaves. Leaves opposite, oval, toothed. Leafstalks on middle leaf pairs longer than those on upper or lower leaves. Flowers small, purplish, snapdragon-like; *in alternating, opposite pairs* in terminal spikes; July–Sept. **WHERE FOUND:** Woods, thickets. Throughout our area. **USES:** Cherokee gargled root tea (or chewed root) for sore throats; drank root tea for rheumatism. Also found in e. Asia,

Maryland Figwort has long stalks with many small flowers with reddish protruding upper petals.

where it is used for fevers, ulcers, ringworm, scabies, and insect bites. Root poulticed for boils, carbuncles, sores, and cancers. Also considered insecticidal; most recent research has focused on larvicidal activity of lignans found in the plant.

FIGWORT, MARYLAND FIGWORT

Leaves, root

Scrophularia marilandica L. Figwort Family (Scrophulariaceae)
[*Scrophularia nodosa* L. var. *marilandica* (L.) Gray]

Perennial; 3–6 ft. Stems angled, grooved. Leaves are oval, rounded, or heart-shaped at base; toothed. Flowers *like a miniature scoop, 4 stamens*, with an additional wide, sterile, purple stamen (yellow in *S. lanceolata*, not shown); June–Oct. **WHERE FOUND:** Rich woods. Sw. ME to n. GA; LA; TX to NE, MN. **USES:** Iroquois used root tea for irregular menses, fevers, hemorrhoids; diuretic, tonic. In colonial New York, the plant had a popular reputation among pregnant woman to allay pain, restlessness, anxiety, and wakefulness; also as a sedative to quiet a fetus, earning the herb the popular name of "Woman's Friend." Poultice a folk remedy for cancer. Also used for sores, wounds, ulcers, and burns; in veterinary practice for sores of cows, and scabies of dogs and

ABOVE: *European Figwort, naturalized in the Northeast, also has reddish protruding petals.*
RIGHT: *European Figwort and other* Scrophularia *spp. are large gangly plants.*

swine. Experimentally analgesic, antispasmodic; contains iridoid glycosides with wound-healing activities, along with antioxidant and anti-inflammatory phenolic acids. Other species used similarly. The Asian species *S. ningpoensis* Hemsl. contains the anti-inflammatory harpagoside, shown to have a protective effect against nerve cell degeneration. *Scrophularia buergeriana* Miq., a Chinese species, shows anti-allergenic, cognitive-enhancing, and antioxidant potential. The European *S. nodosa* L. is sometimes naturalized in s. New England, NY, PA. Used since ancient times in Europe. Lanceleaf Figwort, *S. lanceolata* Pursh, a N. American species used interchangeably with Maryland Figwort.

EUROPEAN VERVAIN

Root, leaves

Verbena officinalis L. — Verbena Family (Verbenaceae)

Mostly smooth, loosely branched annual; 1–3 ft. Leaves paired, with *deeply cut lobes* and sharp teeth. Flowers tiny, purple to pinkish; *in slender spikes*; June–Oct. **WHERE FOUND:** Waste places. Escaped from gardens. Locally established NY, MA to FL, north to WI. Alien (Europe). **USES:** In Europe, plant tea used for obstructions of liver and spleen, headaches, and nervous disorders; also for acute diarrhea, enteritis, and depression. Leaves considered diuretic, milk-inducing; extracts analgesic. In Traditional Chinese Medicine the herb is used for dispersing heat, detoxifying blood, and promoting blood circulation. Used experimentally in China to control malaria symptoms, kill blood flukes (parasites) and germs, stop pain and inflammation. Chinese studies suggest that herbage of this plant is synergistic with prostaglandin E2. Russian studies show adaptogenic activity of the alcoholic extract or tincture. Said to be milder than Blue Vervain (*Verbena hastata*, p. 231) and other species. Animal studies have demonstrated anti-inflammatory, antioxidant, antifungal, cough-suppressing, and milk-stimulating activity; also protects cells of central nervous system, a potential research target for Alzheimer's disease. **WARNING:** Plant suspected of poisoning cattle in Australia.

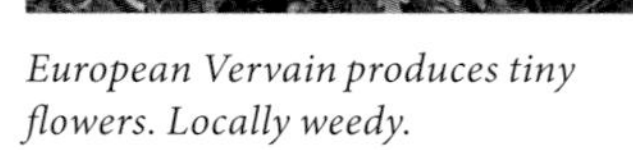

European Vervain produces tiny flowers. Locally weedy.

GROUNDNUT

Root

Apios americana Medik. Pea Family (Fabaceae or Leguminosae)

Twining vine. Leaves with 5–7 oval, sharp-pointed leaflets. *Sweetly fragrant white to maroon or purple-brown flowers*, in crowded clusters in leaf axils; July–Sept. **WHERE FOUND:** Rich, moist thickets. NB to FL; west to TX, north to MN, ND. **USES:** Delicious tubers used as food by Pilgrims during first bleak winters. Favorite Indian food. With 3 times the protein of potatoes, each Groundnut plant, under cultivation, may produce 5 pounds of tubers. The plant has been suggested as a nitrogen-fixing edible ornamental for permaculturists. John Josselyn (1672) suggested a poultice of Groundnut root be used for cancerous conditions known as "proud flesh." Today we learn that, like many legumes, Groundnut contains estrogenic isoflavones (phytoestrogens) with anticancer activity, such as genistein, which can prevent the formation of blood vessels that nourish new tumors. Widely used as food by Native Americans and once cultivated. Surprisingly little researched given its food and medicinal potential.

NAKED-FLOWERED TICK-TREFOIL

Root

Desmodium nudiflorum (L.) DC. Pea Family (Fabaceae or Leguminosae)

Perennial; 18–36 in. Leaves in whorls; 3 oval leaflets, the middle one on a longer stalk. Leafstalk separate from flower stalk. Pinkish red flowers on a leafless stalk; July–Aug. *Pods jointed*; "beggar's ticks" adhere to clothes. **WHERE FOUND:** Woods. ME to FL; TX to MN. **USES:** Cherokee chewed the root for inflammation of the mouth, sore and bleeding

LEFT: *Groundnut has sweetly fragrant flowers with 5–7 leaflets.*
ABOVE: *Groundnut flower close-up.*

Tick-trefoil or Beggar's Lice goes unnoticed until seedpods stick to clothing.

Red Clover is a familiar weed.

gums, and periodontal diseases with pus discharge. Root tea was used as a wash for cramps.

RED CLOVER

Flowering tops

Trifolium pratense L. Pea Family (Fabaceae or Leguminosae)

Familiar biennial or short-lived perennial; to 18 in. Leaves divided into *3 oval leaflets*; leaflets fine-toothed, *with prominent "V" marks*. Flowers *pink to red, in rounded heads*; May–Sept. **WHERE FOUND:** Fields, roadsides. Throughout our area. A weed; found worldwide. **USES:** Historically, flower tea has been used as an antispasmodic, expectorant, mild sedative, blood purifier; for asthma, bronchitis, spasmodic coughs; externally, a wash has been used as a folk remedy for cancer, including the famous Hoxsey treatment, and for athlete's foot, sores, burns, and ulcers. Flowers formerly smoked in anti-asthma cigarettes. Science has not confirmed traditional uses. However, Red Clover contains many biologically active compounds, including flavonoids, saponins, phenolic acids, and phytoestrogenic isoflavones, such as genistein, diadzen, formononetin, and biochanin A, among others. Isoflavone-rich preparations are used as an alternative to hormone-replacement therapies. Phytoestrogens activate estrogen receptors in mammals. Epidemiological studies provide evidence that certain dietary components can have a significant effect on the incidence and location of cancers in humans. A laboratory study found that biochanin A inhibits the activation of cancer. Standardized extracts of Red Clover, produced in Australia, are now sold in the U.S. One tablet contains 40 mg of phytoestrogens, 8 times the amount consumed in the typical American diet. Anti-inflammatory, antioxidant, and potential cancer-preventive activities have been confirmed. **WARNING:** Fall or late-cut hay in large doses can cause frothing, diarrhea, dermatitis, and decreased milk production in cattle. Diseased clover, externally showing no symptoms, may contain the indolizidine alkaloid slaframine, which is much more poisonous than castanospermine, now being studied for anti-AIDS and antidiabetic activity.

FLOWERS IN TIGHT HEADS OR SPIKES

PENNSYLVANIA SMARTWEED

Leaves, tops

Polygonum pensylvanicum L. [*Persicaria pensylvanicum* (L.) M. Gómez]

Buckwheat Family (Polygonaceae)

⚠ Erect or sprawling annual; 1–5 ft. Leaves lance-shaped; *sheaths not fringed* (see Lady's Thumb, below). Flowers rose pink (or white); in crowded, elongate clusters; July–Nov. Flower stalks often have *minute glandular hairs* near top. Highly variable. **WHERE FOUND:** Waste ground. Throughout our area. **USES:** Meskwaki used tea made from whole plant for diarrhea; poulticed leaves for hemorrhoids. Bitter leaf tea used to stop bleeding from mouth. Tops were used in tea for epilepsy. Contains vanicosides A and B, which are protein kinase C inhibitors, suggesting the compounds are an interesting research target for anticancer activity. **WARNING:** Fresh juice is acrid; may cause irritation.

Pennsylvania Smartweed is a highly variable annual.

LADY'S THUMB, HEART'S EASE

Leaves

Polygonum persicaria L.

Buckwheat Family (Polygonaceae)

⚠ Reddish-stemmed, sprawling perennial; 6–24 in. Leaves lance-shaped, often with *a purplish triangular blotch in the middle of leaf*; papery sheath at leaf nodes has fringes. Pinkish flowers in elongate clusters; June–Oct. **WHERE FOUND:** Waste places. Throughout our area. Alien (Europe). **USES:** The Cherokee adopted the leaf tea for heart troubles, stomachaches, and as a diuretic for "gravel" (kidney stones). The whole herb

ABOVE: *Lady's Thumb, Heart's Ease has a purplish triangular blotch in the middle of the leaf.* TOP RIGHT: *Salad Burnet has numerous long stamens.* RIGHT: *Salad Burnet, an alien, is commonly grown in and escaped from gardens.*

was poulticed for pain, rubbed on Poison Ivy rash, and rubbed on horses' backs to keep flies away. Leaf tea used as a foot soak for rheumatic pains of the legs and feet. In European tradition, leaf tea was used for coughs, colds, inflammation, stomachaches, sore throats, wounds, faintness, hemorrhage, and other maladies. Recently antifungal activity confirmed. **RELATED SPECIES:** Other *Polygonum* species in our range, many naturalized from Europe, have been used similarly. **WARNING:** Fresh juice may cause irritation.

GREATER SALAD BURNET

Leaves, root

Sanguisorba officinalis L. — Rose Family (Rosaceae)

Perennial; 1–5 ft. Leaves compound; leaflets 7–15, toothed. Tiny purplish red flowers, in oval or thickly rounded heads, *4 stamens shorter than purple sepals*; May–Oct. **WHERE FOUND:** ME to MN. Escaped elsewhere. Alien. Mostly cultivated in herb gardens. **USES:** In Europe, leaf tea was used for fevers and as a styptic. American soldiers drank tea before battles in the Revolutionary War to prevent bleeding if wounded. Root tea is astringent, allays menstrual bleeding. In China the root tea is used to stop bleeding, "cool" blood; taken for hemorrhoids, uterine bleeding, dysentery; externally for sores, swelling, burns. Experimentally, the plant is antibacterial (in China); stops bleeding and vomiting. Polysaccharides from the root have antioxidant and immunostimulating activity. Terpene glycosides from the plant are responsible for hemostatic effects (stops bleeding). Experimentally, plant extracts have anticancer, blood-thinning, and antioxidant activity. Powdered root used clinically for second- and third-degree burns. **RELATED SPECIES:** American Burnet (*S. canadensis* L.) has *4 stamens much longer than the whitish sepals*, is primarily found in the Northeast and adjacent Canada. Salad Burnet (*S. minor* Scop.) is a European alien, escaped from gardens, with numerous stamens.

Greater Salad Burnet is a European alien with reddish flowers and 4 short stamens.

SQUARE-STEMMED PLANTS WITH LIPPED FLOWERS

AMERICAN DITTANY

Leaves

Cunila origanoides (L.) Britton
(*Cunila mariana* L.)

Mint Family (Lamiaceae or Labiatae)

Wiry-stemmed, branched perennial; 1–2 ft. Leaves oval, to 1 in.; toothed, *oregano-scented*. Small violet to white flowers, in clusters; July–Oct. *Hairy throat; 2 stamens.* **WHERE FOUND:** Dry woods, thickets. Se. NY to FL; TX, OK, MO, IL. **USES:** The Cherokee drank tea for headaches, used it as a cold remedy and to increase perspiration, thus reduce fever. Strong tea used for labor pains; also as a tonic. Leaf tea is a folk remedy for colds, fevers, headaches, snakebites; thought to induce perspiration and menstruation. Used in colonial America for wounds. Francis Porcher, in *Resources of the Southern Fields and Forests* (1863, p. 446), notes that a gentleman in South Carolina said "everybody cured everything with dittany." Porcher went on to write, "Doubtless they took less mercury and drastic purgatives in consequence."

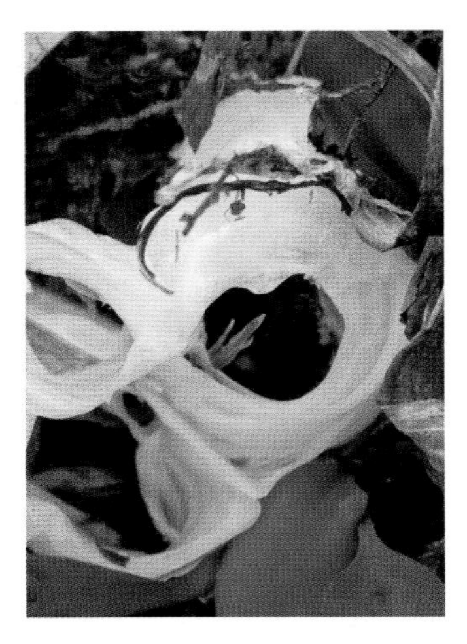

LEFT: *American Dittany produces "frost flowers" from root during the first freezes of autumn.*
ABOVE: *American Dittany* (Cunila origanoides) *blooms in autumn.*

MOTHERWORT

Leaves

Leonurus cardiaca L. — Mint Family (Lamiaceae or Labiatae)

⚠ Square-stemmed perennial; 3–5 ft. Leaves *3-lobed*; lobes toothed. Pinkish flowers in whorls in axils; May–Aug. *Upper lip furry.* **WHERE FOUND:** Much of our area. Alien (Europe). A weed. **USES:** Traditionally, leaf tea is used to promote menstruation, regulate menses, aid in childbirth (hence the common name); also used for asthma and heart palpitations. Said to be sedative; used for insomnia, neuralgia, sciatica, spasms, fevers, and stomachaches. Well known and underutilized for potential mild sedative, hypotensive, and cardiotonic effects. A recent clinical study evaluated Motherwort extracts in hypertension accompanied by sleep disorders and anxiety with positive results and no significant side effects. Scientists have found extracts to have antispasmodic, hypotensive, sedative, cardiotonic, diuretic, antioxidant, immunostimulating, and cancer-preventive activity. Extracts used in European phytomedicine for nervous heart conditions and in the supportive treatment of hyperthyroidism. Experimentally, leonurine, a leaf constituent, is a uterine tonic. **RELATED SPECIES:** Chinese species, well documented with laboratory and clinical reports, have been used similarly. **WARNING:** Rare cases of contact dermatitis reported. Avoid use during pregnancy and lactation.

BEE-BALM, OSWEGO TEA

Leaves

Monarda didyma L. — Mint Family (Lamiaceae or Labiatae)

Perennial; 2–5 ft. Leaves paired. Flowers red, tubular, in crowded heads; June–Sept. Bracts often red or purplish. **WHERE FOUND:** Thickets, streambanks. NY to GA; TN to MI. **USES:** Cherokee used leaf tea

ABOVE: *Motherwort is often found around farmyards.* TOP RIGHT: *Bee-balm, Oswego Tea produces bright red flowers.* RIGHT: *Germander, Wood Sage, Wild Basil, a common adulterant to Skullcap. Sold as "pink skullcap."*

for colic, gas, colds, fevers, stomachaches, nosebleeds, insomnia, heart trouble, measles, and to induce sweating. Believed to reduce muscle spasms, relieve digestive gas, and act as a diuretic. Poultice used for headaches. Introduced into Europe three centuries ago, it has been used for digestive disorders, also as a diuretic, carminative, expectorant, and to induce sweating in fevers. Historically, physicians used leaf to expel worms and gas.

AMERICAN GERMANDER, WOOD SAGE, WILD BASIL — Leaves

Teucrium canadense L. — Mint Family (Lamiaceae or Labiatae)

Variable perennial; 1–3 ft. Leaves oval to lance-shaped, toothed, white-hairy beneath. Flowers purple, pink, or whitish; June–Sept. *Calyx felty, stamens protrude from cleft of upper lip.* **WHERE FOUND:** Woods, thickets. Throughout our area. **USES:** Leaf tea traditionally used to induce menstruation, urination, and sweating. Used like the bugleweeds or water horehounds (*Lycopus*) for lung ailments, worms, hemorrhoids; externally, as a gargle and antiseptic dressing. For well over a quarter-century American Germander has masqueraded in the wholesale herb trade as "Pink Skullcap" (or "Skullcap"), an economic adulterant to supplies of botanical material offered as the herb Skullcap (*Scutellaria lateriflora*). Some have argued that the problem is a simple case of mistaken identity; however, the argument holds no weight since the two plants are not similar in appearance even based on a cursory comparison. **WARNING:** Some species of *Teucrium* have proven to be highly toxic to the liver (resulting in fatalities). *Teucrium canadense* is clearly associated with its liver-toxic compounds and persists as an adulterant to Skullcap.

COMPOSITES WITH FLAT-TOPPED FLOWER CLUSTERS

DAISY FLEABANE (NOT SHOWN)

Whole plant

Erigeron philadelphicus L. — Aster Family (Asteraceae or Compositae)

⚠ Slender, hairy annual or biennial; flowers less than 1 in. across, usually white, sometimes pinkish. See p. 120.

SPOTTED JOE-PYE-WEED

Leaves, root

Eutrochium maculatum (L.) E. E. Lamont (*Eupatorium maculatum* L.) — Aster Family (Asteraceae or Compositae)

⚠ Perennial; 2–6 ft. Stem *purple or purple-spotted*. Leaves lance-shaped, in *whorls of 4–5*. Purple flowers in flat-topped clusters; July–Sept. **WHERE FOUND:** Wet meadows. NS to mtns. of NC; NE to BC. **USES:** Cherokee used tea of whole herb as a diuretic for dropsy, painful urination, gout, kidney infections, rheumatism. Root tea once used for fevers, colds, chills, sore womb after childbirth, diarrhea, liver and kidney ailments; a wash for rheumatism. Its name is derived from "Joe Pye," a nineteenth-century Caucasian "Indian theme promoter" who used the root to induce sweating in typhus fever. **WARNING:** Contains potentially liver-toxic pyrrolizidine alkaloids. Poorly researched.

SWEET JOE-PYE-WEED, GRAVEL ROOT

Leaves, root

Eupatorium purpureum L. — Aster Family (Asteraceae or Compositae)

Similar to *E. maculatum* (above). Perennial; to 12 ft. tall; stems green, *purple at leaf nodes*. Pale pink-purple flowers, in a somewhat rounded cluster; July–Sept. **WHERE FOUND:** Thickets. NH to FL; AR, OK, w. NE to MN. **USES:** Leaf and root tea traditionally used to eliminate stones in urinary tract, to treat urinary incontinence in children, and for dropsy; also for gout, rheumatism, impotence, uterine prolapse, asthma,

Spotted Joe-pye-weed has spotted stems.

Sweet Joe-pye-weed has short leaf-stalks.

chronic coughs. Homeopathically used for gallbladder and urinary ailments. Cistifolin, extracted from the roots, shows potential against rheumatism because of anti-inflammatory activity in experimental antirheumatic models. **REMARKS:** Also known as Queen-of-the-Meadow, a name that is shared with a European species (*Filipendula ulmaria*—see p. 78). **RELATED SPECIES:** German researchers report immunologically active polysaccharides from other *Eupatorium* species, both American and European.

THISTLELIKE, BRISTLY FLOWERS; BURDOCKS AND THISTLES

GREAT BURDOCK — Leaves, root, seeds

Arctium lappa L. — Aster Family (Asteraceae or Compositae)

⚠ Biennial; 2–9 ft. Lower leaves large, rhubarblike. Stalk *solid, celerylike*—grooved above. Reddish purple, thistlelike flowers, 1–1½ in. across; *long-stalked, in flat-topped clusters*; July–Oct. Seedpods (familiar "burs") stick to clothing. **WHERE FOUND:** Waste places. Canada south to PA, NC; west to IL, MI. Local elsewhere. Alien. A widespread Eurasian weed used in traditional medicine in China, Japan, Europe, and N. America. **USES:** Traditionally, root tea (2 ounces dried root in 1 quart of water) used as a blood purifier; diuretic; stimulates bile secretion, digestion, and

ABOVE: *Great Burdock has large leaves with a solid leaf stalk.*
TOP RIGHT: *Burdock roots have cottony tufts at the crown.*
RIGHT: *Great Burdock flowerheads are long stalked.*

sweating; also used for gout, liver and kidney ailments, rheumatism, gonorrhea. Root is high in inulin (up to 50 percent); traditionally used for diabetes. Bitter compounds in roots, particularly arctiopicrin, are antibacterial; also explains use as digestive stimulant. In China, a tea of leafy branches is used for vertigo and rheumatism, and tea mixed with brown sugar is used for measles. Externally, used as a wash for hives, eczema, and other skin eruptions. Juice of fresh plant has been shown to protect against chromosome aberrations. Both flowers and leaves have antibacterial activity. Seeds diuretic; thought to be antiseptic. Seeds used for abscesses, sore throats, insect bites and snakebites, flu, constipation; once used to treat scarlet fever, smallpox, and scrofula. Crushed seeds poulticed on bruises. Leaves poulticed on burns, ulcers, sores. Japanese studies suggest that roots contain compounds that may curb mutations (and hence cancer?). Recent research interest, particularly in Asia, has focused on antioxidant and antidiabetic activity. Seed extracts are used for hypertension, gout, hepatitis, and other disorders related to inflammatory conditions. Ethanolic seed extracts have been found to reduce cholesterol and lower blood pressure through vascular relaxation and suppression of vascular inflammation. Root widely cultivated, particularly in Japan as a vegetable. **WARNING:** Leaf hairs may irritate skin. Do not confuse leaves with the toxic leaves of Rhubarb.

COMMON BURDOCK

Root, seeds, leaves

Arctium minus Bernh. Aster Family (Asteraceae or Compositae)

Smaller than *A. lappa* (see above); 2–5 ft. Leaf stems *hollow, not furrowed.* Flowers smaller—to ¾ in. across, *without stalks or short-stalked in raceme*; not in flat-topped cluster; July–Oct. **WHERE FOUND:** Waste places. Most of our area. Alien. **USES:** Same as for *A. lappa.* Used extensively by Native Americans for uses described for Great Burdock.

CANADA THISTLE

Leaves, root

Cirsium arvense (L.) Scop. Aster Family (Asteraceae or Compositae)

Perennial; 1–5 ft., with vigorous taproots; usually forms colonies. Stems smooth, leafy near top. Leaves oblong to lance-shaped, *margins very prickly.* Flowers small, pink to violet (rarely white), to 3–4 in. across; July–Sept. *Bracts strongly appressed.* **WHERE FOUND:** Fields, pastures, roadsides. Throughout our area. Serious alien weed from Europe. **USES:** Leaf tea "tonic" and diuretic. Once used for tuberculosis; externally, for skin eruptions, skin ulcers, and Poison Ivy rash. Root tea used for dysentery, diarrhea. The Abnaki used root tea as a bowel tonic and dewormer. Contains a milk-clotting enzyme.

LEFT: *Common Burdock has flowerheads on short or no stalks (tightly hugging stem); leaf stems hollow.* ABOVE: *Canada Thistle is a rampant invasive weed (from Europe, not Canada).*

ASIATIC DAYFLOWER

Leaves, root

Commelina communis L.

Spiderwort Family (Commelinaceae)

Sprawling perennial; 1–3 ft. Oval leaves, clasping stem. Flowers with 2 prominent, earlike blue petals and a *smaller whitish petal beneath*; May–Oct. Each Dayflower blooms for only one day, hence the common name. **WHERE FOUND:** Waste places throughout our area. Troublesome weed. Alien (Asia). **USES:** In China, leaf tea gargled for sore throats; used for cooling, detoxifying, and diuretic properties in flu, acute tonsillitis, urinary infections, dysentery, and acute intestinal enteritis. Root used for fevers; leaves were eaten as greens; also reduces pain, used for lung ailments. Science confirms antioxidant, potential antidiabetic activity. Experimentally, offers protection against influenza; alkaloid extract reduces lung lesions and virus loads. **RELATED SPECIES:** Nine species of *Commelina* occur in the U.S., most of which are used similarly and separated on technical characteristics.

CRESTED DWARF IRIS

Root

Iris cristata Ait.

Iris Family (Iridaceae)

⚠ This diminutive iris of southern woods has yellow crests on the downward-curved blue sepals, hence the species name *cristata.* Spreading perennial; 4–8 in. Leaves *short*, lance-shaped; *sheathed* on stem. Blue (rarely white) flowers with yellow crests on down-curved sepals; Apr.–May. **WHERE FOUND:** Wet woods. MD to NC; MS, AR., e. OK to IN. **USES:** Cherokee used root ointment (in animal fats or waxes) on cancerous ulcers. Root tea used for hepatitis. Used as a substitute for *Iris versicolor*

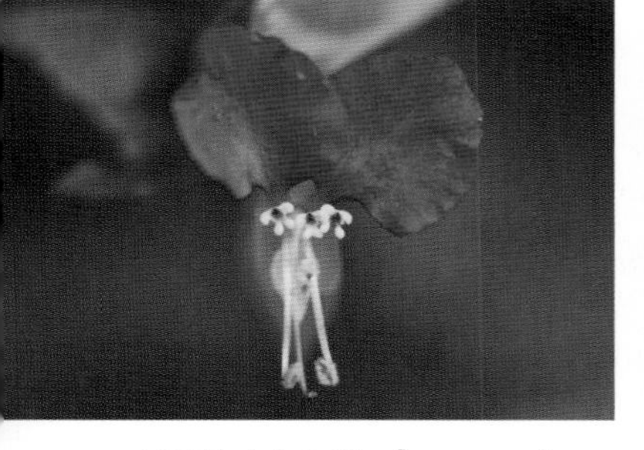

ABOVE: *Asiatic Dayflower, an alien, is one of 9 dayflower species found in N. America.* RIGHT: *Dayflower. All dayflower species are used similarly.*

Crested Dwarf Iris, a small iris common in the Southeast.

Blue Flag is found near water.

in the Midsouth. Fresh roots, sweet at first, then burning like Cayenne, were also used as a cathartic. Leaves used to alleviate thirst. **WARNING:** Same as for Blue Flag (below).

BLUE FLAG

Root

Iris versicolor L.

Iris Family (Iridaceae)

⚠ Perennial; 1–2 ft. Leaves swordlike, similar to those of garden irises. Flowers violet-blue, sepals violet at outer edge, pubescent with greenish to greenish yellow patch with deeply purple veins; with purple on white patch at base; *sheaths papery*; May–July. **WHERE FOUND:** Wet meadows, moist soil, pond edges. NL to VA; OH, WI to MN, MB. **USES:** The Chippewa poulticed roots to swelling and sores; other native groups also poulticed roots for wounds, to relieve pain, inflammation, contusions. Many native groups of the Northeast and upper Midwest used a decoction of the root as a cathartic, for rheumatism and liver disorders, as an emetic, and for blood disorders. In 1830, C. S. Rafinesque observed that many tribes kept a supply of the plant in ponds ready for use. Blue Flag received wide attention in early works on American medicinal plants. Physicians formerly used root of this plant in small, frequent doses to "cleanse" blood and stimulate the bowels, kidneys, and liver. Thought useful for rabies and syphilis. Leaves, with milder effects, used similarly. Seeds used like coffee. In animal experiments root extracts reduced fat tissue and reduced feeding (perhaps more a sign of toxicity than medicinal value). Combined with Button Snakeroot (*Eryngium yuccifolium*) the prescription was considered a strong diuretic. Given the plant's widespread historical use and active components, Blue Flag has been surprisingly neglected by researchers. Primarily used by herbal practitioners for skin eruptions, laxative, diuretic, and to stimulate bile secretion. **WARNING:** Considered poisonous. Fresh root contains furfural, which may nauseate and irritate the gastrointestinal tract and eyes. Can cause headaches and inflammation of the eyes and throat. Iridin, another component, is toxic to humans and livestock. Avoid use during pregnancy and lactation.

SPIDERWORT

Leaves, root, whole plant

Tradescantia virginiana L. Spiderwort Family (Commelinaceae)

Perennial; 1–3 ft. Leaves grass- or irislike; sheathing stem. Purple flowers in a terminal cluster; Apr.–June. Petals 3; many *stamens, with prominent, large-celled hairs.* (Individual stamen hair cells are so large they may be seen with naked eye, or easily with a 10X hand lens.) **WHERE FOUND:** ME to WV, KY, GA, MS; AR to MN. **USES:** Cherokee used root tea for stomachache due to overeating; mild laxative; diuretic; root poultices for skin cancers; leaves mashed and rubbed on insect bites and stings. Cherokee used young shoots as a spring green. Historic reputation as a poultice for spider bites. **RELATED SPECIES:** Thirty species of *Tradescantia* occur in N. America. *T. ohiensis* Raf. always has a bluish film on the smooth stems and leaves; widely distributed in the Midwest and was used similarly to *T. virginiana.*

Ohio Spiderwort has a bluish film on stems.

Virginia Spiderwort. Note the prominent stamen hairs, typical of spiderworts.

SIX FLOWER SEGMENTS; LEAVES NARROW OR ABSENT AT FLOWERING: ALLIUMS

WILD GARLIC, MEADOW GARLIC

Bulb, leaves

Allium canadense L.

Allium Family (Alliaceae)

[Formerly in Lily Family (Liliaceae) or Amaryllis Family (Amaryllidaceae)]

⚠ Herbaceous perennial with 1–4 (or more) clustered bulbs, surrounded by netted membrane; rhizome absent; 7–24 in tall. Leaves 2–6, *solid*, flattened, and channeled, sheathing at base to ¼ up flower scape; green at flowering, and persistent. Flowers 0–60 in loose umbel, usually pink to lavender; Apr.–July. **WHERE FOUND:** Moist meadows, open woods, and roadsides; widespread, NB to FL, west to e. TX, north to IA, SD, and MN. **USES:** Leaves or bulbs with strong characteristic garlic odor, used by Cherokee to treat scurvy, asthma; tincture given to children to prevent worms and colic, applied to chest to treat croup. Used as a carminative and diuretic. Various native groups of the Midwest used bulbs as food, especially as a flavoring for soups and meats. Fresh and dried bulbs used as condiment. During Civil War in South, used as a substitute for garlic; bulbs were commonly pickled. **WARNING:** The strong-smelling bulbs and leaves, or their preparations, can be strongly irritating in contact with skin or mucous membranes. Reported to be poisonous to cattle.

Wild Garlic usually has sparse flowerheads; flowers often replaced by bulbils.

NODDING WILD ONION

Bulb, whole plant

Allium cernuum Roth

Allium Family (Alliaceae)

[Formerly in Lily Family (Liliaceae) or Amaryllis Family (Amaryllidaceae)]

Perennial; 1–2 ft. Leaves soft, flat. *Stem arching at top*. Flowers pink-white; July–Aug. **WHERE FOUND:** Open woods, rocky soil. NY to GA; west to TX; west from MI, MN to BC. **USES:** Cherokee used slender bulbs for colds, colic, croup, and fevers. After a dose of Horsemint (*Monarda punctata*) tea, the juice of this wild onion was taken for "gravel" (kidney stones) and dropsy. Poultice of plant applied to chest for respiratory ailments. Effects probably similar to but weaker than those of Garlic.

ABOVE: *Nodding Wild Onion is easily distinguished by its nodding flowerheads.* RIGHT: *Chives, familiar to gardeners, also occurs in native populations in the far north.*

CHIVES, WILD CHIVES

Leaves

Allium schoenoprasum L.

Allium Family (Alliaceae)
[Formerly in the Lily Family (Liliaceae) or Amaryllis Family (Amaryllidaceae)]

Perennial; 8–24 in. tall; 1 or more bulbs, with short rhizome at base. Leaves, usually 2 per bulb, cylindrical, hollow within, with sheath 1/3–1/2 the length of flower scape. Flowers 30–50 *crowded into a compact, globe-shaped umbel*, pale purple to deep lilac; June–Aug. **WHERE FOUND:** Often cultivated, but native to N. America in moist meadows, rocky streambanks, or lake shores; circumboreal in Northern Hemisphere. Difficult to distinguish between truly native populations (mostly in northern parts of our range) and naturalized plants escaped from cultivation. Throughout our range. **USES:** Familiar as a common garden plant, the fresh leaves with their mild garlic flavor are widely used as a fresh condiment herb for salads, garnish, and flavoring. Flowers also used in salads. Bulbs and leaves widely used by native groups in northern parts of our range as a soup flavoring, meat condiment, or to flavor fish. Health benefits similar to garlic, especially antioxidant activity, which is highest from leaves. Phenolic compounds in flowers may arrest development of some cancer cells. The typical flavor and fragrance is produced by diallyl sulfides, which have antimicrobial activities. **WARNING:** May cause contact dermatitis or irritation.

WILD ONION, AUTUMN ONION, PRAIRIE ONION

Bulb, roots

Allium stellatum Fraser ex Ker Gawl. Allium Family (Alliaceae)
[Formerly in Lily Family (Liliaceae) or Amaryllis Family (Amaryllidaceae)]

Attractive, fall-flowering herbaceous perennial with 1–5 (or more) usually clustered bulbs, and short rhizome; 6–14 in tall. Leaves 3–5, *solid*, flattened, channeled, sheathing just to above ground level; green at flowering, and persistent. Flowers 9–40 in umbel, *distinctly 6-pointed-star-shaped* (stellate, hence the species name), usually deep pink to lavender; July–Oct. **WHERE FOUND:** Open calcareous woods, rocky soils, prairies; TN west to ne. TX, north through eastern portion of Plains states, north to MB, ON, SK. **USES:** Chippewa used the roots as a treatment for congestion due to colds. Used similarly to other wild onion relatives. When leaves or bulb are bruised, they have a rank garlic smell. First described in 1816 as "white Missouri garlick," introduced to English horticulture by Thomas Nuttall (1786–1859), who collected it along the banks of the Missouri River.

Wild Onion has 6-parted, star-shaped flowers.

MISCELLANEOUS NONWOODY VINES WITH VIOLET TO BLUE FLOWERS

PASSION-FLOWER, MAYPOP

Whole plant

Passiflora incarnata L. Passion-flower Family (Passifloraceae)

⚠ Climbing vine; to 30 ft. Tendrils *springlike*. Leaves *cleft with 2–3* slightly toothed lobes. Flowers large, showy, unique, whitish to purplish, *with numerous threads radiating from center*; June–Oct. Fruits fleshy, egg-shaped; Aug.–Nov. **WHERE FOUND:** Sandy soil. PA to FL; e. TX to s. MO. **USES:** Cherokee poulticed root for boils, cuts, earaches, and inflammation. Whole plant has traditionally been used in tea as an antispasmodic, and a sedative for neuralgia, epilepsy, restlessness, painful menses, insomnia, tension headaches. Research shows plant extracts are mildly sedative, slightly reduce blood pressure, increase respiratory rate, and decrease motor activity. Increasing evidence shows value in treatment of anxiety; used in herbal practice for the supportive treatment of alcoholism and withdrawal from opiates. Contains several compounds shown to contribute to sedative activity, including flavonoids, glycosides, alkaloids, and phenolic compounds. Used in European phytomedicine for nervous restlessness, nervous tension; con-

Passion-flower has egg-shaped, edible fruits.

Passion-flower is one of the most beautiful herbaceous vines.

sidered especially useful in sleep disturbances or anxiety arising from restlessness. Fruits edible, delicious. **WARNING:** Potentially harmful in large amounts, may increase effective dose of benzodiazepine-based prescription drugs, therefore should not be used concomitantly.

KUDZU

Root, flowers, seeds, stems, root starch

Pueraria montana var. *lobata* (Willd.) Maesen & S. Almeida, *P. montana* (Lour.) Merr., [*Pueraria lobata* (Willd.) Ohwi]

Pea Family (Fabaceae or Leguminosae)

Noxious, robust trailing or climbing vine. Leaves *palmate, 3-parted*; leaflets entire or palmately lobed. Flowers reddish purple, *grape-scented*; in a loose raceme; July–Sept. **WHERE FOUND:** Waste ground. MA, NY to FL; TX to NE. Asian alien; perhaps the most serious invasive alien weed introduced in the U.S., blanketing everything in its path once it becomes established. **USES:** In China, root tea used for headaches, diarrhea, dysentery, acute intestinal obstruction, gastroenteritis, deafness; to promote measles eruptions, induce sweating. Experimentally, plant extracts lower blood sugar and blood pressure. Flower tea used for stomach acidity; "awakens the spleen," "expels drunkenness." Seeds used for dysentery. Root (*ge-gen*), flowers, and seeds used in China to treat drunkenness or sober an intoxicated person. Animal studies show that pretreatment with Kudzu isoflavones suppressed free choice of ethanol consumption. In families with a negative family history of alcoholism, a clinical study suggested treatment with Kudzu root extract reduced alcohol consumption. Stem poulticed for sores, swellings, mastitis; tea gargled for sore throats. Root starch (used to stimulate production of body fluids) eaten as food. Roots are richer in estrogenic isoflavones, daidzein, and genistein than soybeans. Genistein may prevent development of tumors by preventing the formation of new blood vessels that nourish the tumors. Daidzein and genistein have been shown to inhibit the desire for alcohol and to reduce blood pressure and venous obstruction. An extract of the root was found to have

Kudzu flowers are pealike and strongly grape-scented.

Kudzu produces large, thrice-divided leaves and blankets millions of acres in the Southeast.

100 times the antioxidant activity of vitamin E. Kudzu extracts have also been found to stimulate regeneration of liver tissue while protecting against liver toxins. Recent research suggests Kudzu as a research target to reverse weight gain, fat accumulation, and metabolic syndrome in postmenopausal women.

MISCELLANEOUS PLANTS WITH SHOWY FLOWER SPIKES

TALL BELLFLOWER

Leaves, root

Campanula americana L. — Bellflower Family (Campanulaceae)
[*Campanulastrum americanum* (L.) Small]

Sprawling to erect annual or biennial; to 6 ft. Leaves lance-shaped to oblong-ovate (3–6 in.). Blue flowers in terminal spikes; July–Sept. Star-shaped *petals are fused together*. Note *long, curved style*. **WHERE FOUND:** Moist woods, streambanks. ON to FL; TX to MN. **USES:** The Iroquois used crushed root tea to treat tuberculosis. Leaf tea used by the Meskwaki for tuberculosis. Most *Campanula* species are slightly astringent.

ABOVE: *Tall Bellflower is often found at the edge of woods.* RIGHT: *Foxglove is commonly cultivated and can be deadly toxic.*

FOXGLOVE

Leaves

Digitalis purpurea L.

Plantain Family (Plantaginaceae)
[Formerly in Figwort Family (Scrophulariaceae)]

☠ Biennial; 3–6 ft. Leaves in a *basal rosette*, ovate to lance-shaped, soft-hairy, toothed; to 1 ft. long. Flowers purple to white, *spotted thimbles*, 1¼ in. long, on spikes; in summer of second year. **WHERE FOUND:** Garden escape. New England. Alien. A biennial from Europe that is often cultivated as an ornamental for its showy purple, pink, or white flowers, Foxglove has become naturalized in some areas. **USES:** Dried leaves a source of heart-tonic glycosides. Used in modern medicine to increase force of systolic contractions in congestive heart failure; lowers venous pressure in hypertensive heart ailments; elevates blood pressure in weak heart; diuretic, reduces edema. A famous example of folk use evolving to development of a modern drug, following the 1775 discovery by English physician William Withering testing a folk recipe from one of his patients for dropsy—edema induced by congestive heart failure. The cardiac glycoside digoxin was isolated a hundred years later in 1875. **WARNING:** Lethally toxic. First year's leaf growth (rosette) has been mistaken for leaves of Comfrey (*Symphytum*—see p. 243), with fatal results. Therapeutic dose of *Digitalis* is dangerously close to lethal dose. For use by physicians only. Scottish legal records contain numerous historical reports of children dying from drinking an infusion of the plant.

Purple or Spike Loosestrife is a rampant invasive weed.

Blue Vervain blooms in mid to late summer.

PURPLE OR SPIKE LOOSESTRIFE — Flowering plant

Lythrum salicaria L. — Loosestrife Family (Lythraceae)

Flowers purple-pink, *6-petaled*; in spikes; June–Sept. See p. 207.

BLUE VERVAIN — Leaves, root

Verbena hastata L. — Verbena Family (Verbenaceae)

Perennial; 2–4 ft. Stem *4-angled, grooved*. Leaves mostly lance-shaped, sharp-toothed; base sometimes lobed. Flowers blue-violet, tops branched in *pencil-like spikes*; July–Sept. **WHERE FOUND:** Fields, thickets. Most of our area. **USES:** Cherokee used as a tonic for pain after birth; also for coughs, colds, to induce sweating, strengthen the stomach, and in larger doses as an emetic. Iroquois used for stomach cramps. Root considered more active than leaves. As a folk medicine, valued as a strong bitter to stimulate digestion and as an emetic and expectorant; abandoned by medical practice by the mid-nineteenth century, but popular among folk practitioners for epilepsy, general tonic, an antidote to Poke poisoning; Boneset substitute. Naturalized in West Africa, the Efik ethic group of Southern Nigeria use a Blue Vervain leaf preparation to treat malaria; a recent study finds antimalarial leaf extract as effective as the standard drug chloroquine. Antidiarrheal and gastrointestinal motility-stimulating activity are also confirmed.

COMMON LOW-GROWING WILDFLOWERS OR WEEDS; PETALS IN 4S OR 5S

BACOPA, BRAHMI, WATER-HYSSOP, HERB-OF-GRACE — Whole plant

Bacopa monnieri (L.) Wettst.
[*Bramia monnieri* (L.) Drake;
Lysimachia monnieri L.]

Plantain Family (Plantaginaceae)
[Formerly in the Figwort Family
(Scrophulariaceae)]

⚠ Small, creeping, mat-forming, smooth-stemmed perennial; to 6 in. Leaves oblong-lance- to wedge-oval-shaped, entire; to 1 in. long. Flowers solitary at nodes, mostly purple, pale violet, or white; 5-lobed; *note stamens attached to lower side of flower tube*; June–Sept. **WHERE FOUND:** Wet sandy shores, coastal plains, at the edge of shallow water. MD, VA to FL, west to TX, OK, CA. Alien (Asia). Widespread in tropics. **USES:** In s. India, the herb is known as Brahmi; in n. and e. India, Mandukaparni. The name Brahmi is derived from Brahma, the mythical creator of Hindu tradition. Used in India's Ayurvedic traditional medicine for more than 3,000 years, Bacopa is traditionally considered a nerve tonic, treatment for epilepsy, and used to improve intellect and memory. In recent years it has been widely studied for its neuropharmacological effects for memory enhancement, mental clarity, cognitive enhancement; in short, as a nootropic brain tonic. It contains various active triterpene saponins, notably bacosides A and B, and other active compounds. Recently investigated effects include anxiolytic, antidepressant, anticonvulsive, antioxidant, antiulcer, cardiotonic, analgesic, and other activities. Bacopa is an excellent example of a long-used traditional medicine in which the chemistry, pharmacology, and clinical effects have been extensively studied, leading to worldwide modern use. **WARNING:** May interact with thiazide diuretics.

Bacopa has small 5-petaled flowers and mat-forming, semi-succulent leaves.

SPRING BEAUTY, VIRGINIA SPRING BEAUTY — Roots

Claytonia virginica L.

Montia Family (Montiaceae)
[Formerly in Purslane Family (Portulacaceae)]

Petals lavender to white or pinkish. See p. 57.

LEFT: *Spring Beauty, with white to pink-striped flowers.*
ABOVE: *Bluets are commonly seen on lawns in spring.*

BLUETS

Flowering plant

Houstonia caerulea L.
[*Hedyotis caerulea* (L.) Hook.]

Madder Family (Rubiaceae)

Small perennial; 2–8 in. Leaves narrow, opposite; to ½ in. long. Flowers sky blue to white; *4-parted*, with a *yellow center*; Mar.–July. **WHERE FOUND:** Fields, yards. NS to GA; AR to WI. **USES:** Cherokee used leaf tea to stop bed-wetting. During the Civil War, it was suggested that Bluets and their relatives be investigated as dye plants in the Confederate states, given limited access to imported dye plants at the time. Various weedy bluets, separated by technical characteristics, used similarly.

COMMON SPEEDWELL

Leaves, root

Veronica officinalis L.

Plantain Family (Plantaginaceae)
[Formerly in Figwort Family (Scrophulariaceae)]

Creeping, hairy herb; to 7 in. Leaves elliptical, narrow at base; evenly toothed. Blue-violet flowers in *glandular-haired racemes*; May–Aug. **WHERE FOUND:** Waste places. Much of our area. Alien. Of the more than 20 *Veronica* species that occur in our range, almost all of them are naturalized cosmopolitan weeds from Europe and Asia. From the perspective of the casual herbalist, they are all treated as *Veronica officinalis* despite clear speciation. Commonly found growing on lawns in N. America. **USES:** In Europe, astringent root or leaf lea traditionally used to promote urination, sweating, and menstruation; blood purifier also used for skin and kidney ailments, coughs, asthma, lung diseases, gout, rheumatism, and jaundice. Considered expectorant, diuretic, tonic. Extracts found to prevent and speed healing of ulcers in experiments on

Speedwells, separated on technical characteristics, are a familiar spring lawn weed.

Johnny-jump-up, Heart's Ease flowers are multicolored.

animals; also experimentally lowers cholesterol. **WARNING:** One component, aucubin, though liver-protective, antioxidant, and antiseptic, can be toxic to grazing animals.

THYME-LEAVED SPEEDWELL

Leaves

Veronica serpyllifolia L.

Plantain Family (Plantaginaceae)
[Formerly in Figwort Family (Scrophulariaceae)]

Creeping, much-branched, *smooth* perennial; 2–8 in. Leaves *oval to oblong; short-stalked*. Flowers small, violet-blue (to whitish); 4 petals with pale blue and dark stripes; Apr.–Sept. **WHERE FOUND:** Lawns, roadsides. NL to GA, AR to MN. Alien. **USES:** Leaf juice used by Cherokee for earaches, leaves poulticed for boils; tea used for chills and coughs.

JOHNNY-JUMP-UP, HEART'S EASE

Leaves

Viola tricolor L.

Violet Family (Violaceae)

Angled-stemmed annual; 4–12 in. Leaves toothed, roundish on lower part of plant, oblong above; stipules large, leaflike; *strongly divided*. Pansylike flowers in patterns of purple, white, and yellow; May–Sept. **WHERE FOUND:** Escaped from gardens. Alien. Field weed; much of our area. **USES:** In Europe, leaf tea is a folk medicine for fevers; mild laxative; gargle for sore throats; considered diuretic, expectorant, mild sedative, blood purifier; used for asthma, heart palpitations, and skin eruptions, such as eczema. Contains various saponins and mucilage linked to diuretic activity. Rat experiments confirm possible use for skin eruptions. Contains antioxidant anthocyanidins. Used in European phytomedicine externally in treatment of mild seborrhea and related skin disorders. **WARNING:** Contains saponins; may be toxic in larger doses.

LOW-GROWING PLANTS WITH SHOWY FLOWERS; 5 OR MORE PETALS

PASQUEFLOWER

Whole plant

Anemone patens L. ssp. *multifida* (L.) Pritzel
[*Pulsatilla patens* L (Mill.)]

Buttercup Family (Ranunculaceae)

⚠ Perennial; 2–16 in. Silky leaves arising from root; dissected into linear segments. Showy flowers, 1½–2 in. wide. "Petals" (sepals) purple or white, in *a cup-shaped receptacle*; Mar.–Aug. Seeds with *feathery plumes*. **WHERE FOUND:** Moist meadows, prairies, woods. S. WI, IA to CO, NM, WA, north to AK. **USES:** The Chippewa smelled a handful of crushed leaves to treat headache. The Cheyenne smashed the root and waved it to revive an unconscious person. As few as 5 drops of highly diluted tincture in water used in homeopathic practice for eye ailments, skin eruptions, rheumatism, leukorrhea, obstructed menses, bronchitis, coughs, asthma. **REMARKS:** The homeopathic doses reported here and elsewhere in this plant identification guide are so dilute as to be harmless (i.e., without side effects), if not biologically inactive, by conventional chemical standards. **RELATED SPECIES:** The European species, *Anemone pulsatilla* L. (*Pulsatilla vulgaris* Mill.) and *A. pratensis* L. [*Pulsatilla pratensis* (L.) Mill.], sometimes grown in rock gardens in the

The European Pasqueflower, grown in gardens, is an occasional escapee.

Pasqueflower produces pale violet or white flower petals.

U.S., have been used in German phytomedicine for inflammatory and infectious disease of the skin and mucous membranes, gastrointestinal and urinary tract disorders, migraines, neuralgia, and general restlessness. However, current use is prohibited because the risks outweigh the benefits. **WARNING:** Poisonous. Extremely irritating. Both external and internal use should be avoided.

SHARP-LOBED HEPATICA, LIVERLEAF

Leaves

Hepatica nobilis var. *acuta* (Pursh) Steyerm.
[*Anemone acutiloba* (DC.) G. Lawson, *A. hepatica* var. *acuta* (Pursh) Pritz., *Hepatica acutiloba* DC.]

Buttercup Family
(Ranunculaceae)

Perennial; 4–8 in. Leaves usually evergreen, *3-lobed; lobes pointed.* Flowers lavender, blue, white, or pink; Feb.–June. "Petals" (6–10) are actually sepals. **WHERE FOUND:** Rich woods. W. ME to GA; LA, AR, MO to MN. **USES:** Native Americans used leaf tea for liver ailments, poor digestion, laxative; externally, as a wash for swollen breasts. In folk tradition, tea used for fevers, liver ailments, coughs. Thought to be mildly astringent, demulcent, diuretic. A "liver tonic" boom resulted in the consumption of 450,000 pounds of the dried leaves (domestic and imported) in 1883 alone. Once used in European herbal medicine for liver disorders, as a general appetite stimulant, a tonic, to increase circulation, and many other treatments. Not allowed for use in Germany because of potential risks and lack of scientific substantiation of claimed uses. **REMARKS:** Many modern taxonomic works disagree on the current taxonomic position and placement of *Hepatica.* Therefore, we continue to use the taxonomy from the second edition of this work. **WARNING:** Contains irritating compounds (like many other members of the Buttercup family).

Sharp-lobed Liverleaf flowers range from whitish to pale purple, as shown here.

Sharp-lobed Liverleaf has pointed leaf lobes.

ABOVE: *Sharp-lobed Liverleaf flowers are sometimes deep blue-violet.* RIGHT: *Blue-eyed Grass is not a grass but an iris.*

ROUND-LOBED HEPATICA Leaves

Hepatica nobilis var. *obtusa* (Pursh) Steyerm. Buttercup Family (Ranunculaceae)

[*Hepatica americana* (DC.) Ker., *H. acutiloba* DC., *Anemone americana, A. hepatica* var. *acuta* (Pursh) Pritz.]

Similar to *H. acutiloba,* but leaf *lobes are rounded.* Flowers white or pale blue-violet; Mar.–June. **WHERE FOUND:** Dry woods. NS to GA, AL; MO to MB. **USES:** Same as for *H. acutiloba* (see above).

BLUE-EYED GRASS Root, leaves

Sisyrinchium angustifolium Mill. Iris Family (Iridaceae)

Perennial; 4–18 in. Differs from the other 10 or so *Sisyrinchium* species in our range in that the leaves are narrow (¼ in. wide), much flattened, and deep green. Flowers at tip of *long, flat stalk*; May–July. **WHERE FOUND:** Meadows. Most of our area. **USES:** Cherokee used root tea for diarrhea (in children); plant tea for worms, stomachaches. Several species were used as laxatives. There is no recent science on Blue-Eyed Grass.

BORAGE

Leaves, flowers, seeds

Borago officinalis L. Borage Family (Boraginaceae)

⚠ Coarse annual; 1–4 ft. Stems succulent, hollow. Leaves with cucumber fragrance when crushed; *rough-hairy*, lower leaves broadly ovate and stalked, upper leaves sessile, clasping, 2–5 in. long. Flowers are *brilliant blue stars with prominent black anthers forming a conelike structure* in the center, to 1 in. across; drooping downward; June–Sept. **WHERE FOUND:** Near gardens, sometimes escaped; casual weed. Alien (Europe). Occasional in our range. **USES:** Leaf tea historically used for fevers, jaundice, rheumatism. Considered cooling and diuretic. Externally, a poultice used for wounds; anti-inflammatory. Flowers edible. Extract of leaves with amebicide, antioxidant, and anti-inflammatory activity. Seed oil is a rich source of gamma-linolenic acid (GLA). GLA-rich seed oils are said to alleviate imbalances and abnormalities of essential fatty acids in prostaglandin production. Most research on sources of GLA has involved Evening Primrose, promoted for rheumatoid arthritis, atopic dermatitis, diabetic neuropathy, and menopause symptoms. A small clinical study showed Borage seed oil reduced stress by lowering systolic blood pressure and heart rate. **WARNING:** Like Comfrey leaves, Borage leaves contain potentially liver-toxic and carcinogenic pyrrolizidine alkaloids. Risk of leaf use outweighs benefits. Borage leaves have a casual resemblance to the highly toxic wrinkled, hairy leaves of Foxglove (*Digitalis purpurea*). An unusual Italian case from 2009 alarmingly reported that an entire family was poisoned after eating potato dumplings flavored with "Borage" leaves, which turned out to be misidentified Foxglove leaves.

Borage is usually found in gardens.

MADAGASCAR PERIWINKLE

Leaves

Catharanthus roseus (L.) G. Don Dogbane Family (Apocynaceae)

⚠ Flowers sometimes blue-violet. See p. 56.

VIRGINIA WATERLEAF

Whole plant, root

Hydrophyllum virginianum L. Waterleaf Family (Hydrophyllaceae)

[Modern works place in Borage Family (Boraginaceae)]

Perennial; 1–3 ft. Leaves *deeply divided, with 5–7 lobes*; lower segments 2-parted, with *marks like water stains*. Flowers bell-like with *protruding*

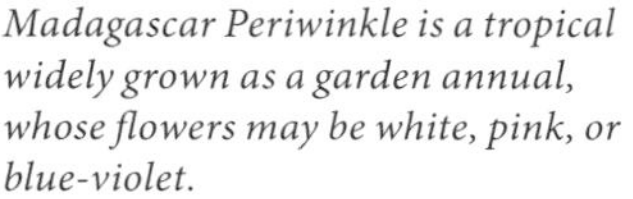

Madagascar Periwinkle is a tropical widely grown as a garden annual, whose flowers may be white, pink, or blue-violet.

Virginia Waterleaf. Note the protruding stamens.

stamens; violet to whitish; May–Aug. **WHERE FOUND:** Rich woods. QC and w. New England to VA; TN, n. AR, e. KS to MB. **USES:** Native Americans used root tea as an astringent for diarrhea, dysentery. Tea (or roots chewed) for cracked lips, mouth sores.

FLAX — Seeds

Linum usitatissimum L. — Flax Family (Linaceae)

Delicate annual; 8–22 in. Leaves linear, 3-veined. Flowers with 5 *slightly overlapping blue petals* (½–¾ in. across); June–Sept. **WHERE FOUND:** Waste places. Throughout our area. Alien. **USES:** Source of linseed oil and linen. Said to be soothing and softening to irritated membranes. Seeds once used for skin and mouth cancers, colds, coughs, lung and urinary ailments, fevers; laxative; poulticed (mixed with lime water) to relieve pain of burns, gout, inflammation, rheumatism, boils. Seeds used in European phytomedicine as a mild, lubricating laxative in constipation, for irritable bowel syndrome, diverticulitis, and for relief of gastritis and enteritis. Flaxseed is also used to correct problems caused from abuse of stimulant laxatives. Oil is a folk remedy used for pleurisy and pneumonia, high in omega-3 fatty acids. A folk remedy for cancer, possibly containing some antitumor compounds found in Mayapple (p. 64). Flaxseed has been suggested as a possible preventive for colon cancer and dietary supplement to improve biochemical parameters associated with the management of diabetes mellitus. Seed oil

Flax has slightly overlapping petals.

Greek Valerian, Jacob's Ladder has "ladderlike" leaves.

antioxidant, said to protect skin against ultraviolet light damage; also has anti-inflammatory, analgesic, and fever reducing effects. **WARNING:** Contains a cyanide-like compound. Oil may be emetic and purgative.

GREEK VALERIAN, JACOB'S LADDER — Root

Polemonium reptans L. — Phlox Family (Polemoniaceae)

Perennial; 8–24 in. Leaves paired; sessile. The name Jacob's Ladder refers to the ladderlike arrangement of the leaves. Flowers are loose clusters of violet-blue bells; *stamens not protruding*; Apr.–June. **WHERE FOUND:** Moist bottomlands. NY to GA, MS, OK to MN. **USES:** Native Americans used root in prescriptions for hemorrhoids, to induce vomiting, treat eczema, enhance action of Mayapple. The Indian name for this plant, which translates as "smells like pine," refers to the root fragrance. Root tea once used to induce sweating, astringent; for pleurisy, fevers, scrofula, snakebites, bowel complaints, and bronchial afflictions. A folk medicine as an expectorant for coughs and to induce sweating; deemed "sweat-root" by one author. **RELATED SPECIES:** *P. vanbruntiae* Britton (not shown) is larger and has protruding stamens.

FLOWERS WITH 5 PETALS; CURLED CLUSTERS: BORAGE FAMILY

HOUND'S TONGUE (NOT SHOWN)

Leaves, root

Cynoglossum officinale L. Borage Family (Boraginaceae)

⚠ Downy biennial; 1–3 ft., with a mousy odor. Leaves lance-shaped. Flowers purplish, enclosed by a soft-hairy calyx, *with leafy bracts beneath inflorescence* (all the way up the flower stalk); Aug.–Sept. Flat fruits covered with soft spines. **WHERE FOUND:** Roadsides, pastures. Much of our area. Alien. **USES:** Leaf and root tea once used to soothe coughs, colds, irritated membranes; astringent in diarrhea, dysentery. Leaf poultice used for insect bites, hemorrhoids. **WARNING:** Contains the potentially carcinogenic alkaloids cynoglossine and consolidine, both CNS-depressant. Pyrrolizidine alkaloids in the plant are responsible for reports of poisoning of livestock, including horses and calves. May cause dermatitis.

WILD COMFREY

Leaves, root

Cynoglossum virginianum L. Borage Family (Boraginaceae)

⚠ Rough, hairy perennial; 1–2 ft. Basal leaves in a rosette; *stalk stem leaves clasping*. Pale violet-blue flowers on spreading racemes, *without leafy bracts beneath flowers*; May–June. Flowers are somewhat like those of Borage. **WHERE FOUND:** Open woods. Sw. CT; NJ; PA to FL; TX to MO; s. IL. **USES:** Cherokee used root tea for "bad memory," cancer, itching of genitals, milky urine. In nineteenth-century texts, authors suggest use as a substitute for comfrey (*Symphytum officinale*); leaves

LEFT: *Wild Comfrey. The stem leaves clasp the stem.* ABOVE: *Wild Comfrey flowers are pale blue with wrinkled petals.*

ABOVE: *Wild Comfrey. Note large basal leaves; no leaf bracts beneath flowers.* RIGHT: *Viper's Bugloss is extremely bristly; its hairs can irritate human skin.*

smoked like Tobacco. **WARNING:** Do not confuse the leaves of either comfrey with those of Foxglove (*Digitalis,* p. 230); fatal poisoning may result. The plant itself is likely actively toxic, because of pyrrolizidine alkaloids and related compounds.

VIPER'S BUGLOSS — Whole plant, root
Echium vulgare L. — Borage Family (Boraginaceae)

Bristly biennial; 1–2½ ft. Leaves lance-shaped. Flowers violet-blue on curled branches. *One flower blooms at a time on each curled branch*; June–Sept. Upper lip longer than lower; *stamens red, protruding*. **WHERE FOUND:** Waste places. Much of our area. Alien (Europe). **USES:** Leaf tea is a folk medicine, used to promote sweating; diuretic, expectorant, soothing; used for fevers, headaches, nervous conditions, and pain from inflammation. Root contains healing allantoin. **WARNING:** Contains a toxic alkaloid. Hairs may cause rash; bees relish the flowers, but the rough hairs of the plant tear their wings. Plant contains toxic pyrrolizidine alkaloids, with at least 8 of them found in the pollen, and recently identified in Viper's Bugloss honey. Analytical methods have been developed to detect these toxic compounds in both honey and pollen supplies intended for human consumption.

MERTENSIA, VIRGINIA COWSLIP, VIRGINIA BLUEBELL — Whole plant
*Mertensia virgini*ca (L.) Pers. ex Link — Borage Family (Boraginaceae)
(*Pulmonaria virginica* L.; *M. pulmonarioides* Roth)

Herbaceous perennial with a stout rootstock, ascending or arched (under the weight of flowers); 9–25 in. tall. Leaves alternate, broadly

Virginia Bluebell has bluish pink, flared, drooping "bells."

Comfrey. Often persistent in old garden sites.

ovate to elliptical, up to 13 in. long, most basal, *with winged stalks*; stem leaves mostly with short (or no) stalks. Flowers blue to pink in dense to elongated cluster; pendant, to 1. in or more long, *tubular with bell-shaped flare*; Apr.–May. **WHERE FOUND:** Rich woods, streambanks. QC, ME, to GA mtns.; AR, KS to MN. A common woodland garden perennial. **USES:** Plant tea used by the Cherokee for treatment of lung ailments, whooping cough, tuberculosis. Iroquois used root decoction for venereal disease. Historically used as a Comfrey substitute; the whole plant high in mucilage, used for coughs, diarrhea, dysentery. **WARNING:** Likely contains toxic pyrrolizidine alkaloids.

COMFREY

Leaves, root

Symphytum officinale L. — Borage Family (Boraginaceae)

⚠ Large-rooted perennial; 1–3 ft. Leaves large, *rough-hairy*; broadly oval to lance-shaped. Bell-like flowers in *furled clusters*; purple-blue, pinkish, or white; May–Sept. **WHERE FOUND:** Widespread alien from Europe. Often cultivated. **USES:** While *Symphytum officinale* is listed as the species used in most herb books, a number of the 25 species of comfrey (*Symphytum*) are cultivated in American gardens. Extremely popular in the 1970s and early '80s, studies showing toxic pyrrolizidine alkaloids, especially in the root, halted the comfrey love affair. Studies show that leaves harvested during the blooming period are very low in alkaloid content. Root tea and weaker leaf tea considered tonic, astringent, demulcent, for diarrhea, dysentery, bronchial irritation, coughs, vomiting of blood, "female maladies"; leaves and root poulticed to "knit bones," promote healing of bruises, wounds, ulcers, sore breasts. Contains allantoin, which promotes healing. In European phytomedicine, external application of the leaf is approved for the treatment of bruises and sprains; root poultice approved for bruising, pulled muscles and ligaments, and sprains. An obvious commercial development has evolved in topical comfrey products (ointments, gels, creams) with 99 percent of toxic alkaloids (see warning below) removed in manufacturing. Clinical studies using these preparations have shown benefits for osteoarthritis in the knees, reduced swelling and pain in ankle sprains, and back pain relief, with confirmed anti-inflammatory and analgesic effects. **RELATED SPECIES:** A widely grown type is Russian Comfrey, *S.* x *uplandicum* Nyman, a hybrid bred as a fodder crop. **WARNING:** Root contains high levels of liver-toxic (or cancer-causing) pyrrolizidine alkaloids. Certain types of leaf tea, although less carcinogenic than beer, were banned in Canada and elsewhere. In Germany, external use of the leaf is limited to 4–6 weeks each year. There is also a danger that the leaves of comfrey (*Symphytum*) may be confused with the first-year leaf rosettes of Foxglove (*Digitalis*), with fatal results. Consult an expert on identification first.

5-PARTED FLOWERS; NIGHTSHADE FAMILY

JIMSONWEED — Leaves, root, seed

Datura stramonium L. — Nightshade Family (Solanaceae)

Annual; 2–5 ft. Leaves *coarse-toothed*. Flowers white to pale violet, 3–5 in.; trumpet-shaped; May–Sept. Seedpods *upright*, shiny, chambered, with prickles; seeds lentil-shaped. **WHERE FOUND:** Waste places. Throughout our area. **USES:** Whole plant contains atropine and other alkaloids, used in eye diseases (to dilate pupils); causes dry mouth, depresses action of bladder muscles, impedes action of parasympathetic nerves; used in Parkinson's disease; also contains scopolamine, used in patches behind ear for vertigo. Leaves once smoked as antispasmodic for asthma. Folk remedy for cancer. **WARNING:** Violently toxic. Causes severe hallucinations. Many fatalities recorded and not rare. Those who collect this plant may end up with swollen eyelids. Licorice (*Glycyrrhiza*) has been suggested as an antidote.

HORSE-NETTLE — Leaves

Solanum carolinense L. — Nightshade Family (Solanaceae)

Flowers *pale* violet to white stars; May–Oct. See p. 60. **WARNING:** Toxic. Fatalities reported in children from ingesting berries.

WOODY NIGHTSHADE, BITTERSWEET — Leaves, stems, berries

Solanum dulcamara L. — Nightshade Family (Solanaceae)

A woody, climbing vine. Leaves oval, often with *1 or 2 prominent lobes at base*. Flowers violet (or rarely white) *stars with yellow protrusions* (stamens); petals *curved back*; May–Sept. Fruits ovoid, red; Sept.–Nov.

LEFT: *Jimsonweed flowers are often pale violet to blue.* ABOVE: *Horse-nettle flowers are mostly whitish or occasionally violet.*

ABOVE: *Woody Nightshade, Bittersweet has shiny red fruits, which should not be eaten.* RIGHT: *Woody Nightshade, Bittersweet, with violet blooms (rarely white).*

WHERE FOUND: Waste places. Throughout our area. Alien (Europe). **USES:** Externally, plant used as a folk remedy for warts, tumors and abcess at the end of a finger. Science confirms significant potential anticancer activity. Used as a starting material for steroids; contains steroidal glycosides. Formerly used as narcotic, diuretic sweat inducer; used for skin eruptions, rheumatism, gout, bronchitis, whooping cough. In Germany, relatively nonpoisonous stems have been approved for use in supportive treatment of acne, eczema, boils, and warts. Stems contain significantly lower amounts of toxic alkaloids than do other parts of the plant. **WARNING:** Toxic. Contains steroids, toxic alkaloids, and glucosides. Will cause vomiting, vertigo, convulsions, weakened heart, paralysis. Green berries can cause diarrhea, dilated pupils, nausea, and vomiting. Lethal dose is estimated to be 200 berries.

LIPPED FLOWERS WITH SPLIT COROLLAS; LOBELIAS

LOBELIA, INDIAN-TOBACCO

Whole plant

Lobelia inflata L. — Bellflower Family (Campanulaceae)

Hairy annual; 6–18 in. Leaves oval, toothed; *hairy beneath*. Inconspicuous white to pale blue flowers; flower throat split at top, in racemes, to ¼ in.; June–Oct. Seedpods *inflated*. **WHERE FOUND:** Fields, waste places, open woods. NS to GA; LA, AR, e. KS to SK. **USES:** Native Americans smoked leaves for asthma, bronchitis, sore throats, coughs. One of the most widely used herbs in nineteenth-century America. Traditionally used to induce vomiting (hence the nickname "Pukeweed") and sweating; sedative; used for asthma, whooping cough, fevers, to enhance or direct action of other herbs. Lobeline, a chemical cousin of nicotine, one of more than 14 alkaloids in the plant, was formerly used in the U.S. in commercial "quit smoking" lozenges, patches, and chewing gums—said to appease physical need for nicotine without addictive effects. Still used in other countries. Lobeline produces dilation of the bronchioles and increased respiration. Lobelane, a minor alkaloid in Lobelia, and lobeline have shown promise in reducing self-selection of methamphetamine in laboratory experiments; a possible research target to thwart stimulant abuse. Multi-drug resistance is a potential limiting factor in chemotherapy treatment for cancers. The alkaloid lobeline

LEFT: *Lobelia. Note inflated seedpods.* ABOVE: *Lobelia, with pale violet flowers.*

Great Lobelia, a showy autumn wildflower.

Pale-spike Lobelia. Flowers are whitish to pale blue.

has recently shown promise as a candidate for development of multidrug resistance reversal agent. **WARNING:** Considered potentially toxic because of its strong emetic, expectorant, and sedative effects.

GREAT LOBELIA

Leaves, roots

Lobelia siphilitica L. — Bellflower Family (Campanulaceae)

Perennial; 1–5 ft. Leaves oval, toothed. Blue-lavender flowers. Corolla throat *white-striped*; Aug.–Oct. **WHERE FOUND:** Moist soil, streambanks. ME to NC; MS, AR, e. KS to MN. Often found growing with Cardinal Flower, this is a common fall wildflower along southern streams, but occurs as far north as MN and ME (where it is rare). **USES:** Early medical writers observed that the Cherokee used the root to treat syphilis, hence the species name *siphilitica;* also against worms. Root tea for syphilis, leaf tea for colds, fevers, "stomach troubles," worms, croup, nosebleeds; gargled leaf tea for coughs; leaves poulticed for headaches, hard-to-heal sores. Formerly used to induce sweating and urination. Considered similar to, but weaker than, *L. inflata,* with a stronger diuretic effect. Little researched; also contains lobeline and lobetyolin, among other alkaloids with unique structures. **WARNING:** May cause gastrointestinal upset and vomiting.

PALE-SPIKE LOBELIA

Leaves

Lobelia spicata Lam. — Bellflower Family (Campanulaceae)

Perennial; 2–4 ft. Stem *smooth* above, *densely hairy* at base. Leaves lance-shaped to slightly oval, barely toothed or without teeth. Flowers pale blue or whitish; June–Aug. **WHERE FOUND:** Fields, glades, meadows, thickets. Most of our area. **USES:** Cherokee used a tea of the plant as an emetic. The Iroquois used a wash made from the stalks for "bad blood" and neck and jaw sores. The root tea was used to treat trembling by applying the tea to scratches made in the affected limb. **WARNING:** Of unknown safety.

MISCELLANEOUS MINT RELATIVES WITH SQUARE STEMS AND PAIRED LEAVES

LEMON BERGAMOT, LEMON BEE-BALM

Leaves, roots

Monarda citriodora Crv. ex Lag. Mint Family (Lamiaceae or Labiatae)

Annual herb; 8–28 in. tall. Leaves lance-shaped, lemon-scented, to 3 in. long; thin, but firm with parallel sides. Flowers in crowded, tiered terminal clusters sometimes branched, purple-pinkish to white, often spotted, *upper petal lip bow-curved*, bracts beneath flowers tinged purple; calyx segments and bracts with *elongate bristle at end*; Apr.–Oct. **WHERE FOUND:** Sandy to rocky soils, meadows, prairies; SC to FL. West to NM, KS, NE; most common in TX. **USES:** Lemon-scented leaves used as a beverage tea and food by Hopi and other native groups. Traditionally used as a folk medicine for headaches, fevers, and stomachache. Considered a pleasant stomachic tea; dried flowers used to induce sneezing to treat headache.

Lemon Bergamot has citrus-scented leaves.

WILD BERGAMOT, PURPLE BEE-BALM

Leaves

Monarda fistulosa L. Mint Family (Lamiaceae or Labiatae)

Highly variable perennial; 2–3 ft. Leaves paired; triangular to oval or lance-shaped. Flowers lavender; *narrow, lipped tubes in crowded heads*; May–Sept. Bracts often slightly purple-tinged. **WHERE FOUND:** Dry wood edges, thickets. QC to GA; LA, e. TX, OK to ND, MN. **USES:** Botanists describe several varieties, with chemists separating different chemotypes based on varied chemistry of the essential oil. The Pawnee noted different variations based on stem morphology and fragrance desirability, with specific names for each. Even the roots were classified according

ABOVE: *Butterflies relish Wild Bergamot, Purple Bee-balm.*
RIGHT: *Perilla is an aggressive, annual Asian weed.*

to specific forms, revealing Native Americans' nuanced relationship with the plant. The Omaha-Ponca made similar distinctions in varietal fragrance variations. The Plains Apache highly valued the leaves as "Indian Perfume," meticulously selecting leaves of individual plants with subtle citrus overtones used as a personal fragrance to attract the opposite sex, and kept leaves among blankets and clothing for fragrance. Once selected, individual plants or populations were carefully conserved to maintain genetic integrity. In Cherokee tradition, leaf tea used for colds, headache, stomach problems, restful sleep, also as a diuretic and diaphoretic; externally, leaf poulticed for headache. Winnebago used a poultice of boiled leaves for skin eruptions and pimples. The Ojibwa used whole plant, concentrating the volatile components to treat bronchial afflictions. Plant preparations were widely used for coughs, colds, to induce sweating, for analgesic effects, and many other maladies. Historically, physicians used leaf tea to expel worms and gas. In one chemotype, the essential oil is high in carvacrol, with anesthetic, worm-expelling, anti-inflammatory, antioxidant, and diuretic activity.

PERILLA — Leaves, seeds

Perilla frutescens (L.) Britt. — Mint Family (Lamiaceae or Labiatae)

⚠ Annual; 1–3 ft. Leaves oval, *wrinkled*, long-toothed; green or often *purple, with a peculiar fragrance.* Flowers whitish to lavender or pale violet in axillary and terminal clusters; July–Sept. **WHERE FOUND:** Moist, open woods. MA to FL; TX to IA. This Asian alien has become an invasive weed in the South. In the Ozarks it is called Rattlesnake Weed because the dried seed cases (the calyx) rattle as one walks by. **USES:** Leaf tea used in Asian traditional medicine for abdominal pains, diarrhea, vomiting, coughs, to "quiet a restless fetus," relieve morning sickness, irritability during pregnancy, fevers, colds. Considered antibacterial, antioxidant (particularly in purple-leaf form), anti-inflammatory, diaphoretic, sedative, and spasmolytic. Dried leaves used in China to treat bronchitis. One Chinese clinical study found application of the fresh leaves (rubbing on infection) for 10 to 15 minutes a day made warts disappear in 2 to 6 days. A favorite culinary herb of Chinese, Japanese, and Korean cultures. Seed oil a rich source of alpha-linolenic acid, an omega-3 essential fatty acid (54–64 percent), and omega-6 (linoleic acid) (around 14 percent). **WARNING:** Avoid during pregnancy. A component in the leaves was found to induce severe lung lesions in mice, rats, and sheep. Cattle have contracted acute pulmonary emphysema from eating the plant. Use in human foods and medicine has been questioned by some, but widely used on a daily basis in Asia. Once used as a fish poison, antidote.

HOARY MOUNTAIN MINT (NOT SHOWN) — Leaves

Pycnanthemum incanum (L.) Michx. — Mint Family (Lamiaceae or Labiatae)

Perennial; 2–6 ft. Leaves oval to lance-shaped, stalked, toothed; *hoary beneath* (upper ones white-haired on both sides). Flowers pale lilac; July–Sept. Calyx lobes *apparently 2-lobed*; lobes lance-shaped. **WHERE FOUND:** Dry thickets. NH to FL, north to TN, s. IL. **USES:** Leaf tea once used for fevers, colds, coughs, colic, stomach cramps; said to induce sweating, relieve gas. Native Americans poulticed leaves for headaches; washed inflamed penis with tea. **REMARKS:** Other *Pycnanthemum* species used similarly.

MAD-DOG SKULLCAP — Leaves

Scutellaria lateriflora L. — Mint Family (Lamiaceae or Labiatae)

Perennial; 1–3 ft. Leaves opposite; oval to lance-shaped, toothed. Flowers small, to ¼ in. long, violet-blue, hooded, lipped; May–Sept. Easily distinguished from other *Scutellaria* species—flowers are *in 1-sided racemes from leaf axils.* **WHERE FOUND:** Rich woods, moist thickets. Much of our area. **USES:** Known as Mad-dog Skullcap because tea was once used as a folk remedy for rabies. A strong tea is traditionally used as a sedative, nerve tonic, and antispasmodic for

Mad-dog Skullcap has small flowers from leaf axils.

all types of nervous conditions, including epilepsy, insomnia, anxiety, and neuralgia. Flavonoids in plant, such as scutellarin, baicalein, and baicalin, have confirmed sedative and antispasmodic qualities. Baicalein and baicalin show promise for reducing factors associated with loss of neuron activity and synaptic function in neurodegenerative diseases such as Parkinson's and Alzheimer's disease. Widely used as an herbal alternative for anxiety, insomnia, and symptoms of depression. Other *Scutellaria* species may have similar properties. **WARNING:** Native Germander or Wood Sage (*Teucrium canadense*) is a widespread adulterant to commercial supplies of Skullcap. Reports of liver toxicity related to Skullcap actually involve Wood Sage, which is traded under the name "pink skullcap." Two cases of presumed Skullcap poisoning, including one fatality, may have involved Wood Sage.

STRONGLY SCENTED MINTS; SQUARE STEMS, PAIRED LEAVES

CANADIAN MINT

Leaves

Mentha canadensis L. Mint Family (Lamiaceae or Labiatae)
(listed as *Mentha arvensis* L. in older texts)

Aromatic perennial with pennyroyal/peppermint scent; 6–25 in. Fine, backward-bending hairs, at least on stem angles. Leaves opposite, round to oval. Pale lavender or whitish flowers in *crowded globular terminal clusters* (or 1–3 clusters below); *calyx hairy*; May–Oct. **WHERE FOUND:** Damp soil. Canada, n. U.S. Our only native *Mentha;* a highly variable species. **USES:** The Flathead Indians of w. Montana used a tea of the leaves for colds, coughs, and fever. Green leaves were packed around an aching tooth to relieve pain and inflammation. Dried leaves used as a condiment, sprinkled on meat and fruit to repel flies; kept in homes for pleasant fragrance. The Cheyenne used the ground leaves in tea to prevent nausea. Other Native American groups used tea for sore throats, gas, colic, indigestion, headaches, diarrhea; in short, same medicinal uses as for Peppermint and Spearmint in Western folk medicine. Science confirms antioxidant activity. **WARNING:** Essential oil of

Canadian Mint is found in moist habitats.

this mint, probably like all essential oils, is antiseptic, but can be toxic to humans in a concentrated form.

PEPPERMINT

Leaves

Mentha x *piperita* L. Mint Family (Lamiaceae or Labiatae)

⚠ Perennial; 12–36 in. Stem *purplish* (not greenish, as in Spearmint); *smooth*, with few hairs. Leaves opposite, stalked; *distinct fragrance of peppermint*. Flowers pale violet; in loose, interrupted terminal spikes; June–frost. Peppermint is a sterile hybrid between Spearmint (*M. spicata*) and Watermint (*M. aquatica*). Native to Europe, it was first grown commercially in England about 1750, also the point at which it begins to appear in herbals. **WHERE FOUND:** Wet places. Escaped from cultivation. Throughout our range. European alien. **USES:** Leaf tea traditionally used for colds, fevers, indigestion, gas, stomachaches, headaches, nervous tension, insomnia. Extracts experimentally effective against herpes simplex, Newcastle disease, and other viruses; antispasmodic, antibacterial, and antioxidant. The oil stops spasms of smooth muscles. Animal experiments show that azulene, a minor component of distilled Peppermint oil residues, is anti-inflammatory and has anti-ulcer activity. Enteric-coated Peppermint capsules are used in Europe for irritable bowel syndrome. Peppermint leaf is used in European phytomedicine for use in muscle spasms of the gastrointestinal tract, as well as spasms of the gallbladder and bile ducts. The essential oil is used externally to treat neuralgia and myalgia. Menthol is an approved ingredient in cough drops. **WARNING:** Oil is toxic if taken internally; causes dermatitis. Menthol, the major chemical component of Peppermint oil, may cause allergic reactions. Infants should never be exposed to menthol-containing products, as they can cause the lungs to collapse. Use should be avoided in cases of gallbladder and bile duct obstruction.

Peppermint. Leaves have a peppermint fragrance, and there is usually more red in the leaf veins.

SPEARMINT

Leaves

Mentha spicata L. — Mint Family (Lamiaceae or Labiatae)

Creeping perennial; 6–36 in. Leaves opposite; *without stalks* (or with very short stalks); with a distinct fragrance of spearmint (think spearmint gum). Flowers pale pink-violet; in slender, elongated spikes; June–frost. **WHERE FOUND:** Wet soil. Much of our area. Escaped. European alien. **USES:** Spearmint is listed in most medieval works on herbs, and was cultivated in convent gardens from the ninth century onward. Used similarly to Peppermint, Spearmint was largely replaced by Peppermint by the late eighteenth century. Spearmint and Spearmint oil are used as carminatives (to relieve gas) and primarily to disguise the flavor of other medicines. Spearmint has been traditionally valued as a stomachic, antiseptic, and antispasmodic. The leaf tea has been used for stomachaches, diarrhea, nausea, colds, headaches, cramps, fevers, and is a folk remedy for cancer. **WARNING:** Oil is toxic if taken internally; causes dermatitis; may cause rare allergic reactions.

Spearmint. Leaves have a spearmint fragrance.

SQUARE-STEMMED PLANTS; LEAVES PAIRED; STRONGLY AROMATIC; NOT MINTY

BLUE GIANT HYSSOP, ANISE-HYSSOP

Leaves, roots

Agastache foeniculum (Pursh) O. Kuntze — Mint Family (Lamiaceae or Labiatae)

Perennial; to 3 ft. Smooth-stemmed, branched above. Leaves *strongly anise-scented*, minutely downy beneath. Bluish flowers in tight whorls; June–Sept. Stamens in *2 protruding pairs*; pairs *crossing*. **WHERE FOUND:** Prairies, dry thickets. ON south to IL, IA; west to CO, SD, WA. Cultivated; escaped eastward. **USES:** The Cheyenne used cold leaf tea to strengthen a weak heart, for chest pains due to cough, for colds, and in steam baths to induce sweating. The Chippewa used leaf tea for fevers and colds. **RELATED SPECIES:** The Chinese use *A. rugosa* (Fisch. & C. A. Mey.) Kuntze leaf tea for angina pains. Root tea used for coughs, lung ailments; antioxidant.

Blue Giant Hyssop, Anise-hyssop has an anise scent.

European Pennyroyal has a distinct, vinyl-like fragrance.

AMERICAN PENNYROYAL (NOT SHOWN)

Leaves

Hedeoma pulegioides (L.) Pers. Mint Family (Lamiaceae or Labiatae)

⚠ Aromatic, *vinyl-scented*, soft-hairy annual; 6–18 in. Leaves small, lance-shaped; toothed or entire. Bluish flowers in leaf axils; July–Oct. Calyx 2-lipped, with 3 short and 2 longer teeth. **WHERE FOUND:** Dry woods. QC to GA; AL to OK; NE to MI. **USES:** Leaf tea traditionally used for colds, fevers, coughs, indigestion, kidney and liver ailments, headaches; to promote sweating, induce menstruation, expectorant; insect repellent. Used essentially for the same purposes as European Pennyroyal *Mentha pulegium* L. (naturalized in NJ, PA, and MD; also Pacific Coast) to stimulate menstruation and as a repellent for insects, thanks to pulegone content. The same warnings apply to both plants, and their use is not recommended. **WARNING:** Pulegone, the active insect repellent compound in essential oil, is absorbed through the skin and converted into a dangerous liver cancer–inducing compound. Ingesting essential oil can be lethal; contact with essential oil (a popular insect repellent) can cause dermatitis. Components of essential oil may be particularly dangerous to epileptics.

HYSSOP

Leaves

Hyssopus officinalis L. Mint Family (Lamiaceae or Labiatae)

Bushy, aromatic perennial; 1–2 ft. Leaves opposite; lance-shaped to linear; *stalkless, entire* (not toothed). Purple, bluish, or pink flowers in whorls of leaf axils, forming small spikes; June–Oct. **WHERE FOUND:** Dry soils. Locally abundant especially in the Northeast and adjacent Canada. Alien (Europe); widely grown in herb gardens. **USES:** Traditionally, leaf tea was gargled for sore throats. Tea thought to relieve gas, stomachaches, loosen phlegm; used with Horehound (p. 98) for bronchitis, coughs, and asthma. The herb has been used externally to treat rheumatism, muscle aches, wounds, and sprains. Experimentally, extracts are useful against herpes simplex; anti-inflammatory. Contains at least 8 antiviral compounds. In 1990 researchers found that a Hyssop extract inhibited replication of human immunodeficiency virus. A polysaccharide was identified in 1995 that inhibited the SF strain of HIV-1 (dose-dependent), preventing replication of the virus. Hyssop contains antioxidant components, such as rosmarinic acid and other compounds, which are associated with antiviral, anti-inflammatory, and antioxidant activity. Therapeutic claims are not permitted in Germany because traditional uses are not substantially researched, though it's certainly a medicinal plant deserving of more research attention.

CALAMINT — Leaves

Clinopodium arkansanum (Nutt.) House — Mint Family
[*Satureja arkansana* (Nutt.) Briq., — (Lamiaceae or Labiatae)
Calamintha arkansana (Nutt.) Shinners,
Clinopodium glabrum (Nutt.) Kuntze, *Hedeoma arkansana* Nutt.]

Botanists can't seem to agree on a single scientific name for this plant. Creeping perennial; 4–8 in. Leaves oval at base of plant; stem leaves linear. Leaves *strongly pennyroyal-scented*. Flowers purplish, 2-lipped;

Hyssop is mostly cultivated, locally naturalized.

Calamint has a pennyroyal fragrance.

Apr.–July. **WHERE FOUND:** Rocky glades. W. NY to AR, TX; north to IL, IN. **USES:** Used as a substitute for American Pennyroyal (see above). Given its localized abundance on glades and rocky outcrops, it was undoubtedly used by Native Americans as a readily available insect repellent. **WARNING:** Contains pulegone in essential oil. The same warning applies as for American and European Pennyroyal.

SQUARE STEMS, PAIRED LEAVES; NO STRONG MINTY AROMA

DOWNY WOODMINT — Leaves

Blephilia ciliata (L.) Benth. — Mint Family (Lamiaceae or Labiatae)

Perennial; 10–26 in. Leaves oblong-oval to lance-shaped, *downy beneath*; stalkless, on flowering stems. Flowers pale bluish purple, in terminal and axillary whorls; June–Aug. Calyx 2-lipped, with bristly teeth; lower lip of corolla narrower than lateral lobes. **WHERE FOUND:** Dry woods, clearings. VT to GA; e. TX to MN. **USES:** Cherokee used poultice of fresh leaves for headaches. Taste bitter, not strongly aromatic; once suggested as a diuretic and antiseptic.

GROUND IVY, GILL-OVER-THE-GROUND — Leaves

Glechoma hederacea L. — Mint Family (Lamiaceae or Labiatae)

Creeping, ivylike perennial. Leaves *scallop-edged*, round to kidney-shaped; sometimes tinged with purple. Two-lipped violet flowers, in whorls of leaf axils; Mar.–July. **WHERE FOUND:** Roadsides, lawns.

LEFT: *Downy Woodmint has whitish to bluish purple flowers.* ABOVE: *Ground Ivy has rounded leaves with scalloped edges.*

Throughout our area. Widespread European alien. **USES:** In English tradition, a well-known folk herb for lung afflictions and fevers. Traditionally, leaf tea used for lung ailments, asthma, jaundice, kidney ailments, blood purifier. Externally, a folk remedy for cancer, backaches, bruises, hemorrhoids. Alcohol extracts are anti-inflammatory and antioxidant and reduce edema. Two components in the plant were found to protect mice from ulcers. Ursolic acid in leaves experimentally anticancer against lymphocytic leukemia and human lung carcinoma. **WARNING:** Reportedly toxic to horses, causing throat irritation and labored breathing. Also reported in humans. In one case the fresh leaves were steeped in ½ cup of hot water for 10 minutes and then drunk. Within 5 minutes tea produced swelling of throat and labored breath, and resulted in difficulty sleeping that night. Symptoms abated in 24 hours.

HEAL-ALL, SELF-HEAL — Whole plant

Prunella vulgaris L. — Mint Family (Lamiaceae or Labiatae)

Low perennial; to 1 ft. Leaves oval to lance-shaped; mostly smooth; opposite, on a weakly squared stem. Purple flowers *crowded on a terminal head*; hooded, with *a fringed lower lip*; May–Sept. **WHERE FOUND:** Waste places, lawns. Throughout our area. A presumed Eurasian alien, evidence suggests some variants represent native genetic material. **USES:** Traditionally, leaf tea was used as a gargle for sore throats and mouth sores, also for fevers, diarrhea; externally, for ulcers, wounds, bruises, sores. Long a folk remedy in Europe for wounds and ulcers, with a reputation for speeding healing. In China, a tea made from the flowering plant is considered cooling. The plant was also used in China to treat heat in the liver and aid in circulation; used for conjunctivitis, boils, and scrofula; diuretic for kidney ailments. Most modern research on the herb has been conducted in relation to use and potential in Traditional Chinese Medicine. Research suggests the plant possesses antibiotic, hypotensive, and antimutagenic qualities. Recent studies suggest possible therapeutic potential in inhibiting diabetes-related vascular conditions, reducing septic conditions of the blood; chemopreventive activity and other effects. Contains the antitumor and diuretic compound ursolic acid. Also rich in natural antioxidant components, containing more rosmarinic acid than Rosemary itself.

Heal-all, Self-heal is a common weed on lawns.

LYRE-LEAVED SAGE, CANCERWEED

Roots, leaves

Salvia lyrata L. — Mint Family (Lamiaceae or Labiatae)

Lyre-leaved Sage, Cancerweed blooms in spring. Leaves are lyre-shaped, hence the name.

Perennial; to 1 ft. Leaves mostly basal, oblong, *cleft* (dandelion-like); *edges rounded*. Purple-blue flowers, to 1 in., in whorled spikes; Apr.–June. **WHERE FOUND:** Sandy soils, lawns. PA to FL; TX to se. KS, IL. **USES:** The Catawba used root in salve for sores. Cherokee used whole-plant infusion for colds, coughs, nervousness, as a mild diaphoretic, and for mild diarrhea. Folk remedy for cancer and warts. In the colonial South, the fresh leaves were bruised and applied to warts, refreshed every 12 hours to destroy them.

LEGUMES (PEA FAMILY)

HOG-PEANUT

Root, beans

Amphicarpaea bracteata (L.) Fern. — Pea Family (Fabaceae or Leguminosae)

Delicate, twining, herbaceous annual vine, with soft-hairy stems. Leaves 3-divided, 1–3¾ in. long, oval to triangular-oval, the *middle leaflet slightly larger than asymmetrical lateral leaflets*; margins entire. Bears two kinds of flowers, above ground and below. Aboveground pealike flowers pale purple to white, 1–15 in drooping racemes in leaf axils; July–Aug. Flattened-oblong seedpods, twisting when dry, with 3–4 lentil-like seeds. *Petalless flowers on subterranean stems* produce a one-seeded fruit suggestive of a lima bean. **WHERE FOUND:** Rich woods and thickets. Throughout our range. **USES:** Cherokee used root tea to treat diarrhea; steam from tea blown on wounds from snakebite. Chippewa used in herbal combination as a laxative. The

Hog-peanut is an herbaceous annual with small, pale purple pea-like flowers.

subterranean fruit (nut) was eaten by various Native American groups. Voles and other rodents store in heaps in winter, up to a pint in volume; collected by the Pawnees and Dakota women, who would replace the rodents' stores with an offering of corn. An important food of prairie native groups. Contains genistein (see Kudzu). Root of an Asian counterpart, *A. bracteata* (L). Fern. var. *edgeworthii* (Benth.) H. Ohashi [*A. edgeworthii* Benth., *A trisperma* (Miq.) Baker], used in Chinese medicine to relieve pain. Root and whole plant are used to clear food stagnation in digestive tract and clear toxins. Root used for indigestion. Whole-plant tea used for night sweats and recurring sores.

PURPLE PRAIRIE-CLOVER

Leaves, root

Dalea purpurea Vent. Pea Family (Fabaceae or Leguminosae)
[*Petalostemon purpureus* (Vent.) Rydb.]

Perennial; 1–2 feet. Leaves numerous, densely crowded on stems; 3–7 (usually 5) leaflets. Flowers purple, tiny, *densely crowded on a conelike or cylindrical* head. **WHERE FOUND:** Dry prairies, glades. IN, KY, AR; TX west to NM. **USES:** The Chippewa used the leaves and flowers in tea to treat heart problems. The Meskwaki used the root to treat measles and pneumonia; the Pawnee, as a general preventive to disease. Several Native American groups chewed the pleasant-tasting root like gum. Strongly antibacterial. Contains various isoflavones.

Purple Prairie-clover has thimblelike flowerheads.

ALFALFA

Flowering plant

Medicago sativa L. Pea Family (Fabaceae or Leguminosae)

⚠ Deep-rooted perennial; 1–3 ft. Leaves cloverlike, but leaflets elongate. Violet-blue flowers in loose heads ¼–½ in. long; Apr.–Oct. Pods *loosely spiral-twisted.* **WHERE FOUND:** Fields, roadsides. Throughout our area. Often cultivated, escaped. Alien (Eurasia). **USES:** Best known as animal fodder; absent from historical herbals; evolved as an herbal health food in the early twentieth century. Nutritious fresh- or dried-leaf tea traditionally used to promote appetite, weight gain; conversely weight loss; diuretic; stops bleeding. Experimentally, antifungal and estrogenic. Unsubstantiated claims include use for cancer, diabetes, alcoholism, and arthritis. A source of commercial chlorophyll and carotene, both with valid health claims. Contains the antioxidant tricin. **WARNING:** Consuming large quantities of Alfalfa saponins may cause breakdown of red

Alfalfa is often found near cultivated fields.

Blue False Indigo, with brilliant blue-violet flowers.

blood cells, causing bloating in livestock. Recent reports suggest that Alfalfa sprouts (or the canavanine therein, especially in seeds) may be associated with lupus (systemic lupus erythematosus), causing recurrence in patients in which the disease had become dormant.

BLUE FALSE INDIGO — Root

Baptisia australis (L.) R. Br. ex Ait. f. Pea Family (Fabaceae or Leguminosae)

Smooth perennial; 3–5 ft. Leaves thrice-divided, cloverlike; leaflets *obovate* (wider at tips). Deep blue to violet flowers, to 1 in. long, on *erect racemes*; Apr.–June. **WHERE FOUND:** Open woods, forest margins, thickets. PA to GA; TX to OK, NE, s. IN. **USES:** A hot tea of the root was used by the Cherokee for toothache, as an emetic and purgative; externally as a poultice to reduce inflammation and treat septic conditions. Cold tea used to alleviate nausea. Like other *Baptisia* species, *B. australis* is currently under investigation as a potential stimulant of the immune system. Contains various saponins. **WARNING:** Considered potentially toxic.

WILD LUPINE — Leaves

Lupinus perennis L. Pea Family (Fabaceae or Leguminosae)

Perennial; 1–2 ft. Leaves long-stalked; *divided into 7–11 oblong, lance-shaped segments*. Flowers blue to violet, pealike; in a showy raceme; Apr.–July. **WHERE FOUND:** Native to dry soils, open woods. Sw. ME, NY

LEFT: *Wild Lupine, with typical blue flowers.* ABOVE: *Wild Lupine often spreads by seed to form large, showy displays.*

to FL; WV, OH, IN, IL. Often cultivated, along with many hybrids. **USES:** Cherokee drank cold leaf tea to treat nausea and internal hemorrhaging. Fed to horses to fatten them and make them "spirited and full of fire." Seeds bitter, inducing flatulence. Adding lupine seed powder to Rainbow Trout feed led to a reduction of infections of the bacterium *Aeromonas hydrophila,* often found in fresh water, sometimes producing gastroenteritis or other diseases in humans. **WARNING:** Seeds can be poisonous; may contain potentially toxic alkaloids. Some lupines are toxic, others are not. Even botanists may have difficulty distinguishing between toxic and nontoxic species.

FLAT-TOPPED CLUSTERS OR SPIKES; ASTER FAMILY

WILD LETTUCE (NOT SHOWN)

Root, stem, leaves

Lactuca biennis (Moench) Fernald

Aster Family
(Asteraceae or Compositae)

⚠ Smooth biennial; 2–15 ft. Leaves *irregularly divided*; coarsely toothed. Flowers bluish to creamy white (rarely yellow); July–Sept. **WHERE FOUND:** Damp thickets. NL to VA mtns.; TN to IA and westward. **USES:** Native Americans used root tea for diarrhea, heart and lung ailments; for bleeding, nausea, pains. Milky stem juice used for skin eruptions. Leaves applied to stings; tea sedative, nerve tonic, diuretic. **REMARKS:** Variable genus; highly technical taxonomy. **WARNING:** May cause dermatitis or internal poisoning.

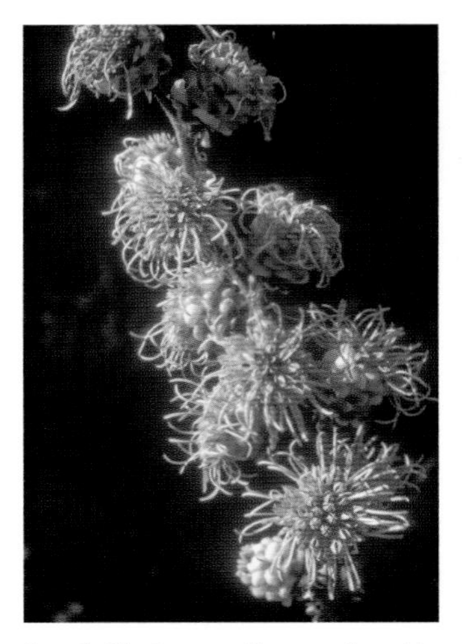

Rough Blazing-star flower spike with tightly packed flowerheads.

Dense Blazing-star has lance-shaped leaves on its flower spikes.

ROUGH BLAZING-STAR — Root

Liatris aspera Michx. — Aster Family (Asteraceae or Compositae)

Perennial; 6–30 in. Leaves alternate, linear. Rose-purple flowers *with 25–30 florets*, in crowded sessile or short-stalked heads on a crowded spike; Aug.–Sept. Note *wide, rounded bracts*. **WHERE FOUND:** Dry soils, prairies. OH to NC; LA, TX to ND. **USES:** Root tea of most *Liatris* species was used as a folk remedy for kidney and bladder ailments, gonorrhea, colic, painful or delayed menses; gargled for sore throats; root used externally in poultice for snakebites. Thought to be diuretic, tonic.

DENSE BLAZING-STAR, DENSE GAYFEATHER, MARSH BLAZING-STAR — Root

Liatris spicata (L.) Willd. — Aster Family (Asteraceae or Compositae)

Smooth-stemmed perennial; 15–43 in. tall with a globe-shaped corm. Lower leaves with 3–5 nerves, narrowly oblong-lanceolate, 4–15 in. long; *very narrow; much reduced above*. Flowers usually tightly crowded, spikelike arrays; bracts tightly surrounding individual flowerheads rounded or *with abrupt tip*; July–Oct. **WHERE FOUND:** Fields, roadsides, fencerows, moist prairies and glades, moist open woods. QC to GA, LA to WI. Our most widely cultivated native *Liatris*, often grown for cut flowers (called Florist's Gayfeather). **USES:** Cherokee used root tea as a warming stimulant, diuretic, expectorant; to induce sweating and relieve pain in colic, backache, dropsy. Menominee used in combination with other herbs for strengthening a weak heart. One of many snakebite remedies. Used in early nineteenth-century America for sore throat,

Ironweeds are challenging to identify because of hybridization.

breast pain, pain following childbirth, and angina. Root has a somewhat spicy, balsamic odor. All tuberous-rooted *Liatris* species were historically considered diuretic. **WARNING:** Reported to cause cross-reactive contact dermatitis.

IRONWEED Root

Vernonia glauca (L.) Willd. Aster Family (Asteraceae or Compositae)

Blue-green perennial; 2–5 ft. Leaves on stem only (not at base); oval to lance-shaped; narrowly sharp-pointed at tip and base. Flowers July–Oct. Seed crowns *yellowish* (brown-purple in other *Vernonia* species). **WHERE FOUND:** Rich woods. NJ to GA; AL to PA. **USES:** Cherokee used the root as a blood tonic, to regulate menses, relieve pain after childbirth; also for bleeding, stomachaches. **RELATED SPECIES:** Many of the 17 N. American species of *Vernonia* traditionally used similarly.

DAISYLIKE FLOWERS OR THISTLES

CHICORY Roots, leaves

Cichorium intybus L. Aster Family (Asteraceae or Compositae)

Biennial or perennial; 2–4 ft. Basal leaves dandelion-like; upper ones reduced. Flowers blue (rarely white or pink), *stalkless; rays square-tipped*; June–Oct. **WHERE FOUND:** Roadsides. Throughout our area. Alien (Europe). **USES:** An ounce of root in 1 pint of water used as a diuretic, laxative. Folk use in jaundice, skin eruptions, fevers. Extract diuretic, cardiotonic; lowers blood sugar, slightly sedative, and mildly laxative. Historically used for liver and gallbladder ailments. Leaf extracts weaker than root extracts. Experimentally, root extracts antibacterial, antioxidant, liver-protectant, and have significant wound-healing activity. In experiments, animals given Chicory root extracts exhibit a slower and weaker heart rate (pulse). Root extracts in alcohol solutions have proven anti-inflammatory effects. Root and leaves used in European phytomedicine for treatment of loss of appetite and dyspepsia. Roasted roots widely used as a coffee substitute and/or additive. **WARNING:** May cause rare allergic reactions.

Chicory has sky blue to violet flowers (rarely white).

MILK THISTLE

Seeds, whole plant

Silybum marianum (L.) Gaertn. Aster Family (Asteraceae or Compositae)

Annual or biennial thistle; to 6 ft. Leaves *mottled or streaked with white veins; sharp-spined,* clasping. Flowers purple tufts; receptacle densely bristle-spined; June–Sept. **WHERE FOUND:** Escaped from cultivation, common in CA. Alien (Europe). **USES:** Young leaves (with spines re-

Milk Thistle has white-mottled, spiny leaves.

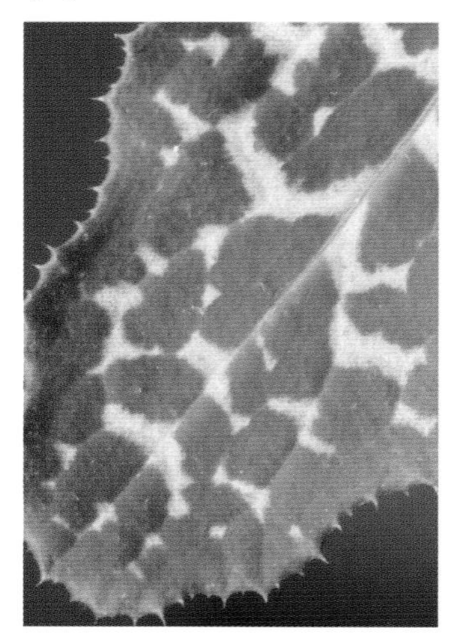

Milk Thistle is a European herb, occasionally naturalized.

moved) eaten as a vegetable. Traditionally, tea made from whole plant used to improve appetite, allay indigestion, and restore liver function. Used for cirrhosis, jaundice, hepatitis, and liver poisoning from chemicals or drug and alcohol abuse. Silymarin, a seed extract, dramatically improves liver regeneration in hepatitis, cirrhosis, mushroom poisoning, and other liver diseases. The science on the pharmacology and therapeutic effects of Milk Thistle is extensive. Components produce structural changes in outer cell membranes of liver cells (competing with toxin receptor points, preventing their entry into cell), have strong antioxidant and free radical scavenging effects, and stimulate the regeneration of liver cells. In the form of an intravenous preparation, silybin, a flavonoid complex of the seed, is clinically useful in Europe in treating severe Amanita mushroom poisoning. While used clinically in Europe, its use in the U.S. is not well known. Oral commercial preparations of the seed extracts are manufactured in Europe and widely available in the U.S. Used in European phytomedicine for the supportive treatment of chronic inflammatory liver disorders such as hepatitis, cirrhosis, and fatty infiltration caused by alcohol or other toxins. In addition to treating liver disease, it also has a preventive effect, helping to prevent liver damage from exposure to toxic chemicals.

NEW ENGLAND ASTER — Root

Symphyotrichum novae-angliae (L.) G. L. Nesom — Aster Family
(*Aster novae-angliae* L.) — (Asteraceae or Compositae)

Hairy-stemmed perennial; 3–7 ft. The showiest wild aster in our area. Leaves lance-shaped, without teeth; clasping stem. Flowers deeper violet than most asters, with up to 100 rays; Aug.–Oct. Bracts sticky.

New England Aster is one of our showiest asters.

WHERE FOUND: Moist meadows, thickets. S. Canada, ME to uplands of NC, AR, KS; CO to ND. **USES:** Cherokee used a root poultice to treat pain; root tea for diarrhea and fevers. Iroquois also used root tea for fevers. The Potawatomi and Meskwaki used plant smudge to revive an unconscious patient. In the early nineteenth century the Shakers of New Lebanon, NY, used a tea of the roots for the treatment of skin eruptions and for the treatment of Poison Sumac [*Toxicodendron vernix* (L.) Shafer (*Rhus vernix* L.)]. Contains various chlorogenic acids associated with anti-inflammatory, antiviral, and antioxidant activity.

NARROW-LEAVED PURPLE CONEFLOWER

Root, whole plant

Echinacea angustifolia DC. Aster Family (Asteraceae or Compositae)

Tap-rooted perennial; 6–20 in. Leaves lance-shaped, stiff-hairy. Flowers with prominent cone-shaped disk surrounded by pale to deep purple spreading rays; June–Sept. *Rays about as long as width of disk* (to 1¼ in.) **WHERE FOUND:** Prairies. TX, w. OK, w. KS, NE; west to e. CO, e. MT, ND, MB, SK. **USES:** Plains Indians used *Echinacea* for more medicinal purposes than any other plant group. Root (chewed or in tea) used for snakebites, spider bites, cancers, toothaches, burns, hard-to-heal sores and wounds, colds, and the flu. Science confirms many traditional uses, plus cortisone-like activity and insecticidal, bactericidal, and immunostimulant activities. Considered a nonspecific immune system stimulant. More than 300 pharmaceutical preparations are made from *Echinacea* plants in Germany, including extracts, salves, and tinctures; used for wounds, herpes sores, canker sores, throat infections; preventive for influenza, colds. A folk remedy for brown recluse spider bites. When introduced into medicine in the late nineteenth century—before the development of antibiotics—*Echinacea angustifolia* preparations were used for sepsis of the blood, gangrene, and other serious bacterial and viral infections for which there were no effective treatments; prescribed by physicians from 1895 to 1925 more than any other American medicinal plant. **REMARKS:** Hybrids occur where the range of this species overlaps that of Pale Purple Coneflower (*E. pallida,* below), though not necessarily hybrids between *E. angustifolia* and *E. pallida.*

PALE PURPLE CONEFLOWER

Root

Echinacea pallida Nutt. Aster Family (Asteraceae or Compositae)

Similar to *E. angustifolia* (above), but larger—to 40 in. Rays *strongly drooping, to 4 in. long.* Flowers Apr.–Aug. **WHERE FOUND:** Prairies,

Narrow-leaved Purple Coneflower produces short ray flowers.

Pale Purple Coneflower. Its pale to deep purple ray flowers grow to 4 in. long.

glades. AR to WI, MN; e. OK, KS, NE. **USES:** Same as for *E. angustifolia,* though some consider this plant less active. Ironically, the root of *E. pallida* is approved for use in Germany as an immunostimulant for preventing and reducing cold and flu symptoms, while *E. angustifolia* is unapproved. It appears that most chemical and pharmacological studies thought to have been conducted on *E. angustifolia* prior to 1988 were actually conducted on *E. pallida,* emphasizing the importance of proper identification of plant material used in scientific research.

PURPLE CONEFLOWER

Root, whole flowering plant

Echinacea purpurea (L.) Moench — Aster Family (Asteraceae or Compositae)

⚠ Perennial; 2–3 ft. Leaves *oval, coarsely toothed.* Bristle tips of flower disks *orange.* Rays typically purple (sometimes white). Flowers June–Sept. **WHERE FOUND:** Open woods, thickets; cultivated in gardens. MI, OH to LA, e. TX, OK. Widely grown as a flower garden perennial. **USES:** Same as for *E. angustifolia.* Widely used in Europe, but not native there. Most commercial *Echinacea* preparations utilize extracts of aboveground parts and/or roots of *E. purpurea.* Extracts enhance particle ingestion capacity of white blood cells and other specialized immune system cells, increasing their ability to attack foreign particles, such as cold or flu viruses. Experimental and clinical studies showed significant immune-system-stimulating activity with orally administered extracts of *E. purpurea, E. angustifolia,* and *E. pallida.* Several clinical studies have revealed that *E. purpurea* reduces severity and duration of cold and flu symptoms, while other clinical studies have shown little or no benefits. Different preparations (water versus alcoholic extracts) may have different active components. Dosage, plant part or species

LEFT: *The leaves of the Purple Coneflower are oval (rather than lance-shaped) and coarsely toothed.* ABOVE: *Purple Coneflower, a widely cultivated native plant, produces orange-tipped spines in the flowerhead.*

used, and type of preparation may all affect the outcome of clinical studies. Cichoric acid, polysaccharides, alkylamides, and other compounds have attributed immunostimulating activity. Tops (not roots) used in various European countries as an immunostimulant for colds and flu; externally for hard-to-heal wounds and sores. In topical preparations, *Echinacea* antioxidant activity reduces degradation of skin when exposed to sunlight. **REMARKS:** Wild Quinine root (*Parthenium integrifolium*) was often used as an adulterant to the root (sold in dried form) of Purple Coneflower (*E. purpurea*). See p. 267. **WARNING:** *Echinacea* and other immunostimulants are not used in cases of autoimmune disease or impaired immune response, including tuberculosis, multiple sclerosis, and HIV infection. Rare allergic reactions have been reported from *Echinacea* use. *Echinacea* is contraindicated in fighting HIV because it appears to increase tumor necrosis factor, which is associated with a poorer prognosis for AIDS patients. *Echinacea* preparations have also been implicated in interfering with enzyme systems that affect the absorption of some prescription drugs.

DRAGON OR GREEN ARUM

Root

Arisaema dracontium (L.) Schott — Arum Family (Araceae)

⚠ Perennial; 1–3 ft. Leaf solitary; divided into 5–15 lance-shaped leaflets along a horseshoe-shaped frond. Spathe sheathlike, *narrow; spadix much longer*. Flowers Apr.–July. **WHERE FOUND:** Rich, moist woods. Sw. QC, VT to FL; TX, e. KS, NE, WI, MI. **USES:** The Menominee dried and aged root for "female disorders"; also an item in sacred bundles. Root considered edible once it has been dried, aged, and elaborately processed. Root used as a substitute for *Arisaema triphyllum* (below). **RELATED SPECIES:** The Chinese use related *Arisaema* species for epilepsy, hemiplegia (paralysis); externally, as a local anesthetic or in ointment for swellings and small tumors. **WARNING:** Whole fresh plant contains intensely burning, irritating calcium oxalate crystals.

JACK-IN-THE-PULPIT

Root

Arisaema triphyllum (L.) Schott — Arum Family (Araceae)
[*Arisaema atrorubens* (Aiton) Raf.]

⚠ Variable perennial; 1–2 ft. Leaves 1–2; 3 leaflets each; green beneath. Spathe *cuplike,* with a *curving flap*; green to purplish brown, often striped. Flowers Apr.–early July. Berries clustered, scarlet. **WHERE FOUND:** Moist woods. Most of our area. **USES:** Cherokee used the dried, aged root for colds, sore throat, dry coughs, and tuberculosis; externally in an ointment (with lard) for ringworm and sores; poulticed for boils. Iroquois used cut root (in whiskey) for colds and bronchitis; antidiarrheal, and to build blood. Externally, the root was poulticed for

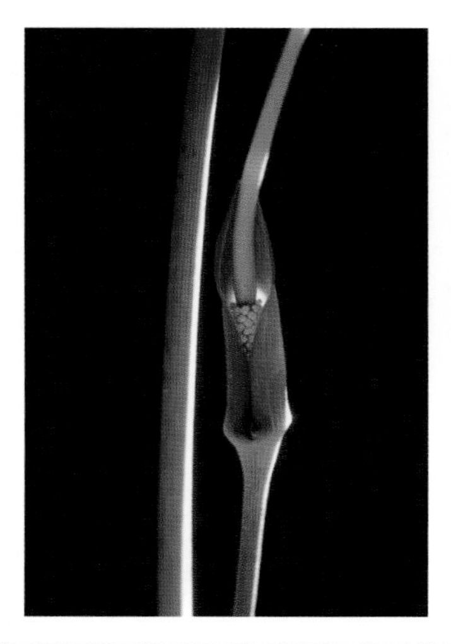

LEFT: *Dragon or Green Arum has a long green flower spathe.* ABOVE: *The leaves of the Dragon or Green Arum grow in a horseshoe pattern.*

ABOVE: *Jack-in-the-pulpit is easily identified when in bloom by its wide, green to reddish spathe.* TOP RIGHT: *Skunk Cabbage. Its unique flowers range from green to red.* RIGHT: *Skunk Cabbage produces large cabbagelike leaves.*

rheumatism, scrofulous sores, boils, abscesses, and ringworm. Roasted seeds were eaten. Dried-root tea traditionally considered expectorant, diaphoretic, and purgative. Historically used for asthma, bronchitis, colds, cough, laryngitis, and headaches. Externally for rheumatism, boils, and swelling from snakebites. **RELATED SPECIES:** The Chinese used related species to treat snakebites. **WARNING:** Intensely irritating. Calcium oxalate crystals found in whole fresh herb; caustic and irritating.

SKUNK CABBAGE

Root

Symplocarpus foetidus (L.) Salisb. ex Nutt. Arum Family (Araceae)

Strongly skunk-scented perennial; 1–2 ft. Leaves broad, oval. Flowers appear before leaves, Feb.–May; *greenish to purple, hooded, sheathing spathe, with a clublike organ within.* **WHERE FOUND:** Wet, rich woods. NS to GA; TN; IL to IA. One of the first spring wildflowers, it often grows in melting snow because of the thermogenesis of salicylic acid and salicylates in the flower, essentially active as a biochemical energy generator. Temperature within the flower spathe is often 60°F higher than the ambient air. **USES:** The Chippewa and other Native American groups used preparations of the root as a cough medicine, for whooping cough, tuberculosis, and for cramps, among other maladies; externally, leaf or root poultice used for pain, wounds, underarm deodorant; leaf poulticed to reduce swelling. Dried root nibbled to stop epileptic seizures. Micmac Indians sniffed the root to relieve migraines. This unusual plant and its claims as a cure for asthma attracted the attention of physicians in colonial America, who used root preparations as antispasmodic for epilepsy, spasmodic coughs, asthma; also used as a diuretic and emetic. Externally root used in lotions to treat itching and rheumatism. **WARNING:** Eating leaves causes burning, inflammation. Roots considered toxic, except in controlled dosages; small amounts may cause vomiting, headache, vertigo, with historic claims of temporary blindness and narcotic effects.

MISCELLANEOUS GREEN-FLOWERED VINES

WILD YAM

Roots

Dioscorea villosa L.

Yam Family (Dioscoreaceae)

Highly variable twining vine; stem smooth. Leaves alternate (lower ones in whorls of 3–8), heart-shaped, hairy beneath; *veins conspicuous*. Flowers not showy; male and female flowers separate; May–Aug. *Three-winged fruit* prominent in autumn. **WHERE FOUND:** Wet woods. CT to TN; TX to MN. **USES:** The Meskwaki used root tea to relieve labor

ABOVE: *Wild Yam leaves have conspicuous veins.* RIGHT: *Note the prominent 3-winged Wild Yam fruit in autumn.*

pains and as an aid in childbirth. Fresh dried root (tea) formerly used by physicians for colic, gastrointestinal irritations, morning sickness, asthma, spasmodic hiccough, rheumatism, and "chronic gastritis of drunkards." Once widely known by the name "Colic Root." Contains diosgenin, formerly used as the starting material to manufacture progesterone and other steroid drugs. Of all plant genera, there is perhaps none with greater impact on modern life but whose dramatic story is as little known as *Dioscorea.* Most of the steroid hormones used in modern medicine, especially those in contraceptives, were developed from elaborately processed chemical components derived from *Dioscorea* species. Drugs made with yam-derived components (diosgenins) relieve asthma, arthritis, eczema; they also regulate metabolism and control fertility. Synthetic products manufactured from diosgenins include human sex hormones (contraceptive pills), drugs to treat menopause, dysmenorrhea, premenstrual syndrome, testicular deficiency, impotency, prostate hypertrophy, and psychosexual problems, as well as high blood pressure, arterial spasms, migraines, and other ailments. Widely prescribed cortisones and hydrocortisones were indirect products of the genus *Dioscorea.* They are used for Addison's disease, some allergies, bursitis, contact dermatitis, psoriasis, rheumatoid arthritis, sciatica, brown recluse spider bites, insect stings, and other diseases and ailments. Wild Yam has appeared in the American market in recent years as a "source" of estrogen or progesterone, prompting some to call this marketing effort the "wild yam scam," since the root does not contain human sex hormones, and some products allegedly simply added progesterone to creams to obtain the desired results. **REMARKS:** Plants formerly classified as *D. quaternata* G. F. Gmel., which occupy the same range as *D. villosa,* are now treated as *D. villosa.* **WARNING:** Fresh plant may induce vomiting and other undesirable side effects.

CINNAMON VINE, CHINESE YAM

Roots, bulblet

Dioscorea polystachya Turcz.
(*Dioscorea batatas* Decne.)

Yam Family (Dioscoreaceae)

Climbing, twining vine, stems turning clockwise; to 15 ft. Leaves on long stems, arrow-shaped, with blunt-rounded lobes at base, heart-shaped at base, with 7–9 primary veins. Flowers axillary, mostly small and inconspicuous; May–Aug. Primarily propagates and spreads by *starchy-round bulblets in leaf axils* ("air potatoes"). **WHERE FOUND:** Thickets, creek bottoms, woods, roadsides, fencerows, waste places. Alien (Asia). PA to n. FL; AR, MO to IL, IN. **USES:** Dried peeled rhizome used in Traditional Chinese Medicine to treat diarrhea, chronic dysentery, diabetes, frequent urination, and cough due to general debility; nutritive tonic. Produces a large potato-like edible root; used as a survival food; cultivated in China as a food plant. Root contains a protein, dioscorin, which is antioxidant and enhances storage of roots. Root extracts found to reduce plasma glucose and insulin levels, reducing fat tissue in animal experiments; anti-inflammatory.

ABOVE: *Cinnamon Vine has starchy bulbils in leaf axils.* TOP RIGHT: *Hops. Its fruiting bodies, or strobiles, produce a yellow crystalline resin within.* RIGHT: *Hops. Note the 3-lobed leaves.*

HOPS

Fruits (strobiles)

Humulus lupulus L.

Hemp Family (Cannabaceae)

Rough-prickly, twining perennial. Leaves mostly with 3–5 lobes; sinuses (notches) rounded; yellow resinous granules beneath. Male and female flowers on separate plants; July–Aug. Fruits (strobiles) *inflated*. **WHERE FOUND:** Waste places. Throughout our area. Four varieties are now recognized in N. America, 3 of which are considered native. The typical European *H. lupulus* var. *lupulus* is cultivated and escaped. **USES:** Tea of fruits (strobiles) traditionally used as sedative, antispasmodic, diuretic; for insomnia, cramps, coughs, fevers; externally, for bruises, boils, inflammation, rheumatism. Experimentally antimicrobial, relieves spasms of smooth muscles; acts as sedative (disputed). Hops contains several sedative and pain-relieving components. Hops-picker fatigue is a condition believed to result from release of the essential oil during harvest. Used to relieve mood disturbances, nervous tension, anxiety, and unrest. Used in European phytomedicine to treat discomfort from restlessness or anxiety and sleep disturbances. Considered calming and helpful in promoting sleep. **RELATED SPECIES:** Japanese hops, *Humulus scandens* (Lour.) Merr. (*H. japonicus* Siebold & Zucc.), is a weedy annual with 5–9 leaf lobes. Leaves much rougher than common hops. Naturalized from New England to NC west to KS. **WARNING:** Handling plant often causes dermatitis. Dislodged hairs may irritate eyes. Crystalline resin in fruits causes rare allergic reactions. Hops are the famous bitter flavoring of beer.

MISCELLANEOUS PLANTS WITH GREEN FLOWERS

GREEN ANTELOPEHORN, SPIDER MILKWEED, GREEN MILKWEED — Root

Asclepias viridis Walter — Milkweed Family (Asclepidaceae)
[Modern works place in Dogbane family (Apocyhaceae)]

⚠ Robust, thick-stemmed, often sprawling, milky-sapped perennial; to 2 ft. tall. Leaves alternate, elliptical-oblong, fleshy, up to 5 in. long and 2 in. wide; margins wavy. Flowers in a large terminal cluster, up to 5 in across; individual flowers, large, up to 1 in. across, with *5 greenish petals, and purple structures within*; May–July. **WHERE FOUND:** Rocky wooded openings, prairies, glades, roadsides. OH, WV, TN to FL; west to TX, north to NE. **USES:** Creek boiled roots for use in kidney problems, such as kidney stones. Despite its showy nature, this milkweed, sometimes called Ozark Milkweed, may have been little used because of its potential toxicity, which was noted to be more powerful than butterflyweed. **WARNING:** Contains toxic components. One report documents 20 sheep that died one at a time after consuming the plant in the summer of 1999.

Green Antelopehorn is a sprawling, large milkweed.

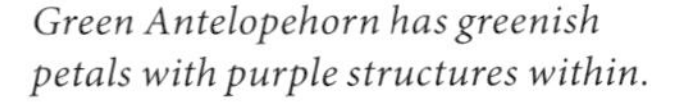

Green Antelopehorn has greenish petals with purple structures within.

CANNABIS, HEMP, MARIJUANA

Leaves, seeds, flowering tops

Cannabis sativa L. — Hemp Family (Cannabaceae)

⚠ Annual weed; 5–14 ft. Leaves *palmate, with 5–7 lobes.* Leaflets lance-shaped, toothed. Male and female flowers on separate plants; female flowers greenish, *sticky and resinous*; Aug.–Sept. **WHERE FOUND:** Escaped or cultivated throughout our area. Alien. **USES:** Leaves smoked or eaten, depending upon local, state, or national laws, as an illegal intoxicant, prescribed as medicine, or enjoyed as a legal high. Legitimate medical use to treat glaucoma; relieve pain and anxiety; alleviate nausea following chemotherapy. Antibiotic for gram-positive bacteria. Many folk uses. Much maligned, but potentially a very useful medicinal plant. We are jointly convinced that the whole leaf is better than the sum of its parts—that whole marijuana is, through synergy, safer, more effective, and cheaper than its isolated silver-bullet active compound, THC, and its various semisynthetic analogs. Further, we believe that legalized marijuana would generate more funds for the government and less organized crime. A legitimate fiber (hemp) and oil-seed plant in many other countries. **WARNING:** Human society has a schizophrenic relationship with this plant. In some modern countries, mere possession of a tiny amount of the plant can result in life imprisonment or a death sentence, whereas in other countries, one can openly buy and smoke the plant in public. In the U.S. legality and enforcement vary from state to state, even municipality to municipality. We are entering a period in which society is slowly rethinking attitudes toward this plant, resulting in decriminalizing the possession and use of marijuana in some states. The laws banning its use and unwarranted negative social attitudes related to the plant are far more dangerous than the effects of the plant itself.

Marijuana, with its iconic leaves, grown openly among sunflowers in a front yard in the Swiss Alps.

BLUE COHOSH

Root

Caulophyllum thalictroides (L.) Michx. — Barberry Family (Berberidaceae)

⚠ Perennial; 1–2 ft. Smooth-stemmed; stem and leaves covered with bluish film. Leaves divided into 3 (occasionally 5) *leaflets with 2–3 lobes.* Flowers greenish yellow, in terminal clusters; Apr.–June, before leaves expand. **WHERE FOUND:** Moist, rich woods. NB to SC; AR to ND, MB. Like Black Cohosh, coming under increasing collecting pressures. **USES:** Root tea used by Cherokee to aid labor, for nervous conditions and colic, to reduce inflammation, and to treat rheumatism. The Chippewa used a root decoction for bleeding from the lungs, as an emetic, and to treat digestive cramps. The Iroquois used root for fevers and as a tonic. The Menominee and the Meskwaki used a root tea to treat profuse menstruation. A folk remedy for rheumatism, cramps, epilepsy, and inflammation of the uterus. Championed by C. S. Rafinesque in his 1830 *Medical Flora,* with the poignant point that it was widely used by the Indians and their Indian doctor imitators, referring to a class of itinerant folk physicians who claimed to learn their trade from Native Americans. Historically prescribed by physicians for chronic uterine diseases. Said to cause abortion by stimulating uterine contractions. Roots possess estrogenic activity and check muscle spasms. Studies by scientists in India suggest the root may possess some contraceptive potential. Extracts shown to be anti-inflammatory (in rats). An alkaloid in the root, methylcytisine, has effects similar to those of nicotine, increasing blood pressure, stimulating the small intestine, and causing hyperglycemia. It also contains glycosides, which are believed to be responsible for its uterine-stimulant activity and to constrict coronary blood vessels. Herbalists and midwives continue to use the plant to induce labor, but safety is questioned. **REMARKS:** Most northern populations from s. QC to WV mtns., north to e. MI, are now recognized as *C. giganteum* (Farw.) H. Loconte & W. H. Blackw., which as the name implies is generally a larger plant, with distinctly purple to red to greenish flowers. **WARNING:** Root powder strongly irritating to mucous membranes. Avoid during pregnancy.

Blue Cohosh produces green-yellow flowers.

WILD IPECAC (NOT SHOWN)

Leaves, root

Euphorbia ipecacuanhae L. Spurge Family (Euphorbiaceae)

⚠ Large-rooted perennial with underground stems; 3–12 in. tall. Stems smooth, succulent. Leaves inserted at joints; rounded to linear, green to purple. Solitary flowers on long stalks; "cups" have 5 glands, with narrow white, yellow, green, or purple appendages; Apr.–May. **WHERE FOUND:** Sandy soil. Mostly coastal. NJ to FL. **USES:** Native Americans used leaf tea for diabetes; root tea as a strong laxative and emetic, for pinworms, rheumatism; poulticed root on snakebites. **WARNING:** Extremely strong laxative. Juice from fresh plant may cause blistering.

COLUMBO ROOT

Root

Frasera caroliniensis Walt.
[*Swertia caroliniensis* (Walt.) Ktze.]

Gentian Family (Gentianaceae)

Smooth biennial; 3–8 ft. Reduced leaves in 4s on stem; with large lance-shaped or oblong basal leaves. Flowers *greenish yellow with brown-purple dots*; 4-parted, with a *large, glandular, greenish dot* on each division; June–July. **WHERE FOUND:** Limey slopes, rich woods. NY to GA; LA to WI. **USES:** Root tea formerly used for colic, cramps, dysentery, diarrhea, stomachaches, lack of appetite, nausea; general tonic. **RELATED SPECIES:** Asian species have been used similarly.

CASTOR-OIL-PLANT, CASTOR BEAN

Seed oil

Ricinus communis L.

Spurge Family (Euphorbiaceae)

Large annual or perennial (in tropics); 5–12 ft. Leaves large, *palmate with 5–11 lobes.* Flowers in clusters—female ones above, male ones below; July–Sept. Seed capsule with soft spines; seeds mottled. **WHERE FOUND:** Escaped exotic cultivar, alien; widespread particularly in subtropics and tropics. **USES:** Seed oil famous since ancient Egyptian time as a purgative or laxative, folk remedy used to induce labor. Nauseous taste may induce vomiting. Oil used as a laxative to treat food poisoning and before X-ray diagnosis of bowels. Used externally for ringworm, itch, hemorrhoids, sores, abscesses, and hairwash for dandruff. Oil even suggested as a renewable energy resource. Poulticed boiled leaves are a folk remedy to produce milk flow. Jim Duke wishes his

Columbo Root produces distinctive, 4-parted, purple-spotted flowers with narrow sepals beneath.

Castor Bean seeds are mottled with brown streaks and are highly toxic.

Castor-oil-plant, Castor Bean is a large-leaved tropical annual with green to red seed capsules.

mother had heeded the warning "not to be administered to children under twelve years." His mother believed that castor oil cut with orange juice was a panacea. Ricin, the deadly poison found in the seeds, can be bioengineered to attach to monoclonal antibodies that attack only cancer cells, a technique reportedly tried in 1,000 cancer patients. The AIDS virus can infect an immune cell by locking onto its cell receptor protein, CD4. By genetically affixing the deadly poison ricin to genetically engineered CD4 proteins, one obtains CD4-ricin, which will lock on the external viruses of infected cells 1,000 times more often than on healthy cells. Using such techniques, one might possibly kill enough infected cells to prevent the disease from spreading and causing life-threatening symptoms. Ricinoleic acid has served as a component in contraceptive jellies. In some African countries, seed is used as a contraceptive. **WARNING:** Seeds are a deadly poison—1 seed may be fatal to a child. After oil is squeezed from seeds, the toxic protein ricin, which is deadly, remains in seed cake. Oil is used in industrial lubricants, varnishes, and plastics. May induce dermatitis. Ricin is a restricted chemical given its harmful potential.

AMERICAN WHITE OR FALSE HELLEBORE — Root

Veratrum viride Ait. — Melanthium Family (Melanthiaceae)
[Formerly placed in Lily Family (Liliaceae)]

Perennial; 2–8 ft. Leaves large, broadly oval, *strongly ribbed and pleated.* Flowers yellowish, turning dull green; small, *star*-shaped, in a many-flowered panicle; Apr.–July. **WHERE FOUND:** Wet wood edges, swamps. New England to GA mtns. TN to WI. **USES:** Historically valued as an analgesic for pain, epilepsy, convulsions, pneumonia; heart sedative; weak tea for sore throats, tonsillitis. The Shakers of NY manufactured a cardiac product from the plant prescribed by physicians into the 1930s. Used in pharmaceutical drugs to slow heart rate, lower blood pressure; for arteriosclerosis forms of nephritis. Powdered root used in insecticides. **WARNING:** All parts, especially root, are highly or fatally toxic. Leaves have been mistaken for Pokeweed and Marsh-marigold and eaten, with fatal results. The pleated basal leaves being mistaken for those of Ramps has caused poisonings in several cases. Recent medical literature on the plant describes cases of poisoning from mistaking the leaves for those of a wild edible, which highlights caution necessary in wild plant identification when intended for consumption.

Highly toxic American White or False Hellebore produces greenish flowers and pleated leaves.

FLOWERS IN SLENDER TERMINAL CLUSTERS

HORSEWEED, CANADA FLEABANE

Whole plant

Conyza canadensis (L.) Cronq. (*Erigeron canadensis* L.)

Aster Family (Asteraceae or Compositae)

Highly variable bristly annual or biennial weed; 1–7 ft. Leaves numerous, lance-shaped. Tiny (to ¼ in.) *greenish white flowers on many branches from leaf axils*; disk yellow, with short rays; July–Nov. **WHERE FOUND:** Waste places, roadsides. Throughout our area; has become a weed on other continents. **USES:** Plant tea used as a folk diuretic, astringent for diarrhea; treatment for "gravel" (kidney stones), diabetes, painful urination, hemorrhages of stomach, bowels, bladder, and kidneys; also used for nosebleeds, fevers, bronchitis, tumors, hemorrhoids, coughs. Africans used it for eczema and ringworm. Essential oil traditionally used for bronchial ailments and cystitis. Contains pain-relieving, antioxidant, spasm-relieving, antibacterial, antifungal, anticancer, and insecticidal components. **WARNING:** May cause contact dermatitis.

Horseweed is weedy with greenish flowers.

Alumroot produces small greenish white flowers.

ALUMROOT

Root, leaves

Heuchera americana L. — Saxifrage Family (Saxifragaceae)

Variable perennial; 1–2 ft. Leaves toothed, roundish to somewhat maple-shaped; base heart-shaped. Flowers small, greenish white; short-stalked on long wispy spike; Apr.–June. **WHERE FOUND:** Woods, shaded rocks. S. ON; CT to GA; OK to MI. **USES:** Cherokee used root tea for bowel complaints such as dysentery, menstrual problems, hemorrhoids; with honey for thrush, mouth sores; powered root used externally on hard-to-heal sores. Roots considered a strong astringent; similar to alum; styptic. Leaf tea used for diarrhea, dysentery, hemorrhoids; gargled for sore throats. Root poulticed on wounds, sores, abrasions. Other *Heuchera* species are used similarly.

FIGWORT, MARYLAND FIGWORT

Leaves, root

Scrophularia marilandica L. — Figwort Family (Scrophulariaceae)
[*Scrophularia nodosa* L. var. *marilandica* (L.) Gray]

Flowers like a *miniature scoop with reddish brown upper lip, greenish toward base*; June–Oct. **WHERE FOUND:** Rich woods. Sw. ME to n. GA; LA; TX to NE, MN. See p. 209.

Maryland Figwort produces tiny flowers that have a protruding upper petal.

FLOWERS IN AXILS; PLANTS WITH STINGING OR BRISTLY HAIRS

CANADIAN WOOD-NETTLE

Leaves

Laportea canadensis (L.) Wedd.
(*Urtica canadensis* L.)

Nettle Family (Urticaceae)

⚠ Herbaceous perennial; 15–40 in. tall. Stems and leaves with *stinging hairs*, often mistaken for stinging nettle without regard to identity. Leaves alternate, *broadly oval* in outline 3–6 in. long, on *long stalks*, coarsely toothed and hairy. Flowers small, whitish greenish, in loose, branching, elongate cluster in leaf axils, mostly longer then leaf petioles; June–Aug. **WHERE FOUND:** Rich, moist woods, often in large colonies in bottomlands. NS to FL; OK to MB. **USES:** Iroquois used an infusion of the roots as an emetic; treatment for tuberculosis; aid in childbirth. The Meskwaki and Ojibwa used root as a diuretic, also for urinary incontinence. Traditional use possibly mirrors use of Stinging Nettle root (see below) for treatment of benign prostatic hyperplasia in men. Leaves of a *Laportea* species used in Trinidad for treating diabetes. **WARNING:** Like Stinging Nettle, Wood-nettle has stinging, irritating hairs. Most species of *Laportea* are tropical from Cen. & S. Am., Africa, and Pacific Islands.

Wood-nettle, with its large leaves on long stalks, is common in bottomlands in the South.

Stinging Nettle is seldom noticed until the plant is touched.

Dwarf Nettle is a small annual with deeply sharp-toothed small leaves.

STINGING NETTLE

Whole plant

Urtica dioica L.

Nettle Family (Urticaceae)

⚠ Perennial; 12–50 in. *Stiff, stinging hairs.* Leaves opposite, mostly oval; bases sometimes barely heart-shaped. Flowers greenish, in branched clusters; June–Sept. Male and female flowers on separate plants or branches. The N. American genetic material, designated *U. dioica* ssp. *gracilis* (Ait.) Seland., with 6 varieties, differs from the European *U. dioica* ssp. *dioica* in that the European material has male and female flowers on separate plants. **WHERE FOUND:** Waste places, moist soils. Most of N. America; NL to AK and southward. European subspecies occasionally naturalized in our range. **USES:** Traditionally, leaf tea used in Europe as a blood purifier, blood builder, diuretic, astringent; for anemia, gout, glandular diseases, rheumatism, poor circulation, enlarged spleen, mucous discharges of lungs, internal bleeding, diarrhea, dysentery. Its effect involves the action of white blood cells, aiding coagulation and formation of hemoglobin in red blood corpuscles. Iron-rich leaves have been cooked as a potherb. Studies suggest CNS-depressant, antibacterial, and mitogenic activity; inhibits effects of adrenaline. This plant should be studied further for possible uses against kidney and urinary system ailments. Root used in European phytomedicine in treatments for prostate cancer. Russians are using the leaves in alcohol for cholecystitis (inflammation of the gallbladder) and hepatitis. Some people keep potted Stinging Nettle in the kitchen window, alongside an aloe plant, in the belief that an occasional sting alleviates arthritis. Leaves used in European phytomedicine for supportive treatment of rheumatism and kidney infections. Root preparations approved for symptomatic relief of urinary difficulties associated with early stages of benign prostatic hyperplasia, which affects a majority of men over

50 years of age. **RELATED SPECIES:** Other *Urtica* species occurring in N. America are said to be used interchangeably, such as Dwarf Nettle (*Urtica urens* L.), a small European annual with *deeply sharp-toothed small leaves*, occasional in much of N. America, except for prairie states. **WARNING:** Fresh plants sting. Dried plant (used in tea) does not sting. One fatality has been attributed, rightly or wrongly, to the sting of a larger tropical nettle. Histamine, acetylcholine, 5-hydroxytryptamine, small amounts of formic acid, leukotrienes, and other unknown compounds act together to produce the sting. Some of these compounds are neurotransmitters in the human brain.

COCKLEBUR

Leaves, root

Xanthium strumarium L. — Aster Family (Asteraceae or Compositae)

⚠ Variable weedy annual; to 5 ft. Leaves oval to heart-shaped, somewhat lobed or toothed, on long stalks. Flowers inconspicuous, green. Fruits oval, with crowded, *hooked prickles*; Sept.–Nov. **WHERE FOUND:** Waste places. Scattered. **USES:** Root historically used for scrofulous tumors (strumae—hence the species name). This plant and the related species *X. spinosum* L. (not shown) were formerly used for rabies, fevers, malaria; experimentally antioxidant, anti-inflammatory, diuretic, fever-reducing, sedative. Native Americans used leaf tea for kidney disease, rheumatism, tuberculosis, and diarrhea; also as a blood tonic. Chinese used it similarly. **REMARKS:** Taxonomy of *Xanthium* species is confusing. **WARNING:** Most Cocklebur species are toxic to grazing animals and are usually avoided by them. Seeds contain toxins, but seed oil has served as lamp fuel.

Cocklebur produces familiar characteristic burrs.

SLENDER, MOSTLY TERMINAL FLOWER CLUSTERS; BUCKWHEAT FAMILY

COMMON SMARTWEED, MILD WATER PEPPER (NOT SHOWN)

Leaves

Polygonum hydropiper L. — Buckwheat Family (Polygonaceae)
[*Persicaria hydropiper* (L.) Delarbre]

Reddish-stemmed annual; to 2 ft. Leaves lance-shaped, lacking sheath bristle; very acrid and *peppery to taste*; margin wavy. Greenish flowers in arching clusters (most *Polygonum* species have pink flowers); June–

Nov. **WHERE FOUND:** Moist soils, shores. Much of our area. **USES:** The Cherokee used leaf tea as a diuretic for painful or bloody urination, diarrhea, fevers, chills; poulticed leaves for pain, hemorrhoids; rubbed them on a child's thumb to prevent sucking. Leaf tea a folk remedy for internal bleeding and menstrual or uterine disorders. Leaves contain rutin, which helps strengthen fragile capillaries and thus helps prevent bleeding. Contains pain-relieving compounds and polygodial, which numbs a toothache. Also contains hot pungent compounds, which may explain the name "smartweed." **RELATED SPECIES:** Many other *Polygonum* species have been used in American, European, and Asian folk and traditional medicine. **WARNING:** Plant can irritate skin.

SHEEP-SORREL

Leaves, root

Rumex acetosella L. — Buckwheat Family (Polygonaceae)

Slender, smooth, sour-tasting perennial; 4–12 in. Leaves *arrow-shaped*. Tiny flowers in green heads, interrupted on stalk; turning reddish or yellowish; Apr.–Sept. **WHERE FOUND:** Acid soils. Throughout our area. **USES:** Leaf tea of this common European alien traditionally used for fevers, inflammation, scurvy. Fresh leaves considered cooling, diuretic; leaves poulticed (after roasting) for tumors, wens (sebaceous cysts); folk remedy for cancer. Root tea used for diarrhea, excessive menstrual bleeding. Also used for antispasmodic and astringent effects. Has become popular in recent years as a component of the reputedly anticancer essiac formula and Ojibwa teas. Sheep-sorrel is rich in cancer-preventive vitamins; also includes four antimutagenic and four antioxidant compounds, perhaps laying the foundation for reported anticancer (or cancer-preventing) folk uses. Ethanolic extracts of this and other *Rumex* species, used as folk cancer remedies, have recently shown the ability to affect cancer cell death by preventing their reproduction. Still much more research is necessary. **RELATED SPECIES:** *Rumex hastatulus* is very similar and likely confused with *R. acetosella,* separated on highly technical characteristics. **WARNING:** May cause poisoning in large doses, because of high oxalic acid and tannin content.

YELLOW OR CURLY DOCK

Roots, leaves

Rumex crispus L. — Buckwheat Family (Polygonaceae)

Perennial; 1–5 ft. Leaves large, lance-shaped; margins distinctly *wavy* ("crisped," hence the species name). Flowers green, on spikes; Apr.–Sept. Winged, heart-shaped seeds; June–Sept. Roots yellowish in cross-section. **WHERE FOUND:** Waste ground. Throughout our area. **USES:** Herbalists consider dried-root tea an excellent blood purifier. Used to treat "bad blood," chronic skin diseases, chronic enlarged lymph glands, skin sores, rheumatism, liver ailments, sore throats. May cause or relieve diarrhea, depending on dose, harvest time, and concentrations of anthraquinones (laxative) and/or tannins (antidiarrheal). Anthraquinones can arrest growth of ringworm and other fungi. Contains compounds that are antioxidant, antifungal, antibacterial.

Sheep-sorrel produces reddish fruits and often covers infertile fields.

The leaves of Yellow or Curly Dock have prominently wavy margins.

Rumex obtusifolius L. used similarly. **WARNING:** Large doses may cause gastric disturbance, nausea, and diarrhea.

FLOWERS IN TERMINAL CLUSTERS AND IN UPPER AXILS: AMARANTH AND GOOSEFOOT FAMILIES

SMOOTH PIGWEED — Leaves, roots

Amaranthus hybridus L. — Amaranth Family (Amaranthaceae)

Highly variable, *smooth-stemmed* annual; 1–6 ft. Leaves to 6 in. long, hairy. Flower spikes green (or red-tinged); *lateral spikes erect or ascending*; Aug.–Oct. **WHERE FOUND:** Throughout our area. Alien weed, first found along riverbanks, and now spread throughout agricultural lands with a nearly worldwide distribution. **USES:** Cherokee used leaves to relieve profuse menstruation. Seeds collected and used as food by various western Native American groups. Leaf tea astringent, stops bleeding; used in dysentery, diarrhea, ulcers, intestinal bleeding. Reduces swelling and pain. Recent research confirms a significant pain-reducing effect from root extracts. Many members of the Pigweed (*Amaranth*) family and Goosefoot family serve as potherbs and/or cereal grains. The National Academy of Sciences is vigorously investigating both grain amaranths and the goosefoot relatives as food crops.

Smooth Pigweed is a highly variable annual with smooth stems.

Green Amaranth, Pigweed is an annual with blunt flower spikes interspersed with bristly bracts.

GREEN AMARANTH, PIGWEED

Leaves

Amaranthus retroflexus L.

Amaranth Family (Amaranthaceae)

Grayish, downy annual; 6–24 in. Leaves oval, stout-stalked. Flower spikes to 2½ in. long; blunt, *chaffy, interspersed with bristly bracts*; July–Oct. **WHERE FOUND:** Throughout our area. Native to cen. and e. N. America, but successfully invasive on most continents. **USES:** Cherokee used leaves like *A. hybridus* to treat profuse menstruation; leaf tea used by the Mohegan to treat sore throat. Astringent. Used for diarrhea, excessive menstrual flow, hemorrhages, hoarseness.

SPINY AMARANTH

Leaves, herb

Amaranthus spinosus L.

Amaranth Family (Amaranthaceae)

Branched, erect annual; to 3 ft. tall. Leaves ovate to lance-ovate, 1¼–2½ in. long, with a *pair of curved spines in the leaf axils*. Flower spikes numerous, 2–6 in. long; rough and spongy toward top; June–August. **WHERE FOUND:** Waste places, fields. NY to MO southward. Weed found throughout the tropics, spreading into N. America, originating in S. American lowland tropics. **USES:** Leaves astringent. Adopted by Native American groups for the treatment of profuse menstruation. Many Amaranth species valued for astringency; most often used to stop bleeding, both internal and external.

Spiny Amaranth. Note spines in leaf axils.

The leaves of Lamb's-quarters, Pigweed have a mealy surface.

LAMB'S-QUARTERS, PIGWEED

Leaves

Chenopodium album L.

Amaranth Family (Amaranthaceae)
[Formerly in Goosefoot Family (Chenopodiaceae)]

Highly variable annual weed; 1–3 ft. Stem often mealy, red-streaked. Leaves somewhat diamond-shaped, coarsely toothed; *mealy white beneath.* Flowers greenish, inconspicuous; in clusters; June–Oct. **WHERE FOUND:** Gardens, fields, waste places. Throughout our area. Alien, probably from Europe; one of the most successful cosmopolitan weeds in the world. **USES:** The Cherokee ate leaves to treat stomachaches and prevent scurvy. Cold tea used for diarrhea; leaf poultice used for burns. A folk remedy for vitiligo, a skin disorder. Leaves considered edible.

MEXICAN TEA, AMERICAN WORMSEED

Seeds, essential oil

Chenopodium ambrosioides L.
[*Dysphania ambrosioides* (L.) Mosyakin & Clemants]

Amaranth Family (Amaranthaceae)
[Formerly in Goosefoot Family (Chenopodiaceae)]

⚠ Stout, *sharply aromatic* herb; 1–5 ft. Leaves wavy-toothed. Flowers greenish in spikes, among leaves; Aug.–Nov. Seeds *glandular-dotted.* **WHERE FOUND:** Waste places. Throughout our area. Native to N. America and S. America; widespread in tropical regions throughout the world. **USES:** For centuries the herb has been used in infusions to treat worm infections. Until recently, the essential oil distilled from flowering and fruiting plant was used against roundworms, hookworms, dwarf (not large) tapeworms, intestinal amoeba. Now largely replaced by synthetics, though still widely used in developing and tropical countries as a folk medicine for worms and infectious disease. The component in the essential oil with potent worm-expelling effects is the toxic compound ascaridole. **WARNING:** Oil is highly toxic. Still, a dash of the leaves is added as a culinary herb to Mexican bean dishes in the belief that it may reduce gas. May cause dermatitis or an allergic reaction. Steven Foster has experienced vertigo from contact with essential oil released during harvest. Ingesting just 10 ml of the essential oil has proven fatal (much less in children). The active compound ascaridole in the essential oil may be partially dissipated in hot water.

Mexican Tea, American Wormseed. Its deeply lobed leaves are strongly aromatic.

NODDING, INCONSPICUOUS FLOWERS WITH YELLOW POLLEN; RAGWEEDS

COMMON RAGWEED

Leaves, root

Ambrosia artemisiifolia L. Aster Family (Asteraceae or Compositae)

⚠ Annual; 1–5 ft. Leaves dissected, artemisia-like; highly variable—as a rule alternate, but opposite as well. Drooping, *inconspicuous green flowerheads on conspicuous erect spikes*; July–Oct. **WHERE FOUND:** Waste ground throughout our area. Noxious weed. **USES:** Cherokee and other Native American groups rubbed leaves on insect bites, infected toes, minor skin eruptions, and hives; topically applied to prevent infections. Tea used for fevers, nausea, mucous discharges, intestinal cramping; very astringent, emetic. Root tea used for menstrual problems and stroke. Formerly used in fevers as a substitute for quinine. **WARNING:** Pollen causes allergies. Ingesting or touching plant may cause allergic reactions. Pollen from the genus *Ambrosia* is responsible for approximately 90 percent of pollen-induced allergies in the U.S. Goldenrods (*Solidago* species) are often pointed to as the source of late-summer allergies, but at the same time the showy goldenrods are blooming, the inconspicuous flowers of ragweeds are really guilty.

Common Ragweed is a major culprit in summer allergies.

Giant Ragweed. Note its large, 3-lobed leaves.

GIANT RAGWEED

Leaves, root

Ambrosia trifida L.

Aster Family (Asteraceae or Compositae)

⚠ Annual; to 6–15 ft. Stems and leaves with stiff hairs, rough to the touch. Leaves opposite, *deeply 3-lobed* (sometimes 5-lobed or without lobes); tips pointed. Lower leaves are most uniform in appearance. Flowers similar to those of Common Ragweed. **WHERE FOUND:** Alluvial waste places, sometimes forming vast, pure, pollen-producing stands. Much of our area. **USES:** Astringent; stops bleeding. Leaf tea formerly used for prolapsed uterus, leukorrhea, fevers, diarrhea, dysentery, nosebleeds; gargled for mouth sores; said to stop salivation. Cherokee used the crushed leaves on insect bites. The root was chewed to allay fear at night. The pollen of both this and Common Ragweed is

Giant Ragweed flower close-up. Note upside-down flowerheads.

harvested commercially, then manufactured into pharmaceutical preparations for the treatment of ragweed allergies. Essential oil contains antibacterial and antifungal components. **WARNING:** Pollen causes allergies. Ingesting or touching plant may cause allergic reactions. The Aster family, to which the ragweeds and the allergy-inducing artemisias (see following pages) belong, is the worst family as far as pollinosis is concerned.

FLOWERS IN TERMINAL CLUSTERS; LEAVES SILVER-HAIRY, AT LEAST BENEATH; ARTEMISIAS

WORMWOOD

Leaves

Artemisia absinthium L. Aster Family (Asteraceae or Compositae)

⚠ Aromatic perennial; 1–4 ft. Leaves silver to gray-green, strongly divided; segments blunt, with *silky silver hairs on both sides*. Flowers are tiny, 1/8 in. across, greenish yellow, drooping; July–Sept. **WHERE FOUND:** Waste ground. Escaped from cultivation in n. U.S. Alien. **USES:** Extremely bitter leaves are nibbled to stimulate appetite. Tea is a folk remedy for delayed menses, fevers, worm expellent, and liver and gallbladder ailments. Used as a bitter tonic to stimulate digestion to treat dyspepsia, particularly for those who drank too much alcohol. The famous flavoring basis of absinthe; banned in Europe and the U.S. in 1915, then re-legalized in the European Union and the U.S., by 2007, absinthe is widely available again (with thujone content limited). Whether absinthe's dark reputation is warranted or not, absinthe became the poster child of the prohibition movement in the late nineteenth and early twentieth centuries. New research suggests that the thujone content of vintage absinthe was much lower than previously estimated, and the condition once called "absinthism" cannot be distinguished medically from alcoholism. The toxic principle is thujone, which occurs only in minute amounts in water extracts (such as teas). Intoxication from absinthe liqueurs was historically likened to that induced by marijuana. It is theorized that the active component of both plants may react with the same receptors of the central nervous system. Despite potential toxicity, the herb is approved in some European countries to stimulate appetite and treat dyspepsia. Wormwood extracts have recently been shown to have antioxidant, antimicrobial, neuroprotective, and liver-protective effects. **WARNING:** Relatively small doses allegedly causes nervous disorders, convulsions, insomnia, nightmares, and other symptoms. Flowers of

Wormwood has extremely bitter gray-green leaves.

artemisias may induce allergic reactions. Approved as a food additive (flavoring) with thujone removed.

WESTERN MUGWORT, WHITE SAGE, CUDWEED — Leaves

Artemisia ludoviciana Nutt. — Aster Family (Asteraceae or Compositae)

⚠ Highly variable aromatic perennial; to 3 ft. Leaves *white-felty beneath*; lance-shaped, entire. Flowers in dense panicles; July–Sept. **WHERE FOUND:** Waste ground. MI to s. IL, TX; north to MT and westward. Naturalized east to New England. A weed. **USES:** Much used by Native Americans as an astringent, to induce sweating, curb pain and diarrhea. Weak tea used for stomachaches, menstrual disorders. Leaf snuff used for sinus ailments, headaches, nosebleeds. Externally, wash used for itching, rashes, skin eruptions, swelling, boils, sores. Compress for fevers. Used in steam baths for rheumatism, fevers, colds, and flu. Extracts have antibacterial, antispasmodic, antioxidant, and potential antimalarial activity. **WARNING:** May cause allergies.

MUGWORT — Leaves

Artemisia vulgaris L. — Aster Family (Asteraceae or Compositae)

⚠ Aromatic; 2–5 ft. Leaves *deeply cut, silvery-woolly beneath*. Flowerheads erect; July–Aug. **WHERE FOUND:** Waste ground. S. Canada to GA; KS, MI; occasional westward. Variable alien weed from Eurasia. **USES:** Leaf tea diuretic, induces sweating; checks menstrual irregularity, promotes appetite; "tonic" to nerves. Historically with a reputation for treating epilepsy. Used for bronchitis, colds, colic, epilepsy, fevers, kidney ailments, sciatica. Experimentally, lowers blood sugar. Components have

ABOVE: *Mugwort has become a rampant weed.* LEFT: *Western Mugwort, White Sage, Cudweed leaves are white and felty beneath.*

antibacterial, antifungal, antispasmodic, bronchodilator, pain-relieving, and liver-protectant activity. Dried leaves used as "burning stick" (moxa), famous in Chinese medicine to stimulate acupuncture points and treat rheumatism. Clinically proven to lower incidence of breached birth presentations. **WARNING:** May cause dermatitis. Reported to cause abortion and allergic reactions.

FLOWERS IN TERMINAL CLUSTERS; LEAVES NOT SILVER-HAIRY; ARTEMISIAS

ANNUAL WORMWOOD, SWEET ANNIE, QINGHAO — Leaves, seeds

Artemisia annua L. — Aster Family (Asteraceae or Compositae)

⚠ Sweet-scented, bushy annual; 1–9 ft. Leaves thrice-divided, *fernlike* segments oblong to lance-shaped, sharp-toothed or cleft. Tiny green-yellow flowers, in clusters; July–Oct. **WHERE FOUND:** Waste ground. Throughout our area; becoming much more common, escaping from cultivation. Alien. **USES:** Leaf tea (gather before flowering) used for colds, flu, malarial fevers, dysentery, diarrhea. Externally, poulticed on abscesses and boils. For quinine- and/or chlorquinine-resistant malaria; clinical use of derivative compounds in China shows near 100 percent efficacy. Now semisynthetic derivatives from the sesquiterpene lactone artemisinin are widely used in malaria-infected tropical countries as a preventive and treatment for malaria. Seeds traditionally used for night sweats, indigestion, flatulence. The compound responsible for the antimalarial activity also demonstrates marked herbicidal activity. Contains six or more antiviral compounds, some proven synergetic. Jim Duke's assistant is allergic to nonflowering material, even in a sealed envelope. **WARNING:** May cause allergic reactions or dermatitis.

Wormwod close-up of flowerheads.

WILD TARRAGON — Leaves, roots

Artemisia dracunculus L. — Aster Family (Asteraceae or Compositae)
[*A. redowskii* Led.; *A. glauca* var. *draculina* (S. Wat.) Fern.]

⚠ Variable—aromatic to odorless; to 5 ft. Leaves lance-shaped to linear (sometimes divided), without teeth, dark green. Whitish green flowers in loose, spreading clusters; July–Oct. **WHERE FOUND:** Prairies, dry soil. WI, MO, TX, and westward. Rare eastward to New England. **USES:**

ABOVE: *Annual Wormwood, Sweet Annie is a tall, fast-growing annual.* TOP RIGHT: *Annual Wormwood, Sweet Annie, with finely divided, aromatic leaves.* RIGHT: *Wild Tarragon usually has typical fragrance of the culinary herb tarragon, though sometimes is without fragrance.*

Native Americans used leaf or root tea for colds, dysentery, diarrhea, headaches, difficult childbirth. Promotes appetite. Leaves poulticed for wounds, bruises. Sometimes substituted for the cooking herb French Tarragon, which must be propagated vegetatively; it does not produce viable seed. French Tarragon smells strongly of anise; Wild Tarragon may be sweet-scented because of estragole, or odorless and flavorless. **WARNING:** Allergic reactions may result from use.

MISCELLANEOUS PLANTS WITH GREEN-BROWN FLOWERS

VIRGINIA SNAKEROOT

Root

Aristolochia serpentaria L. Birthwort Family (Aristolochiaceae)

⚠ Delicate; 8–20 in. Leaves elongate, *strongly arrow-shaped*. Flowers calabash-pipelike, purplish brown; at base of plant, often under leaf litter; May–July. **WHERE FOUND:** Woods. Sw. CT to FL; TX to MO, IL; OH. Too rare to harvest. **USES:** Famous as an American medicinal plant introduced in the 1636 Thomas Johnson edition of *Gerard's Herball.* By 1650 it was included in the *London Pharmacopoeia.* It was primarily extolled as a stimulating tonic and diaphoretic, and further as a cure for venomous snakebites (hence the name "*serpentaria*") and the bites of rabid dogs. Thomas Jefferson listed it among the 20 most important medicinal plants of VA. Aromatic root nibbled (in minute doses) or in weak tea (1 teaspoon dried root in 1 cup of water) promotes sweating, appetite; expectorant. Used for fevers, stomachaches, indigestion, suppressed menses, and snakebites. Tea gargled for sore throats. **REMARKS:** Recent high prices following increasing demand and decreasing supplies may suggest this as a prospect for cultivation in the forest—further justification for saving our forests. Likely harvested in tonnage by the eighteenth century, reducing populations to the localized scattered occurrence seen today. **WARNING:** Irritating in large doses; probably contains toxic aristolochic acid.

Virginia Snakeroot is a small plant with a camphorous root. Flower is beneath leaf litter.

DUTCHMAN'S-PIPE

Leaves

Aristolochia tomentosa Sims Birthwort Family (Aristolochiaceae)

⚠ A climbing, woody vine. Leaves *heart-shaped*; blunt-tipped, lower surface has dense, soft white hairs. Flowers pipe-shaped; calyx yellowish; May–June. **WHERE FOUND:** Rich riverbanks. NC, FL, and TX; north to e. KS, MO, s. IL to s. IN. **USES:** Like those of *A. serpentaria* (above), but much weaker in effect. Little used. **RELATED SPECIES:** *A. macrophylla* (not shown) has nearly smooth, sharp-pointed leaves. Flowers brown-

Dutchman's-pipe produces a woody, twining vine.

Skunk Cabbage, with dark red to brown flowers.

purple. Ironically, Virginia farmers spray *A. macrophylla* as a weed. It contains the antiseptic, antitumor compound and mutagenic toxic compound aristolochic acid. **WARNING:** Potentially irritating in large doses. Aristolochic acid is considered an insidious toxin, inducing mutations.

SKUNK CABBAGE — Root

Symplocarpus foetidus (L.) Salisb. ex Nutt. — Arum Family (Araceae)

⚠ Strongly skunk-scented perennial. Flowers appear before leaves, Feb.–May. Greenish to purple, *hooded, sheathing spathe with a clublike organ within*. Root toxic. See p. 270.

COMMON JUNIPER

Fruits, leaves (needles)

Juniperus communis L.

Cypress Family (Cupressaceae)

⚠ Shrub or small tree; 2–20 ft. Bark reddish brown, shredding off in papery peels. Leaves (needles) taper to a spiny tip, in *whorls of 3s with 2 white bands above* (or 1 white band sometimes divided by a green midrib, broader than green margin). Fruits (technically cones) on short stalk; round to broadly oval, bluish black, usually with 3 seeds. **WHERE FOUND:** Rocky, infertile soils. Canada to AK, south to mtns. of GA; e. TN north to IL; MN west to NM, CA; N. America, Europe, and Asia. N. American forms are small and shrublike, whereas European forms are more treelike. **USES:** Fruits used to flavor gin and other alcoholic beverages; also used commercially in some diuretic and laxative products. Widely used by Native American groups for a broad range of purposes. The Cheyenne used the leaves as a ceremonial fumigant. Leafy boughs and fruits used in steam baths or as tea for colds, fevers, sedative, and for sore throat and tonsillitis. Chippewa use decoction of leafy boughs for asthma. Most extensively used by First Nation peoples in Canada and Alaska. Juniper berries are one of the most widely used herbal diuretics. Used in European phytomedicine in teas for stomach complaints and to simulate appetite. Science confirms anti-inflammatory, spasm-reducing, and diuretic activity. Eaten raw or as tea, berries a folk diuretic and urinary antiseptic for cystitis, carminative for flatulence, antiseptic for intestinal infections. Once used for colic, coughs, stomachaches, colds, and bronchitis. Externally, used for sores, aches, rheumatism, arthritis, snakebites, and cancer. Components of the essential oil, as well as flavonoids and glycosides, are responsible for diuretic, intestinal antiseptic, antimicrobial, and antioxidant activity. Diuretic activity results from irritation of renal tissue. **WARNING:** Potentially toxic. Large or frequent doses cause kidney failure, convulsions, and digestive irritation. In Germany, use limited to four weeks. Avoid during pregnancy. Oil may cause blistering.

Common Juniper, with typical juniper "berries," actually cones.

ENGLISH YEW

Leaves (needles), bark

Taxus baccata L.

Yew Family (Taxaceae)

☠ Often grown as ornamental shrub, but the tree will grow to 40 ft. Bark is red-brown, peeling in plates. Leaves arranged in spirals—narrow, glossy, dark green above, lighter beneath, to 1¼ in. long. Fruit (aril) *globose, red*, to ½ in. in diameter. **WHERE FOUND:** Sometimes naturalized near old plantings; widely cultivated as an ornamental evergreen shrub or hedge throughout, represented by more than 250 cultivated varieties. **USES:** Yew toxins were used in Old World cultures as poison arrow tips to kill fish and animals. The pitch of yew trees was mixed with clarified butter and used for the treatment of cancer, portending the use by modern Western societies. The bark and leaves contain paclitaxel (once popularly called taxol, which is now a registered trademark of Bristol-Myers). Paclitaxel was first isolated in 1969, and its structure determined by 1971. The name was first applied by its discoverer, Monroe Wall. At first commercially derived from the bark of the Pacific Yew *Taxus brevifolia* Nutt., most supply now comes from *T. baccata* and its relatives in Europe and India. Today it is used in chemotherapy for the treatment of certain forms of breast cancer and ovarian cancer. Sales of the drug exceed $1 billion per year, making it one of the most important natural products ever. **WARNING:** All plant parts (except perhaps the red aril) of this and other yews contain highly toxic components and are considered poisonous, possibly fatally so.

AMERICAN YEW, CANADA YEW (NOT SHOWN) — Leaves (needles)

Taxus canadensis Marsh. — Yew Family (Taxaceae)

☠ Straggling evergreen shrub; rarely to 7 ft. Twigs *smooth*, green; reddish brown on older branches. Needles 2-ranked, ⅜–1 in. long, narrowing into abrupt fine points. Needles are *green on both sides* with light green bands below; they often develop a reddish tint in winter. Female plants produce juicy, cuplike, red arils (pulp) surrounding ½ in. fruits. Seeds stony. **WHERE FOUND:** Rich woods. NL to w. VA; ne. KY to IA, MB. **USES:** Native Americans such as the Abnaki, Algonquins, Chippewa, and Menominee used minute amounts of toxic leaf tea internally and externally for rheumatism, bowel ailments, fevers, colds, scurvy; to expel

English Yew is commonly cultivated as an ornamental hedge.

Florida Yew is a federally listed endangered species; leaves similar to those of Canada Yew.

afterbirth, dispel clots; as a diuretic. Twigs used as fumigant in steam baths for rheumatism. Leaves (needles) said to be anti-rheumatic and hypotensive. **RELATED SPECIES:** Florida Yew, *Taxus floridana* Nutt. ex Chapm., which has been investigated as a source of novel taxanes (anticancer components), is a federally listed endangered species that occurs only in a small area along the Apalachicola River in n. FL. **WARNING:** All plant parts (except perhaps the red aril) of this and other yews contain highly toxic components and are considered poisonous. Ingesting as few as 50 leaves (needles) has resulted in fatalities.

EVERGREEN SHRUBS WITH SWORDLIKE LEAVES; PALMS AND YUCCAS

SAW PALMETTO

Fruits

Serenoa repens (W. Bartram) Small
[*Sabal serrulata* (Michx.) Schult.f.]

Palm Family (Arecaceae or Palmae)

Shrubby palm; 3–9 ft., with horizontal creeping stems aboveground. Leafstalks armed with *sawlike teeth*. Leaves fanlike, with sword-shaped leaf blades radiating from a central point. Flowers whitish green, with 3–5 petals; May–July. Fruits black, fleshy, 1 in. long, surrounding 1 large seed; in large, branched clusters; Aug.–Nov. **WHERE FOUND:** Low

TOP LEFT: *Saw Palmetto, a small palm, covers millions of acres in FL and GA.* BOTTOM LEFT: *One day's intake of fresh, wild-harvested Saw Palmetto berries at a drying facility in FL.* ABOVE: *Saw Palmetto. Freshly harvested fruits.*

pine woods, savannas, thickets. SC, GA, FL to AL, MS. **USES:** Fruits once used as a staple food of Indian groups of FL. Early European settlers found it less palatable; one described their flavor as that of "rotten cheese steeped in tobacco juice." Introduced into medicine in 1879, then suggested as gynecological aid that could enlarge breasts. Fruit extracts, tablets, and tincture subsequently used to treat prostate enlargement and inflammation. Also used for colds, coughs, irritated mucous membranes, tickling feeling in throat, asthma, chronic bronchitis, head colds, and migraine. A suppository of the powdered fruits in cocoa butter was used as a uterine and vaginal tonic. Considered expectorant, sedative, diuretic. Pharmacological and clinical studies show that the fruits are useful in the treatment of prostate disorders. In the 1990s, Saw Palmetto berry extracts emerged as the most important natural treatment for benign prostatic hyperplasia, a nonmalignant enlargement of the prostate that affects a majority of men over 50 years of age. Evaluated in several thousand men in dozens of controlled clinical trials, alone or in comparison with conventional drugs, Saw Palmetto preparations have been found as effective as the conventional drug in relieving symptoms of benign prostatic hyperplasia while producing fewer side effects in some studies; other studies have indicated no benefit. Saw Palmetto fruit preparations are approved in Germany, France, Italy, and other countries for treatment of symptoms related to benign prostatic hyperplasia, and are often the treatment of choice by physicians in those countries, usually taken for six months before results are evident. Thousands of tons of Saw Palmetto berries are harvested from wild habitats in FL and adjacent GA each year. **WARNING:** Benign prostatic hyperplasia can be diagnosed only by a physician.

YUCCA, ADAM'S NEEDLES

Roots

Yucca filamentosa L. — Asparagus Family (Asparagaceae)

[Formerly in the Lily Family (Liliaceae) or Agave Family (Agavaceae)]

Perennial; to 9 ft. in flower. Leaves in a rosette; stiff, spine-tipped, oblong to lance-shaped, with *fraying twisted threads on margins.* Flowers whitish green bells, on *smooth,* branched stalks; June–Sept. **WHERE FOUND:** Sandy soils. S. NJ to GA. **USES:** Native Americans used root in salves or poultices for sores, skin diseases, and sprains. Pounded roots were put in water to stupefy corralled fish so they would float to the surface for easy harvest. Could be used as yet another starting material for steroids. **WARNING:** Root compounds toxic to lower life forms. See p. 25.

YUCCA, SOAPWEED

Roots

Yucca glauca Nutt. — Asparagus Family (Asparagaceae)

[Formerly in the Lily Family (Liliaceae) or Agave Family (Agavaceae)]

Blue-green perennial; 2–4 ft. Leaves in a rosette; stiff, swordlike, rounded on back, margins rolled in. Flowers are whitish bells *on nonbranching stalk*; May–July. **WHERE FOUND:** Dry soils. IA, ND to MO, TX. **USES:** Numerous Native American groups used plant for medicine,

Flowers of Adam's Needle Yucca grow on smooth-branched stalks.

Flowers of the Soapweed Yucca grow on a single stalk. More than 30 species of yucca occur in N. America, most in the desert Southwest.

food, and fiber. Poulticed root on inflammations, used to stop bleeding; also in steam bath for sprains and broken limbs; hair wash for dandruff and baldness. Leaf juice used to make poison arrows. Antifungal, antitumor, and antiarthritic activities have been suggested by research. Water extracts have shown antitumor activity against B16 melanoma in mice. One human clinical study suggests that saponin extracts of yucca root were effective in the treatment of arthritis, but the findings have been disputed. Sometimes confused with or considered synonymous to *Y. arkansana* Trel. **WARNING:** Same as for *Yucca filamentosa.*

EVERGREEN SHRUBS WITH 5-PARTED FLOWERS; HEATH FAMILY

SHEEP LAUREL, LAMBKILL

Twigs, leaves, flowers

Kalmia angustifolia L.

Heath Family (Ericaceae)

Slender shrub; 3–5 ft. Leaves opposite, leathery, elliptical to lance-shaped. Flowers deep rose pink (or white), to ½ in. across in *clusters on sides of twigs*; May–July. **WHERE FOUND:** Dry soils. NL to VA, GA mtns.; north to MI. **USES:** Considered more poisonous than medicinal. Native Americans used minute amounts of flower, leaf, and twig tea for bowel ailments. Tiny amounts of leaf tea used for colds, backaches, stomach ailments; externally, for swelling, pain, and sprains. Contains grayano-

LEFT: *Sheep Laurel. Note flowers in leaf axils.* ABOVE: *Mountain Laurel. Flowers terminal.*

toxins, which affect the heart and cause neurological symptoms. **WARNING:** Highly toxic. Do not ingest.

MOUNTAIN LAUREL — Leaves

Kalmia latifolia L. — Heath Family (Ericaceae)

☠ Shrub or small tree; 5–30 ft. Leaves evergreen, leathery, ovate, without teeth. Flowers pink (rose or white), about 1 in. wide; *in terminal clusters*; May–July. **WHERE FOUND:** Rocky woods, clearings. New England; NY to FL, LA to OH, IN. **USES:** The Cherokee used leaf tea as an external wash for pain, rheumatism, in liniments for vermin. The bristles of up to a dozen leaves were brushed across the skin to treat rheumatism; crushed leaves used to relieve thorn scratches. Historically, herbalists used minute doses to treat syphilis, fever, jaundice, heart conditions, neuralgia, and inflammation. Contains grayanotoxins, which affect the heart and cause neurological symptoms. **WARNING:** Plant is highly toxic; even the honey from flowers is reportedly toxic. Avoid use.

LABRADOR TEA — Leaves

Rhododendron groenlandicum (Oeder) Kron & Judd (*Ledum groenlandicum* Oeder) — Heath Family (Ericaceae)

⚠ Shrub; to 3 ft. Leaves fragrant, oblong or linear-oblong, *white to rusty-woolly* beneath; *edges turned under*. Small white flowers in terminal clusters; May–July. **WHERE FOUND:** Peat soils, bogs. NL to NJ, PA, OH, MI, WI, MN, across Canada to AK. **USES:** Widely used by Native Americans in the n. U.S. and Canada; leaf tea for asthma, colds, stomachaches, kidney ailments, scurvy, fevers, rheumatism; blood purifier;

LEFT: *Labrador Tea. Like its relatives, it can be toxic.* ABOVE: *Great Rhododendron. A showy native shrub.*

externally, as a wash for burns, ulcers, stings, chafing, Poison Ivy rash. Subsequently adopted by European settlers and promoted by itinerant self-taught physicians as a remedy for coughs, lung ailments, dysentery, indigestion; used externally for leprosy, itching, and to kill lice. **WARNING:** Reportedly toxic. Contains grayanotoxins, which affect the heart and cause neurological symptoms.

GREAT RHODODENDRON — Leaves

Rhododendron maximum L. — Heath Family (Ericaceae)
(*Hymenanthes maxima* (L.) H. F. Copel.)

⚠ Thicket-forming evergreen shrub or small tree; 10–14 ft. Leaves large, *leathery*, without teeth, edges *rolled under*. Rose pink (white) spotted flowers in very showy clusters; June–July. **WHERE FOUND:** Damp woods. S. ME to GA; AL to OH. **USES:** Cherokee poulticed leaves to relieve arthritis pain, headaches; taken internally in controlled dosage for heart ailments. **WARNING:** Leaves toxic. Ingestion may cause convulsions and coma. Avoid use. Contains grayanotoxins, which affect the heart and cause neurological symptoms.

SEMI-EVERGREEN SHRUBS; LEAVES LEATHERY

BEARBERRY, UVA-URSI — Leaves

Arctostaphylos uva-ursi (L.) Spreng — Heath Family (Ericaceae)

⚠ Trailing shrub; to 1 ft. Bark finely hairy. Leaves shiny, *leathery, spatula-shaped*. Flowers white, *urn-shaped*; May–July. Fruit is a dry red berry. **WHERE FOUND:** Sandy soil, rocks. Arctic to n. U.S. See p. 35.

LEFT: *Bearberry, Uva-ursi in fruit.* ABOVE: *Bearberry, Uva-ursi in flower.*

Wintergreen, Teaberry in fruit.

Wintergreen, Teaberry in flower.

WINTERGREEN, TEABERRY, CHECKERBERRY — Leaves

Gaultheria procumbens L. — Heath Family (Ericaceae)

Wintergreen-scented; to 6 in. Leaves oval, glossy. Flowers *waxy*, drooping bells; July–Aug. Fruit is a dry berry. See p. 36.

YAUPON HOLLY, CASSINA, BLACK DRINK — Leaves, berries

Ilex vomitoria Ait. — Holly Family (Aquifoliaceae)

Evergreen shrub or small tree; 6–15 ft. Leaves to 2 in. long; elliptical, leathery, *round-toothed*. Berries in clusters, red (rarely yellow); Sept.–Nov. Calyx segments *rounded*, with few hairs on margins. **WHERE FOUND:** Sandy woods. Se. VA to FL; TX, AR. **USES:** The most under-exploited

Yaupon Holly, N. America's only native caffeine-producing plant.

Yaupon Holly, showing flowers and persistent fruit from the previous season.

natural resource of the American South. Yaupon Holly is the only caffeine-containing plant in N. America with viable commercial potential. Recent studies reveal that Yaupon Holly contains appreciable amounts of caffeine and theobromine, along with antioxidant, anti-inflammatory, and chemopreventive polyphenols, comparable to coffee and tea. Historical commercial development was limited to the Confederacy during the Civil War, when "yaupon factories" roasted leaves as a coffee and tea substitute. If you were a sixteenth- to eighteenth-century European explorer entering a Native American village along the Gulf Coast, elders would greet you with an offering of Yaupon Holly tea, the leaves of which were grown in semiwild plantations, and served up as a sacred or ceremonial beverage. It was called "black drink" as the leaves were usually decocted down to an espresso-like extract. Native Americans called it "cassine." In a 1542 travel journal, Álvar Núñez Cabeza de Vaca (1490–1558) provides the first published observations on its use. In addition to a ceremonial beverage, Native Americans made a thick brew of black tea, drunk before going into battle or counsel, as both a stimulant and a ceremonial cleanser imbibed until it induced vomiting. For better or worse, honoring this tradition, in 1769 English botanist William Aiton named the plant *Ilex vomitoria*, a scientific name that will live in infamy, and which perhaps thwarted the shrub's commercial development. **RELATED SPECIES:** The S. American species *Ilex paraguariensis* A. St. Hil. is the source of the S. American beverage maté or yerba maté, available wherever herbal teas are sold, and the caffeinated beverage of choice in Argentina, Paraguay, and Uruguay.

PARTRIDGEBERRY, SQUAW-VINE — Leaves

Mitchella repens L. — Madder Family (Rubiaceae)

Leaves opposite, rounded. Flowers white (or pink), *4-parted*; terminal, *paired*; May–July. Each flower produces a single dry red berry, lasting over the winter. See p. 37.

LEFT: *Partridgeberry, Squaw-vine in flower.* ABOVE: *Partridgeberry, Squaw-vine in fruit.*

THYME

Leaves

Thymus vulgaris L. Mint Family (Lamiaceae or Labiatae)

Prostrate perennial subshrub; to 6 in. Short-stalked; leaves small (to 3/8 in. long), oval, *entire* (not toothed). Flowers small, purple, or (rarely) white; clustered at ends of branches; July–Aug. **WHERE FOUND:** Scattered to rare. NS to NC; OH to IN. European alien; escaped from cultivation. **USES:** In European folk tradition, Thyme leaf tea has been used for nervous disorders, angina pectoris, flu, coughs, stomachaches; blood purifier; also to relieve cramps. Widely used in modern European herbal medicine to treat spasmodic coughs, bronchitis, whooping cough, emphysema, and asthma. Experimentally, oil of Thyme is antispasmodic, expectorant, antimicrobial; lowers arterial pressure, increases heart rhythms and respiratory volume; lowers blood pressure, alleviates toothaches. Eriodictyol, a flavonoid component of the leaf, is strongly antioxidant. Widely used in European phytomedicine for the treatment of

Thyme, mostly grown in gardens, sometimes escaped.

bronchitis, whooping cough, and inflammation of the upper respiratory tract. **WARNING:** Oil is toxic and highly irritating to skin.

CRANBERRY

Fruits

Vaccinium macrocarpon Aiton — Heath Family (Ericaceae)

Trailing evergreen shrub with slender stems; to 6 in. tall. Leaves leathery, somewhat spatula-shaped, rounded at end, stalkless, ¼ to ½ in. long. Flowers, 3–6 in clusters in leaf axils, white to pink, to ½ in. long with *4 reflexed lobes*; June–Aug. Fruits are the familiar red, tart berries of cranberry sauce. **WHERE FOUND:** Bogs. NL to MB, south to VA; OH south to mtns. of NC and TN. Cultivated commercially in bog habitats. **USES:** Native Americans used the fruits as food; given to early European sailors to treat scurvy and dysentery. Regarded as a urinary antiseptic throughout the twentieth century. Contrary to popular belief, Cranberry does not serve as a urinary antiseptic by acidifying the urine. Recent studies show that Cranberry contains compounds that prevent the adhesion of bacteria to linings of the bladder and gut, therefore preventing infection. Recent clinical studies have shown that Cranberry juice cocktail and dried Cranberry juice extracts prevent urinary tract infections and act as a urinary deodorant. Proanthocyanins in the fruit have antibacterial, antioxidant, chemopreventive, antifungal (against *Candida*), and other potential health benefits. **RELATED SPECIES:** Small Cranberry, *Vaccinium oxycoccos* L. (not shown), occurs throughout boreal regions of the Northern Hemisphere (south to NJ, MN, and n. IN in the U.S.); in Europe, used as a source of the berries.

Cranberry, the familiar cranberry of commerce.

SHRUBS WITH WHITE, 5-PETALED FLOWERS; ROSE FAMILY

NINEBARK

Bark

Physocarpus opulifolius (L.) Maxim.
(*Opulaster opulifolius* (L.) Kuntze)

Rose Family (Rosaceae)

Shrub; to 9 ft. Bark *peels in thin strips or layers*. Leaves oval to obovate; irregularly toothed, with star-shaped hairs. Flowers white; May–July. Seedpods inflated, 2-valved; usually 3 pods per cluster. **WHERE FOUND:** Streambanks. QC to SC; AL, AR to MN. Often cultivated and then escaped. The bark on older branches separates, peeling into several layers, hence the common name Ninebark. The species name *opulifolius* refers to the resemblance of the leaves to those of Crampbark, *Viburnum opulus*. **USES:** Menominee used inner-bark tea for "female maladies," gonorrhea, tuberculosis; to enhance fertility; Chippewa, as an emetic, laxative. **WARNING:** Potentially toxic. The western N. American species *Physocarpus capitatus* (Pursh) Kuntze, contains potentially toxic cucurbitacins, which probably explains traditional laxative and emetic effects.

Ninebark has delicate, fragrant flowerheads.

LARGE-HIP, RUGOSA, OR WRINKLED ROSE

Fruit, flowers, roots

Rosa rugosa Thunb.

Rose Family (Rosaceae)

Coarse, *bristly stemmed* shrub; 2–6 ft. Leaves *strongly wrinkled*; 5–9 leaflets. Large rose (or white) flowers to 3¼ in. across; June–Sept. Fruits (hips) large, red, to 1 in., crowned with sepals. **WHERE FOUND:** Seaside, sand dunes. N. U.S., Canada. Asian alien. Common, often in large thickets along coastal beaches and dunes, this Asian introduction has larger fruits (rose hips) than any of our native roses. **USES:** The Chinese use flower tea to "regulate vital energy," promote blood circulation; also for stomachaches, liver pains, mastitis, dysentery, leukorrhea, rheumatic pains; also thought to "soothe a restless fetus." Fruits (rose hips) make a pleasant, somewhat tart tea. High in vitamin C, the fruits have been used to treat scurvy (a disease caused by deficiency of vitamin C); also strong antioxidant activity. Petals contain antioxidant polyphenols. Roots are used in e. Asia as a treatment for diabetes mellitus, pain, folk remedy for cancer, and chronic inflammatory disease. Flower extracts have been shown useful against allergic diseases such as atopic dermatitis. Euscaphic acid isolated from the roots has anti-inflammatory activity.

Large-hip, Rugosa, Wrinkled Rose produces our largest rose hip.

Large-hip, Rugosa, Wrinkled Rose flowers.

RED RASPBERRY

Leaves, root, fruits

Rubus idaeus L.

Rose Family (Rosaceae)

Upright shrub, 3–6 ft.; *canes do not root at tips. Smooth* bristly stem, with or without hooked prickles; 3–7 oval leaflets. Flowers white; June–Oct. Drupelets not *separated by bands of hairs.* **WHERE FOUND:** Cultivated throughout our area. European alien. **USES:** Astringent leaf tea a folk remedy for diarrhea, dysentery; used to strengthen pregnant women, aid in childbirth. Root also used. Animal studies suggest efficacy in childbirth, painful menstrual cramps. Therapeutic use of leaf not approved in Germany because of lack of scientific support of claimed activities as a uterine tonic. Active compound relaxes and stimulates the uterus. Fruit syrup (juice boiled in sugar) gargled for inflamed tonsils. Polyphenols in fruit with strong antioxidant activity. Polyphenols in leaf extract recently showed potential as an antioxidant with chemopreventive activity against human laryngeal and colon cancer cell lines. Leaves used in Middle East as a folk remedy for kidney stones, recently shown to reduce formation of kidney stones in laboratory experiments.

BLACK RASPBERRY

Root, leaves, fruits

Rubus occidentalis L.

Rose Family (Rosaceae)

Shrub with *arching canes* that *root at tips* to 6 ft. in height. *Stem glaucous* with *curved prickles.* Leaves whitened beneath; sharply double-toothed. Flowers white; Apr.–July. Fruits purple-black; July–Sept. Rows of *white hairs between drupelets.* **WHERE FOUND:** Throughout our area; often

Red Raspberry leaves are used in herbal medicines.

Red Raspberry fruits.

cultivated. **USES:** Astringent root tea traditionally used for diarrhea, dysentery, stomach pain, gonorrhea, back pain; a "female tonic"; blood tonic for boils. Leaf tea a wash for sores, ulcers, boils. Leaf tea used in European phytomedicine for treatment of diarrhea and mild inflammation of the mouth and throat. It is astringent, because of the tannins in both leaf and root. Anthocyanins in fruit show potential to inhibit cell proliferation of oral, esophageal, and colon cancer cell lines. **RELATED SPECIES:** The same parts of most blackberry plants (other *Rubus* species) have been used similarly.

Black Raspberry. The roots are the primary herbal ingredient.

MISCELLANEOUS SHRUBS WITH THORNY BRANCHES

AMERICAN OR ALLEGHENY BARBERRY (NOT SHOWN)
Root bark

Berberis canadensis P. Mill. Barberry Family (Berberidaceae)

Shrub; 10–25 in. *Brownish to dull purple branches*; spines usually 3-parted. Leaves spatula-shaped, sparsely toothed; grayish white beneath, without prominent veins. Flowers bright yellow, 5–10 per raceme; May. Petals *notched*. Berries red, *round*. **WHERE FOUND:** Rocky woods, mountains. VA to GA, AL; MO to IN. **USES:** Root tea used for fevers. A Cherokee remedy for diarrhea. See Common Barberry (below).

COMMON BARBERRY
Root bark, fruit, leaves

Berberis vulgaris L. Barberry Family (Berberidaceae)

Branching shrub; to 9 ft. *Grayish branches; spines 3-parted.* Leaves alternate or in rosettes from previous year's leaf axils, spatula-shaped, with *numerous spiny teeth; veins beneath prominent.* Flowers 10–20 per raceme; Apr.–June. Petals not notched. Fruits red, *elliptical*. Root bark yellow. **WHERE FOUND:** Widely planted and escaped; gone wild in s. New England. Alien (Europe). **USES:** Berry tea used to reduce fevers, promote appetite; diuretic, expectorant, laxative; also relieves itching. Root-bark tea promotes sweating; astringent, antiseptic, blood purifier; used for jaundice, hepatitis (stimulates bile production), fevers, hemorrhage, diarrhea. Leaf tea for coughs. Root-bark tincture used for arthritis, rheumatism, sciatica. Contains berberine, which has a wide spectrum of biological activity, including antibacterial activity; useful against infection. Root bark experimentally liver protective, potential benefit in preventing kidney stones; antioxidant, anti-inflammatory, antiarrhythmic, and sedative effects. Contains berbamine, which increases white blood cell and platelet counts. Barberry is used in Traditional Chinese Medicine for treating lowered white blood cell counts following chemo-

Typical bright yellow inner bark of Barberry.

Japanese Barberry, an Asian alien, is becoming increasingly common.

LEFT: *Common Barberry. Its fruits turn bright red when they ripen.* ABOVE: *English Hawthorn is commonly planted and naturalized.*

therapy or radiation therapy in cancer patients. **RELATED SPECIES:** Of the 22 *Berberis* species in N. America, several Asian species, including Japanese Barberry (*Berberis thunbergii* DC.) and Beale's Barberry (*Berberis bealei* Fortune), are increasingly becoming seriously invasive aliens. **WARNING:** Large doses potentially harmful.

HAWTHORNS

Flowers, fruits, leaves

Crataegus species

Rose Family (Rosaceae)

⚠ Very complex plant group; 100–1,000 species in N. America. Highly variable; hybridize readily—species identification is difficult even for the specialist. Spiny shrubs. Leaves simple, toothed; cut or lobed. Flowers mostly white, usually with 5 petals; calyx tube bell-shaped, 5-parted; spring–early summer. Fruits are dry red berries; each berry has 1–5 hard seeds. **WHERE FOUND:** Most abundant in e. and cen. U.S. **COMMENT:** The most commonly used species in phytomedicine include European species English Hawthorn, *C. laevigata* (Poir.) DC., naturalized in WI, MI, and adjacent Canada; and One-seeded Hawthorn, *C. monogyna* Jacq., naturalized from QC to NC, TN, north to ON. Chinese Hawthorn *C. pinnatifida* Bunge sometimes grown as a specimen in U.S. **USES:** Fruits and flowers famous in herbal folk medicine (Native American, Chinese, European) as a heart tonic. Studies confirm use in hypertension with weak heart, angina pectoris, arteriosclerosis, and high cholesterol. Dilates coronary vessels, reducing blood pressure; acts as direct and mild heart tonic; antioxidant. Prolonged use necessary for efficacy. Tea or tincture used. Hawthorn products are very popular in Europe and China. Hawthorn leaf and flower (though not the fruits) are approved in Germany for treating early stages of congestive heart failure, characterized by diminished cardiac function, a sensation of pressure or anxi-

One-seeded Hawthorn. A common European native naturalized in our flora.

One-seeded Hawthorn flowers are used in European herbal medicine.

ety in the heart area, age-related heart disorders that do not require digitalis, and mild arrhythmias. Use confirmed by numerous controlled clinical studies. **WARNING:** Eye scratches from thorns can cause blindness. Contains heart-affecting compounds that may affect blood pressure and heart rate.

MATRIMONY VINE, WOLFBERRY, GOJI

Fruits, bark

Lycium barbarum L.
(*Lycium halimifolium* Mill.)

Nightshade Family (Solanaceae)

Smooth, weak-stemmed, trailing, climbing, or sprawling; *sparsely thorny* shrub; to 6 ft. or more. Leaves elliptical, lance-shaped, or oval, to 2½ in. long, an inch or more wide. Flowers pale to deep purple, *tubular, with 4–5 lobes*; June–Oct. Stamens exceed petals. Fruits fleshy, red, shiny, *elliptical to ovoid*, less than an inch long; Sept.–Nov. **WHERE FOUND:** Increasingly spreading in N. America, forest edges, open areas, waste ground. Naturalized in nearly all lower 48 states and adjacent Canada. Alien from China, where it is widely grown for the fruits; used as a tonic food. **USES:** In the past 15 years, sale of dried fruits proliferated in Western countries under the name "goji berries," a bastardized marketing name derived from the Chinese name *guo qi zi*. In Traditional Chinese Medicine, the fruits are used as a tonic to improve qi (vital essence) and for general debility,

Freshly harvested fruits of Matrimony Vine are widely sold in health food stores as "goji berries."

Matrimony Vine has light to deep purple flowers; stamens protrude beyond petals.

Matrimony Vine fruits are glossy red with papery calyx at base.

manifested by symptoms such as dizziness, tinnitus, diabetes, anemia, and impaired vision. Contains polysaccharides, steroidal glycosides, and flavonoids with studies suggesting chemopreventive effect against human cervical cancer. Suggested useful as an anti-aging tonic and preventive for retinal decline, also against ischemic stroke, nerve inflammation; antioxidant, anti-inflammatory, immunostimulant. Chinese clinical studies suggest an overall improvement of neurological and psychological factors such as fatigue, sleep quality, mental acuity, calmness, and well-being. **REMARKS:** *Lycium chinense* Mill. is also grown commercially in China and escaped in the U.S. Most "goji" in Western markets is *L. barbarum.* **WARNING:** Some individuals may experience allergic reactions; may interact with prescription drugs such as warfarin (effectively increasing the prescribed dose).

SHRUBS OR SMALL TREES WITH COMPOUND LEAVES; BRANCHES OR TRUNKS ARMED WITH SHORT SPINES

DEVIL'S WALKING-STICK, ANGELICA TREE — Root, berries

Aralia spinosa L. — Ginseng Family (Araliaceae)

Woody; 6–30 ft. Main stem and leafstalks with many sharp *(often stout) spines.* Leaves large (to 6 ft. long), twice-divided; leaflets numerous, oval, toothed. Tiny white flowers in umbels, in a very large panicle;

Devil's Walking-Stick, Angelica Tree in fruit.

Devil's Walking-Stick's spine-studded bark.

July–Sept. **WHERE FOUND:** Rich woods, alluvial soils. S. New England (cultivated) to FL; TX north to MI. **USES:** Cherokee used an infusion of the pounded, roasted roots to induce vomiting; also for fevers, tonic, carminative; for rheumatism, venereal disease, colic, and aching teeth. Root salve used to treat hard-to-heal sores. In folk tradition, fresh bark strongly emetic, purgative; thought to cause salivation. Tincture of berries used for toothaches, rheumatic pain. Tincture of wood used for colic. Root poulticed for boils, skin eruptions, swelling. Diabetic Koreans in the Washington, D.C., area take the plant to lower their insulin requirements. **WARNING:** Handling roots may cause dermatitis. Large amount of berries poisonous.

NORTHERN PRICKLY-ASH — Bark, berries

Zanthoxylum americanum P. Mill. — Rue Family (Rutaceae)
(Spelling variant *Xanthoxylum americanum* P. Mill.)

Aromatic shrub, 4–25 ft., with paired short spines. Compound leaves with 5–11 leaflets; oval, toothed, *lemon-scented* when crushed. Tiny green-yellow flowers; Apr.–May, before leaves. Fruits are red-greenish berries covered with lemon-scented dots; Aug.–Oct. **WHERE FOUND:** Moist woods, thickets. QC to FL; OK to MN. **USES:** Bark tea or tincture historically used by numerous Native American groups as well as herbalists for chronic rheumatism, dyspepsia, dysentery, kidney trouble, heart trouble, colds, coughs, lung ailments, uterine cramps, and nervous debility. When chewed, bark induces copious salivation. Once popular to stimulate mucous surfaces, bile, and pancreas activity.

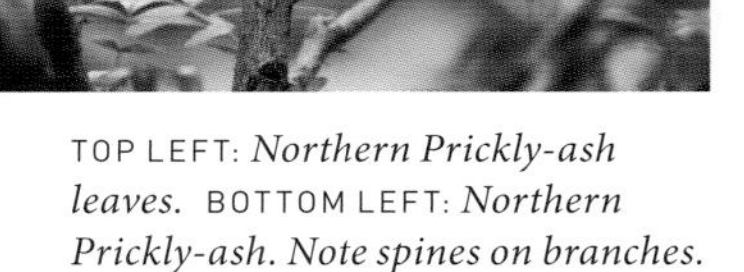

TOP LEFT: *Northern Prickly-ash leaves.* BOTTOM LEFT: *Northern Prickly-ash. Note spines on branches.*

TOP RIGHT: *Southern Prickly-ash leaves.* BOTTOM RIGHT: *Southern Prickly-ash, with large triangular spikes on trunk.*

Bark chewed for toothaches. Berry tea used for sore throats, tonsillitis; also used as a diuretic. Dried, aromatic fruit used as a condiment by the Menominee to season other medicines. Asian species such as *Z. simulans,* also known as Sichuan Pepper, widely used in China. Extracts of all plant parts and the furanocoumarins contained therein show significant antifungal activity.

SOUTHERN PRICKLY-ASH Bark, berries

Zanthoxylum clava-herculis L. Rue Family (Rutaceae)

Small tree or shrub; much larger than *Z. americanum*—to 30 ft. Bark with large, *triangular, corky knobs.* Fruits Aug.–Oct. **WHERE FOUND:** Poor soils. S. VA to FL; TX to s. AR, se. OK. **USES:** Similar to *Z. americanum;* also a folk remedy for cancer. Contains the alkaloid chelerythrine, with antibacterial and anti-inflammatory activity. Affects muscle contractility by blocking or stimulating neuromuscular transmissions, rather than through a direct effect on smooth muscle tissue. Used in European herbal medicine for treating rheumatic conditions, Raynaud's disease, and as a circulatory stimulant in intermittent claudication. An alkylamide, neoherculin, produces a localized numbing effect. Most research on the genus relates to Asian or African species.

COMPOUND LEAVES; SHRUBS WITHOUT SPINES

ELDERBERRY

Flowers, berries, inner bark, leaves

Sambucus canadensis L.

Moschatel Family (Adoxaceae)

[Formerly in Honeysuckle Family (Caprifoliaceae)]

⚠ Shrub; 3–12 ft. Stem with white pith. Leaves opposite (paired), compound, *usually 7* (sometimes 5–11) elliptical to lance-shaped leaflets, sharply toothed. Fragrant white flowers in flat, umbrella-like clusters; June–July. Fruits purplish black; July–Sept. **WHERE FOUND:** Rich soils. NS to GA, TX to MB. **USES:** Iroquois used bark in tea for measles, headache; strong laxative, diuretic, spring cleansing emetic; poulticed for cuts. Cherokee used berry tea for boils, rheumatism; also diuretic, cathartic, emetic; externally in salve for burns and skin eruptions. Leaves poulticed on bruises and on cuts to stop bleeding. Bark tea was formerly used as a wash for eczema, old ulcers, skin eruptions. A tea with Peppermint in water is a folk remedy for colds; induces sweating and nausea. Considered a mild stimulant, carminative, and diaphoretic. In WV, concentrated fruit syrup is made as a wintertime remedy for colds and flu. Various research groups have conducted clinical studies on the use of a fruit extract of European Elder (*Sambucus nigra* L.) with positive results for colds and flu. Flowers and fruit extracts used in European phytomedicine in treating colds, as they reduce fever while increasing bronchial secretions. Polyphenolics and anthocyanins in the fruit have antioxidant

Elderberry fruiting-heads, drooping under their own weight.

LEFT: *Elderberry in flower.*
ABOVE: *Chaste Tree produces showy flowers.*

activity. **REMARKS:** Treated as a variety of the closely related European black elder [as *S. nigra* subsp. *canadensis* (L.) Bolli] in a 1994 taxonomic revision with a broad circumscription of the genus. *S. nigra* usually has 5 leaflets, whereas *S. canadensis* usually has 7 leaflets. The two species are used interchangeably, though the fruits contain different, but similar, polyphenolic compounds. **WARNING:** Bark, root, leaves, and unripe berries toxic; said to cause cyanide poisoning, severe diarrhea. Fruits edible when cooked. Flowers not thought to be toxic; eaten in pancakes and fritters.

CHASTE TREE

Fruits

Vitex agnus-castus L.

Mint Family (Lamiaceae or Labiatae)
[Formerly in Verbena Family (Verbenaceae)]

Aromatic shrub or low tree; 3–15 ft. Twigs with dense short hairs, resinous. Leaves palmate; 5–9 (rarely 3) narrowly elliptical leaflets, pointed on both ends; to 4½ in. long, ½–¾ in. wide; entire or wavy margined; densely white hairy beneath, slightly hairy above. Leaflets unequal in size; *only the three largest with leafstalks*. Flowers small lavender to lilac (rarely white), tubular, 5-lobed, ¼ in. long; mostly in terminal, large, pyramid-shaped panicles to 1 ft. tall; Apr.–Oct. **WHERE FOUND:** Dry to moist soils. Alien. Native to w. Asia and sw. Europe. Widely naturalized in the South; MD to FL; west to TX, OK. **USES:** Seeds (fruits) used for more than 2,500 years for menstrual difficulties. In medieval Europe, seeds were thought to allay sexual desire, hence the names Chaste Tree and Monk's Pepper. In the late nineteenth century, American physicians used tincture to increase milk secretion and treat menstrual irregularities. Today it is widely prescribed by European gynecologists for treatment of premenstrual syndrome, heavy or too frequent periods, acyclic bleeding, infertility, suppressed menses, and to stimulate milk flow. Fruit preparations approved in Germany and elsewhere for menstrual disorders, pressure and swelling in the breasts, and premenstrual syndrome. A recent study showed that a fruit extract may reduce severity and duration of migraines associated with premenstrual syndrome. **WARNING:** May cause rare dermatitis. Avoid during pregnancy or hormone replacement therapy.

YELLOWROOT — Root

Xanthorhiza simplicissima Marsh. — Buttercup Family (Ranunculaceae)

Small shrub; 1–3 ft. Thick, deep yellow root with yellowish bark. The erect, unbranched woody stem, usually 2–3 ft. high, bears leaves and flowers only on the upper portion, which is marked with the scars of the previous year's leaves. Leaves usually divided into *5 leaflets*, on long stalks; leaflets cleft, toothed. Flowers small, *brown-purple in drooping racemes*, 5 petals, 2-lobed, with glandlike organs on a short claw; Apr.–May. **WHERE FOUND:** Moist woods, thickets, and streambanks. NY to FL; AL to KY. **USES:** Cherokee used root tea as an astringent and tonic for hemorrhoids, cramps, nerves; stem chewed for sore mouth and throat. Other Native American groups used root tea for stomach ulcers, colds, jaundice, cramps, sore mouth or throat, menstrual disorders; blood tonic, astringent; externally for cancers. A folk remedy used in the South for diabetes and hypertension. Contains berberine—anti-inflammatory, astringent, hemostatic, antimicrobial, anticonvulsant, immunostimulant, uterotonic; also produces a transient drop in blood

ABOVE: *Yellowroot has a bright yellow root.* RIGHT: *Yellowroot, with chocolate-colored flowers.*

pressure. Berberine stimulates the secretion of bile and bilirubin and may be useful in correcting high tyramine levels in patients with liver cirrhosis. Yellowroot was formerly used as an adulterant to or substitute for Goldenseal, though nineteenth-century physicians believed its medicinal action was quite different than that of Goldenseal. An Alabama herbalist, the late Tommie Bass, used to sell "sticks" of the root for smokers to chew on in efforts to quit smoking. **WARNING:** Yellowroot may be potentially toxic in large doses.

LEAVES OPPOSITE; FLOWERS SHOWY; MISCELLANEOUS SHRUBS

CAROLINA ALLSPICE

Root, bark

Calycanthus floridus L. — Calycanthus Family (Calycanthaceae)

⚠ Aromatic shrub; 3–9 ft. Leaves oval, opposite, entire (not toothed), with *soft fuzz beneath.* Flowers terminal, maroon-brown, about 2 in. across; on an urn-shaped receptacle; Apr.–Aug. Flowers have a *strong, strawberry-like fragrance when crushed.* **WHERE FOUND:** Rich woods. VA to FL, AL to WV. **USES:** Cherokee used root or bark tea as a strong emetic, diuretic for kidney and bladder ailments. Cold tea used as eye drops for failing sight. Settlers used tea as a calming tonic for malaria. Widely cultivated in gardens; one of the most aromatic plants of the South; bark once used as a substitute for Cinnamon; spicy fruit (seldom set) used as an Allspice (*Pimenta dioica* (L.) Merr.) substitute. **WARNING:** Grazing cattle have been reported to have a toxic reaction to eating this plant. Historically, dogs and wolves said to be poisoned by the plant.

Carolina Allspice. Note its deep maroon flowers.

BUTTONBUSH

Bark, root, flowers, leaves

Cephalanthus occidentalis L. — Madder Family (Rubiaceae)

⚠ Shrub; 9–20 ft. Leaves *oblong*-ovate; essentially smooth. White flowers in a *globe-shaped* cluster; July–Aug. *Stamens strongly protruding.* **WHERE FOUND:** Streambanks, moist soils. NB; New England to FL and Mex.; north to WI and west to CA. **USES:** Kiowa used a root decoction to treat internal bleeding. Meskwaki used inner bark as an emetic. Choctaw used bark for dysentery, fevers; wash for sore eyes, and chewed bark to relieve toothaches. Thought to be tonic, diuretic, astringent;

promotes sweating. Bitter bark tea used for difficult-to-treat coughs. Syrup of the flowers and leaves once used as a mild laxative and tonic. Leaf tea once used for fevers, coughs, "gravel" (kidney stones), malaria, palsy, pleurisy, and toothaches. Interestingly, this plant, which superficially resembles a diminutive Cinchona bush (source of quinine), belongs to the same plant family and has a folk reputation, as Dogwood does, for relieving fever and malaria. Contains 30 or more triterpenoid saponins. **WARNING:** Contains the glucosides cephalanthin and cephalin. The leaves have caused poisoning in grazing animals.

WILD HYDRANGEA

Root, bark

Hydrangea arborescens L.

Hydrangea Family (Hydrangeaceae)
[Formerly in Saxifrage Family (Saxifragaceae)]

⚠ Shrub; to 9 ft. Leaves opposite, mostly ovate; toothed, pointed. Flowers in flat to round clusters, often with papery, *white, sterile, petal-like calyx lobes on outer edge*; June–Aug. Like cultivated hydrangeas, the flowerhead of our native Wild Hydrangea is surrounded by sterile, papery, white, flowerlike structures that attract pollinators to the inconspicuous (tiny) fertile blooms. **WHERE FOUND:** Rich woods. NY to n. FL, LA; OK to IN, OH. **USES:** A missionary among the Cherokee in the early nineteenth century observed that a syrup or infusion of the root given in teaspoonful doses three times a day relieved excruciating pain from passage of kidney stones. Cherokee used inner bark and leaves as antiseptic, for ulcers, to stop vomiting, treat high blood pressure; externally poulticed for burns, sore or swollen muscles. Root traditionally used for kidney stones, mucous irritation of bladder, bronchial afflictions, stomach problems. The root bark was formerly marketed under the name Gravel Root, referring to its use for kidney stones. Compounds in

ABOVE: *Buttonbush has a spherical flowerhead.* RIGHT: *Wild Hydrangea is common in rich woods.*

plant have anti-inflammatory and diuretic activity; inhibit tumor formation. **WARNING:** Experimentally, causes bloody diarrhea, painful gastroenteritis, cyanide-like poisoning. Little research has been conducted on the plant.

LEAVES OPPOSITE; FRUITS RED TO PURPLE

AMERICAN BEAUTY BUSH, FRENCH MULBERRY

Leaves, roots, berries

Callicarpa americana L.

Mint Family (Lamiaceae or Labiatae)
[Formerly in the Verbena Family (Verbenaceae)]

Shrub; 3–6 ft. Leaves ovate-oblong, toothed; woolly beneath. Tiny whitish blue flowers in whorl-like cymes; June–Aug. *Rich blue-violet berries in clusters*, in leaf axils; Oct.–Nov. **WHERE FOUND:** Rich thickets. N. MD to FL; north to AR, OK. Occurring only in the southern part of our range (to n. AR), though Asian species of this genus are grown as ornamentals as far north as Boston. Easily recognized by the sticky, aromatic, opposite leaves and whorls of magenta fruits. **USES:** Native Americans used root and leaf tea in steam baths for rheumatism, fevers, and malaria. Root tea used for dysentery, stomachaches. Root and berry tea used for colic. Formerly used in the South for dropsy and as a blood purifier in skin diseases. Leaf tea used as a diuretic. Sweetish, subastringent, acid fruits once eaten (but not in quantity); purple berries used to dye wool purple (with alum). **RELATED SPECIES:** The Chinese use the leaves of a related *Callicarpa* species as a hemostat (to stop bleeding of wounds). It is also used to treat flu in children, and for menstrual disorders.

American Beauty Bush, French Mulberry, with purplish berries in leaf axils.

STRAWBERRY BUSH

Stem and root bark, seeds

Euonymus americanus L.

Staff-tree Family (Celastraceae)

Erect or straggling, deciduous or nearly evergreen shrub; 3–6 ft. Stalks green, 4-angled. Leaves rather thick, lustrous, sessile; tips sharp-pointed. Flowers greenish purple; petals stalked; May–June. Fruits *scarlet, warty.* **WHERE FOUND:** Rich woods. Se. NY, PA to FL; TX, OK to IL. **USES:** Cherokee used an infusion of the bark with other herbs for urinary problems; bark also a cathartic and irritant to gastrointestinal system. Native Americans used root tea for uterine prolapse, vomiting of blood, stomachaches, painful urination; wash for swellings. Bark formerly used by physicians as tonic, laxative, diuretic, and expectorant. Tea used for malaria, indigestion, liver congestion, constipation, lung afflictions. Powdered bark applied to scalp was thought to eliminate dandruff. Seeds strongly laxative. **WARNING:** Fruit, seeds, and bark may be poisonous. Do not ingest—fruits may cause vomiting, diarrhea, and unconsciousness.

WAHOO

Stem and root bark, seeds

Euonymus atropurpureus Jacq. — Staff-tree Family (Celastraceae)

⚠ Shrub or small tree; 6–25 ft. Leaves hairy beneath, oblong-oval, stalked. Flowers purplish; June–July. Fruits purplish, *smooth*; seeds covered with scarlet pulp. **WHERE FOUND:** Rich woods. ON to TN, AL, AR; OK to ND. **USES:** Essentially the same as for *E. americanus* (see above). Historically, the bark was considered tonic, laxative, diuretic, and expectorant. Extracts, syrups, and tea were used for fevers, upset stomach, constipation, dropsy, lung ailments, liver congestion, and heart medicines. The seeds were considered emetic and strongly laxative. Bark and root contain digitalis-like compounds. Seeds once used to kill head lice. **WARNING:** Fruit, seeds, and bark are considered poisonous. Leaves have poisoned sheep and other animals feeding on them.

Strawberry Bush. Note its warty fruit.

Wahoo fruit is smooth.

POSSUMHAW, SOUTHERN WILD-RAISIN (NOT SHOWN) — Bark

Viburnum nudum L. — Moschatel Family (Adoxaceae)
[Formerly in Honeysuckle Family (Caprifoliaceae)]

Deciduous shrub; to 12 ft. Leaves *glossy, leathery*; oval, wavy-edged, or slightly toothed; those below flowers wedgelike, widest near middle. Small white flowers in flat clusters; Apr.–June. **WHERE FOUND:** Bogs, low woods. MD to FL; AR to KY. **USES:** Cherokee used bark tea as a diuretic, tonic, uterine sedative, antispasmodic; for diabetes, often in combination with other herbs. According to Ed Croom, Lumbees boiled bark for 12 hours to reduce liquid to ⅓ original amount. A 1-ounce dose was taken 3 times per day for 4 days, then dosage was reduced to ½ ounce, taken twice a day.

CRAMPBARK, GUELDER ROSE, HIGHBUSH CRANBERRY — Bark, fruit, leaves

Viburnum opulus L. — Moschatel Family (Adoxaceae)
[Formerly in Honeysuckle Family (Caprifoliaceae)]

Shrub; to 12 ft. Leaves maplelike, with 3–5 lobes; *hairy beneath*. Leafstalks with a narrow groove and a *disk-shaped gland*. White flowers in a rounded head, to 4 in. across; Apr.–June. Berries bright red to orange. **WHERE FOUND:** Ornamental from Europe. Sometimes escaped in urban areas; occasionally abundant such as in Chicago area. Widespread range in N. America, N. Asia, Europe, extending into N. Africa. **USES:** Montagnais used wash of boiled berries for eye inflammation. In Europe, bark tea has been used to relieve all types of spasms, including menstrual cramps; astringent, uterine sedative. Bark used as a calmative and to relieve pain associated with menstrual disorders. Science confirms antispasmodic activity. In China, leaves and fruit are used as an emetic, laxative, and antiscorbutic. Folk remedy for cancer. Proanthocyanidins in berries have significant antimicrobial and antioxidant activity; research target for early stages of colon cancer. **WARNING:** Bluish black berries are considered edible but potentially poisonous; they contain

Crampbark, Guelder Rose, Highbush Cranberry produces maplelike leaves and bright red fruits.

chlorogenic acid, beta-sitosterol, and ursolic acid, at least when they are unripe.

BLACKHAW

Bark, leaves, twigs

Viburnum prunifolium L. Moschatel Family (Adoxaceae)
[Formerly in Honeysuckle Family (Caprifoliaceae)]

Large shrub to small tree; 6–30 ft. Leaves elliptic to ovate; *finely toothed*; mostly smooth, *dull* (not shiny). White flowers in *flat clusters*; Mar.–May. Fruits black (bluish at first). **WHERE FOUND:** Bogs, low woods. CT to FL; TX to e. KS. **USES:** Cherokee used infusion of root bark to reduce recurring muscle spasms and fevers; as a tonic. Leaf poultice used for inflamed skin cancers. Bark smoked like tobacco. In the early nineteenth century, bark was largely supplied and promoted by the Shakers as a diuretic. Root- or stem-bark tea used by Native Americans, then adopted by Europeans for painful menses, to prevent miscarriage, relieve spasms after childbirth. Astringent tea of young twigs a folk remedy for diarrhea. Considered uterine tonic, sedative, antispasmodic, and nervine. Also used for asthma. Research has confirmed uterine-sedative, pain-relieving (like willows, it contains salicin), anti-inflammatory, and spasm-reducing properties. Iridoid glucosides likely responsible for antibacterial activity. **WARNING:** Berries may produce nausea and other discomforting symptoms, though they were also eaten in small amounts.

Blackhaw in flower. Most of the dozen or so species in our range are used similarly.

Blackhaw in fruit.

MISCELLANEOUS SHRUBS WITH ALTERNATE LEAVES

NEW JERSEY TEA, RED ROOT

Leaves, root

Ceanothus americanus L.

Buckthorn Family (Rhamnaceae)

Small shrub with red root; 1–2 ft. Leaves oval, toothed, to 2 in. long, with *3 prominent parallel veins*. White flowers in showy, puffy clusters on herbaceous (nonwoody) flower stalks; Apr.–Sept. **WHERE FOUND:** Dry, gravelly banks, open woods. ME to FL; OK to MN. **USES:** The Alabama used a root decoction (boiled for 2 or 3 hours), then cooled to bathe injured feet or legs, repeated three times a day for up to a month until injury healed. Cherokee used the root tea for bowel complaints (mild laxative); infusion held in mouth to treat toothache. Chippewa used as a laxative; root infusion for lung complaints and shortness of breath. Leaf tea used as a beverage by Native Americans; later adopted and popular among settlers, especially during Revolutionary War. During the Civil War, in South Carolina the leaf tea was promoted as a substitute for "foreign tea." Also root tea for colds, fevers, snakebites, stomachaches; blood tonic. Historically touted as a syphilis and gonorrhea cure. Root strongly astringent (8 percent tannin content), expectorant, sedative. Tea gargled for mouth sores and sore throat. Triterpenes and flavonoids from root extracts have antimicrobial effect against several oral pathogens. Root tea was once used for dysentery, asthma, sore throats, bronchitis, whooping cough, and spleen inflammation or pain. Alkaloid in root is mildly hypotensive (lowers blood pressure).

New Jersey Tea, Red Root.

BLACK CURRANT

Root bark, fruit, seeds, leaves

Ribes nigrum L.

Currant Family (Grossulariaceae)
[Formerly in the Saxifrage Family (Saxifragaceae)]

Prickly-thorned shrub, 3–6 feet. *Leaves 3–5 divided, maplelike*, base cordate; both sides shiny with *yellow glandular dots* (use hand lens). Flowers white to reddish, in a drooping raceme; Apr.–June. Fruits black, smooth. One of about 16 *Ribes* species in our range. **WHERE FOUND:** Native to Europe. Locally established NL to MD, west to IL, MN, ON, QC. Cultivated elsewhere. **USES:** Traditionally an infusion of the root or bark was used for fevers, dysentery; fruits and jelly for sore throat. Fruits considered diuretic, useful in dysentery, fevers, soothing to stomach. Fruit used in cordials for stomachic qualities. In recent years black cur-

Currants. The seeds of Ribes nigrum *have become a commercial seed oil source.*

Steeplebush, Hardhack produces woolly stems and leaves that have tawny, woolly undersides.

rant seed oil, high in gamma-linolenic acid, has been sold as a dietary supplement. Berry extract antioxidant, antibacterial, antiviral, anti-inflammatory, and chemopreventive. Black Currants have been suggested for hypertension and other cardiovascular disease, neurodegenerative disorders, kidney disorders, and diabetic neuropathy. Leaves used as a folk medicine for gout, rheumatism, diarrhea, spasmodic coughs; externally for wounds. Contains various proanthocyanidins.

STEEPLEBUSH, HARDHACK

Leaves, flowers

Spiraea tomentosa L. Rose Family (Rosaceae)

Small shrub; 2–4 ft. Stems woolly. Leaves *very white or tawny-woolly beneath*; oval-oblong, saw-toothed. Flowers rose (or white), in a *steeple-shaped* raceme; July–Sept. **WHERE FOUND:** Fields, pastures. NS to NC; AR to ON. **USES:** Native Americans used leaf tea for diarrhea, dysentery; flower and leaf tea for morning sickness. Leaves and flowers were once used to stop bleeding; also for leukorrhea. Other *Spiraea* species and the related genus *Ulmaria* were used similarly. Leaves and bark, as well as the root, were widely available in pharmacies in the nineteenth century, and the best form was said to be an extract prepared by the Shakers. Mostly used as a treatment for diarrhea, bleeding, and noted among other astringents as one that seldom caused stomach upset. Little researched.

LATE LOWBUSH BLUEBERRY

Leaves, fruit

Vaccinium angustifolium Ait.

Heath Family (Ericaceae)

Shrub; 3–24 in. Leaves narrowly *lance-shaped*, with *tiny stiff teeth* that are green and hairless on both sides. Flowers white (or pink-tinged); urn-shaped, 5-lobed; May–June. Fruits (blueberries) Aug.–Sept. **WHERE FOUND:** Sandy or acid soils. NL to MD; n. IA to MN. **USES:** Wild Lowbush Blueberries are primarily harvested commercially in eastern ME. Ojibwa used leaf tea as a blood purifier; also used for colic, labor pains, and as a tonic after miscarriage; Chippewa used fumes of burning dried blueberry leaves as a fumigant to treat craziness. Leaves used as an astringent folk tea to treat diarrhea; gargled for sore throats. Many Native American groups collected this and other species of wild blueberries, dried them and used as food, often stored for winter use. Certainly wild blueberries are among the most relished wild fruits from America, leading to development of commercial hybrids forming the familiar large blueberries of commerce. Today the proanthocyanidins that give blueberries their color are widely researched and valued for antioxidant and potential cancer-preventive activities. As with other common food items in the past decade, including various fruits, coffee, tea, chocolate, red wine, and so on, new science of plant phenolic compounds has led to a growing body of literature exploring the health benefits of compounds once thought of only as the color components of food.

Late Lowbush Blueberry, our most common blueberry. Note its serrated leaves.

SHRUBS WITH ALTERNATE COMPOUND LEAVES; SUMACS

FRAGRANT OR STINKING SUMAC

All parts

Rhus aromatica Ait.

Cashew Family (Anacardiaceae)

Bush or shrub; 2–7 ft. Leaves *3-parted, fragrant*, blunt-toothed; *end leaflet not stalked*. Flowers small, yellow, *appearing before leaves* in early spring. Fruits very oily to touch; hairy, red; May–Aug. Highly variable. **WHERE FOUND:** Dry soil. W. VT to nw. FL; TX to SD and westward. **USES:** Native Americans used leaves for colds, bleeding; chewed leaves for stomachaches; diuretic. The patient chewed the bark for colds and slowly swallowed juice. Fruits chewed for toothaches, stomachaches, and grippe. Physicians formerly used astringent root bark to

Fragrant or Stinking Sumac, with small yellow flowers in early spring.

Fragrant or Stinking Sumac produces fragrant leaves and oily, hairy fruits.

treat irritated urethra, leukorrhea, diarrhea, dysentery, bronchitis, laryngitis, and bed-wetting in children and elderly. Contraindicated if inflammation is present. A root/stem bark extract was found to interact with viral and host cell surfaces, thwarting the ability of herpes simplex viruses 1 and 2 to infect cells, suggesting a strong antiviral effect. **WARNING:** May cause dermatitis.

WINGED OR DWARF SUMAC

Berries, bark, leaves, root

Rhus copallinum L.

Cashew Family (Anacardiaceae)

[*Rhus copallina* L.; *Schmaltzia copallinum* (L.) Small]

⚠ Shrub or small tree; to 30 ft. Leaves divided into 9–31 shiny, mostly toothless leaflets, with a *prominent wing along midrib*. Fruits red, short-hairy, Oct.–Nov. **WHERE FOUND:** Dry woods, clearings. S. ME to FL; e. TX to n. IL. **USES:** Cherokee used this and other sumacs similarly, as bark tea to stimulate milk flow; wash for blisters. Berries chewed to treat bedwetting and mouth sores. Root tea used for dysentery. Stems used to make flutes. Various components extracted from the fruits show experimental anticancer activity. **WARNING:** Do not confuse this sumac with Poison Sumac, *Toxicodendron vernix* (L.) Kuntze (*Rhus vernix* L.), which has white fruits and toothless leaves and grows in or near swamps.

SMOOTH SUMAC

Fruits, bark, leaves, root

Rhus glabra L.

Cashew Family (Anacardiaceae)

⚠ Shrub; 3–20 ft. Twigs and leafstalks *smooth, without hairs*. Leaves with 11–31 *toothed* leaflets. Fruits red, with short, appressed hairs; June–Oct. **WHERE FOUND:** Fields and openings. Throughout our area. **USES:** Chippewa used a root infusion for colds and as an emetic. Native Americans used berries to stop bedwetting. Leaves smoked for asthma; leaf tea used for asthma, diarrhea, stomatosis (mouth diseases), dysentery. Root tea emetic, diuretic, antidiarrheal. Bark tea formerly used for di-

Winged or Dwarf Sumac. Note wings on stems.

Smooth Sumac has smooth leaves and stems.

arrhea, dysentery, fevers, scrofula, general debility from sweating; also for mouth or throat ulcers, leukorrhea, and anal and uterine prolapse; astringent, tonic, antiseptic. Powdered bark, boiled in equal parts of milk and water, mixed in flour to make a plaster, once used for burn treatment, and claimed to leave no scars (don't try this at home). Fruit tea famous as a cooling drink; gargle for inflammation and ulceration of the throat. Inner-bark tea also a gargle for sore throat caused by mercurial salivation (during a time when physicians routinely prescribed and poisoned patients with mercury treatments for a wide range of disease conditions). Self-described "Indian doctors" used a root decoction to treat gonorrhea and gleet; a wash for ulcers. Gallotannins and other polyphenols from the seeds a source of energy-producing seed oils and a tanning material. Of 100 medicinal plants screened for antibiotic activity, this species was most active, attributed to content of gallic acid, 4-methoxygallic acid, and methyl gallate. Alcoholic extracts had the strongest activity. **WARNING:** Do not confuse this sumac with Poison Sumac *Toxicodendron vernix* (L.) Kuntze (*Rhus vernix* L.), which has white fruits and toothless leaves and grows in or near swamps.

STAGHORN SUMAC

Leaves, berries, bark, root

Rhus typhina L. [*Rhus hirta* (L.) Sudworth] Cashew Family (Anacardiaceae)

Shrub or small tree; 4–15 ft. Similar to *R. glabra* (above), but *twigs and leafstalks strongly hairy*. Fruits long-hairy; June–Sept. **WHERE FOUND:** Dry, rocky soil. NS to NC, SC, GA; IL to MN. **USES:** Similar to those for *R. glabra.* Native Americans used berries in cough syrups. Berry tea used for "female disorders," lung ailments. Gargled for sore throats, worms. Fruit tea of this and other *Rhus* species once widely used as a folk medicine for a cooling effect in fevers and inflammatory condition; gargle for sore throat. Leaf tea used for sore throats, tonsillitis. Root or bark tea astringent; used for bleeding. **WARNING:** Do not confuse this sumac with Poison Sumac *Toxicodendron vernix* (L.) Kuntze (*Rhus vernix* L.), which has white fruits and toothless leaves and grows in or near swamps.

Staghorn Sumac has strongly hairy leaves and twigs.

SHRUBS WITH SIMPLE ALTERNATE LEAVES; NOT TOOTHED

LEATHERWOOD

Bark

Dirca palustris L.

Mezereum Family (Thymelaeaceae)

⚠ Branched shrub; 1–9 ft. Branchlets *pliable*, smooth, jointed; bark very tough. Leaves oval to obovate, on short stalks. Yellowish bell-like flowers appear before leaves; Apr.–May. **WHERE FOUND:** Rich woods, along streams. NB to FL; LA to MN. **USES:** Chippewa and other Native American groups used bark tea as a laxative and for lung ailments. Minute doses cause burning of tongue, salivation. Folk remedy for toothaches, facial neuralgia, paralysis of tongue. **WARNING:** Poisonous. Causes severe dermatitis, with redness, blistering, and sores.

SPICEBUSH

Leaves, bark, berries, twigs

Lindera benzoin (L.) Blume

Laurel Family (Lauraceae)

Shrub; 4–15 ft. Leaves aromatic, ovate, without teeth. Tiny yellow flowers in axillary clusters *appear before leaves*; Mar.–Apr. Fruits highly *aromatic, glossy, scarlet*, with a single large seed; Sept.–Nov. **WHERE FOUND:** Moist, rich soils. Common along streambanks in rich, moist woods from ME to FL, TX to MI. The scarlet, strongly spice-scented berries and the pleasant-scented, entire (without teeth) leaves distinguish this plant. The berries often persist on branches after leaves drop in autumn. Male and female flowers on separate plants (sex may change from year to year). **USES:** Cherokee and other Native Americans used all plant parts as diaphoretic for treating colds, coughs, cramps, delayed menses, croup, measles; bark tea used as a blood purifier and for sweating, colds, rheumatism, anemia. Settlers used berries as an Allspice substitute. Medicinally, the berries were used as a carminative

TOP LEFT: *Leatherwood has pliable stems, hence the name Leatherwood.* TOP RIGHT: *Spicebush has bright red, spicy-aromatic berries that remain after leaves drop.* BOTTOM LEFT: *Spicebush blooms in early spring.*

ABOVE: *Staggerbush is a poisonous shrub.*

for flatulence and colic. The oil from the fruits was applied to bruises and muscles or joints (for chronic rheumatism). Twig tea was popular for colds, fevers, worms, gas, and colic. The bark tea was once used to expel worms, for typhoid fevers, and as a diaphoretic for other forms of fevers. An extract of the stem bark has been found to strongly inhibit yeast (*Candida albicans*), much better than any of the other 53 species studied. Thirty-nine components have been identified from the essential oils of different plant parts of Spicebush. Contains various biologically active lactones, similar to those found in the Chinese species *L. obtusiloba* Blume.

STAGGERBUSH Leaves

Lyonia mariana (L.) D. Don. Heath Family (Ericaceae)

Slender, deciduous shrub; to 7 ft. Leaves thin, oblong to oval. White or pinkish flowers in *umbrella-like racemes, in clusters on old leafless*

branches; Apr.–June. **WHERE FOUND:** Sandy, acid pine thickets. Southern RI, CT, NY to FL; e. TX to AR. **USES:** Cherokee used leaf tea externally for itching, ulcers. Benjamin Smith Barton, in his classic *Collections for an Essay Towards a Materia Medica of the United States* (1801), wrote that leaf tea was used as a wash for "disagreeable ulceration of the feet, which is not uncommon among the slave, & c., in the southern states." **WARNING:** Poisonous; produces "staggers" in livestock, hence the common name.

ALTERNATE, FRAGRANT, LEATHERY LEAVES

SWEETFERN

Leaves

Comptonia peregrina (L.) Coult. — Wax-myrtle Family (Myricaceae)
(*Myrica aspleniifolia* L.)

Strongly aromatic, deciduous shrub; 2–5 ft. Leaves soft-hairy, lance-shaped; 3–6 in. long, with *prominent rounded teeth*. Flowers inconspicuous. Fruits *burlike*; Sept.–Oct. **WHERE FOUND:** Dry soil. NS to VA, GA mtns.; OH, NE, IL to MN, MB. The distinctly shaped, leathery, aromatic leaves of *Comptonia* give the plant a feathery or fernlike appearance. While often found in infertile soils near shores, it is a common weedy shrub of dry roadsides, gravel banks, and woodland clearings. The species name *peregrina* means "foreign." This is a misnomer from an American perspective—it was foreign to the European botanist who first named it. **USES:** Cherokee used tea of leaves for roundworms; Delaware used a wash to treat blisters; also to purify blood, as an expectorant, and for tuberculosis; in combination with other herbs to treat bladder inflammation. The Menominee valued the leaves as a seasoning and aid in childbirth; placed among berries to avoid spoilage. Micmac used leaf tea as a tonic beverage; wash for Poison Ivy rash. Malecite used for sprains, swelling, and tuberculosis. Leaf tea astringent; folk remedy for diarrhea, dysentery, leukorrhea, rheumatism, vomiting of blood. In colonial America used as a substitute for imported balsams to treat asthma and bronchitis. Contains stilbenes, synthetic analogs of which have been produced, which

Sweetfern leaves, suggestive of fern leaves, are sweetly fragrant.

show research promise against the organisms that cause anthrax and tuberculosis. Essential oil components show experimental promise against several cancer cell lines.

WAX-MYRTLE, CANDLEBERRY

Leaves, fruit, root bark

Myrica cerifera L. [*Morella cerifera* (L.) Small]

Wax-myrtle Family (Myricaceae)

Coarse shrub or small tree; to 26 ft. Young branchlets waxy. Leaves oblong to lance-shaped; *leathery*, evergreen, with *waxy globules*. Fruits 1/8 in. across; Mar.–June. The hard seeds are covered in a white or gray wax, long used in candle making. **WHERE FOUND:** Swamp thickets. S. NJ to FL; TX to AR. **USES:** Candle wax is produced from fruits. Root bark formerly used in tea as an astringent and emetic for chronic gastritis, diarrhea, dysentery, leukorrhea, "catarrhal states of the alimentary tracts," jaundice, scrofula, and indolent (hard to heal) ulcers. Leaf tea was used for fevers, externally as a wash for itching. Powdered root bark was an ingredient in "composition powder," once a widely used home remedy for colds and chills. One component in the plant, myricitrin, has anti-inflammatory, antimutagenic, diuretic, and antibacterial activity. Myricitrin, myricetin, and a recently isolated compound, myricanol, show research interest in targeting a protein associated with Alzheimer's disease development. **WARNING:** Wax is irritating. Constituents of the wax are reportedly carcinogenic.

Wax-myrtle, Candleberry can grow into a small tree.

SWEET GALE (NOT SHOWN)

Berries, root bark, leaves

Myrica gale L.

Wax-myrtle Family (Myricaceae)

Fragrant deciduous shrub; 2–6 ft. Leaves gray; oblong to lance-shaped. Flowers in clusters, at ends of previous year's branchlets; Apr.–June. Fruit with 2 winglike bracts; July–Aug. **WHERE FOUND:** Swamps, shallow water. NL to mtns. of NC, TN to MI; WI, MN; AK, Canada; circumboreal also in n. Europe and Asia. **USES:** Potawatomi used burning leaves as a smudge to repel insects; leaves stored with berries to prevent spoilage. Branch tea once used as a diuretic for gonorrhea. In Europe a folk medicine for gout, fevers, itching, and to repel insects. Inner bark or tea and leaves used for tuberculosis. Berries historically used in England to flavor malt beverages. This plant, like legumes, takes nitrogen from the atmosphere and locks it into the soils as a natural fertilizer. Myrigalone B, an extract from the fruit exudates, is a potent antioxidant that inhibits

Bayberry fruits are often used to scent candle wax.

free radical damage from the liver, therefore possibly having liver-protectant activity. Myrigalone A from the leaves has a novel mode of action in inhibiting seed germination of competing plant species; also herbicidal activity. Antiviral activity has been reported for aqueous extracts of the fresh plant. **WARNING:** Essential oil is reportedly toxic; inhibits growth of various bacteria; also with potential insect repellent or insecticidal activity.

BAYBERRY — Leaves, bark, fruits

Myrica pensylvanica Mirb. — Wax-myrtle Family (Myricaceae)

[*Morella pensylvanica* (Mirb.) Kartesz]

Stout shrub; 3–12 ft. *Branches grayish white*. Leaves elliptic to obovate (widened at tips). Flowers in clusters *below leafy tips*; Apr.–July. Young fruits very hairy. **WHERE FOUND:** Sterile soils near coast. Canadian coast to VA, NC (rare). **USES:** Same as for *Myrica cerifera* (above). Micmac used leaf snuff for headaches, leaf tea as a stimulant; poulticed root bark for inflammation. **WARNING:** Wax is considered toxic.

ALTERNATE, TOOTHED, OVAL LEAVES

SPECKLED ALDER; TAG ALDER — Stem bark, leaves

Alnus incana (L.) Moench ssp. *rugosa* (Du Roi) Clausen — Birch Family (Betulaceae)

Open, spreading shrub; to 18 ft. Leaves ovate to elliptical, tip acute or blunt, base wedge-shaped, coarsely double toothed. Bark *smooth*, grayish to reddish brown, with prominent *white horizontal speckles*. Flowers (catkins) Feb.–May. "Cones" woody, erect, persistent. Hybridizes with *A. serrulata* (Ait.) Willd. **WHERE FOUND:** Forms thickets along waterways. PE, QC, ME to MD, west to ND; adjacent Canada. **USES:** Native Americans used bark tea for diarrhea, pain of childbirth, coughs, toothaches, sore mouth, and as a blood purifier; diuretic, purgative, emetic. Externally, as an eyewash, and a wash for hives, Poison Ivy rash, hemorrhoids, swellings, and sprains. Leaves poulticed as a folk medicine to stop bleeding, relieve pain. Used in 1800s for malaria and

Speckled Alder, Tag Alder forms thickets along waterways.

American Hazelnut produces distinctive fringed fruit.

syphilis. Alders have traditionally been used in the treatment of hepatitis, cancers, rheumatism, dysentery, diarrhea, and stomachaches, among others. They contain a wide range of biologically active compounds, including antioxidant and anti-inflammatory diarylheptanoids, flavonoids, phenols, and tannins. The glycosides oregonin and salicortin show experimental potential in mitigating the progression of obesity. **RELATED SPECIES:** Native Americans across N. America used other alders similarly; also European and Asian species in their respective cultures.

AMERICAN HAZELNUT — Inner bark, twig hairs, nuts

Corylus americana Walt. — Birch Family (Betulaceae)

Shrub; to 15 ft. Bark light gray, smooth. Stems and leafstalks with *bristly, glandular hairs*. Leaves with offset edge at base to heart-shaped, broadly oval with nearly parallel sides (appearing square in overall outline), double-toothed, to 5 in. long. Flowers Apr.–May. Fruits with edible nuts encased in beaked, toothed bracts. **WHERE FOUND:** Thickets. ME to GA; AR, OK to SK. **USES:** Cherokee drank bark tea for hives, fevers; astringent; also emetic. Bark poultice used to close cuts and wounds and to treat tumors, old sores, and skin cancers. Twig hairs were used by Native Americans and historically by physicians to expel worms and treat nephritis. Nuts, similar to but inferior in size and flavor to filberts and hazelnuts, have a low-quality seed oil once used for obstinate coughs.

WITCH-HAZEL — Bark, leaves, twigs

Hamamelis virginiana L. — Witch-hazel Family (Hamamelidaceae)

Deciduous shrub or small tree; to 15 ft. Leaves obovate, *scalloped margins* (large wavy teeth), with *uneven, wedge-shaped* bases; end buds distinctly scalpel-shaped. Flowers yellow, in axillary clusters; petals *very slender*, to 1 in.; flowers bloom after leaves drop; Sept.–Dec. **WHERE**

ABOVE: *Witch-hazel usually flowers after leaves drop.* RIGHT: *Witch-hazel's leaves have scalloped edges. Note fruits in leaf axils, maturing a year after flowering.*

FOUND: Woods. NS, QC to FL; TX to MN. **USES:** One of the most important American medicinal plants through history and today. Cherokee used leaf tea for colds, fevers, sore throat; as a wash to bathe sores and skinned knees; bruised leaves poulticed on scratches; bark tea for tuberculosis. Chippewa used bark infusion as a wash for sore eyes and skin eruptions. Menominee used a twig decoction to treat a lame back, also rubbed on the legs of game participants to keep limber. The hard brown to black seeds were sacred beads. Widely used today (in distilled extracts, ointments, eyewashes) as an astringent for hemorrhoids, toning skin, suppressing profuse menstrual flow, eye ailments. Used commercially in preparations to treat hemorrhoids, irritations, minor pain, and itching. Tannins (hamamelitannin and proanthocyanidins) in the leaves and bark are thought to be responsible for astringent and hemostatic properties, antioxidant activity. In the U.S., approved as a nonprescription drug for use in external analgesic and skin protectant products, and as an external anorectal, primarily used for symptomatic relief of hemorrhoids, irritation, minor pain, and itching. Products are available wherever over-the-counter drugs are sold. Used in European phytomedicine for treatment of burns, dermatitis, hemorrhoids, local inflammation of mucous membranes, minor skin injuries, varicose veins, and venous conditions, among others. Bottled Witch-hazel water, widely available, is a steam distillate that may not contain the same levels of polyphenols as other extraction methods. Light breaks down active components, therefore Witch-hazel preparations should be stored in a dark place. Hamamelitannins and related compounds of promise in research for colon cancer prevention. **RELATED SPECIES:** Vernal Witch-hazel, *Hamamelis vernalis* Sarg., with a range centered in the Ozarks, blooms Dec.–Mar. Leaves and bark indiscriminately harvested

Vernal Witch-hazel, with red-tinted flowers, blooms from winter to early spring.

Hop-hornbeam fruits superficially resemble hops.

as Witch-hazel, without distinguishing species. Three additional *Hamamelis* species also occur in Japan and China.

HOP-HORNBEAM

Bark, heartwood

Ostrya virginiana (Miller) K. Koch — Birch Family (Betulaceae)

Small tree, often shrublike; usually to 20 ft. (rarely to 60 ft.). Leaves narrow to broad ovate or oblong, margins sharply (often doubly) serrated, 3–5 in. long. Catkins are short, cylindrical, producing flattened oval seed to 1 in. across, with bristly hairs at base. Fruit in *overlapping clusters resembling hops*. **WHERE FOUND:** Moist and dry woods, streambanks. NS to FL, west to TX, north to MB. **USES:** Cherokee used bark tea as blood builder, to bathe sore muscles. Chippewa used wood steam baths to treat rheumatism. Iroquois used bark decoction for rectal cancer. Decoction of the heartwood used as folk medicine to treat fevers; tonic; and blood purifier for malaria, nervous conditions, and dyspepsia.

CONIFERS WITH FLAT NEEDLES IN SPRAYS

BALSAM FIR Resin, leaves

Abies balsamea (L.) Mill. Pine Family (Pinaceae)

Spire-shaped tree; to 60 ft. Flattish needles, to 1¼ in. long, in flattened sprays; stalkless. Needles *rounded at base*, each with *2 white lines beneath*. Cones 1–4 in. long, erect; purple to green, scales mostly twice as long as broad. Bark *smooth*, with numerous resin pockets. Widely used in Northeast as a Christmas tree. **WHERE FOUND:** Moist woods. Canada, south through New England and along mtns. to VA and WV; west through n. OH to ne. IA, MI. **USES:** Canada balsam, an oleoresin, is collected by cutting bark blisters or pockets in wood, July–Aug. Used as an antiseptic, in creams and ointments for hemorrhoids, wounds, and as a root-canal sealer. Diuretic (may irritate mucous membranes). The Penobscot applied resin as an analgesic antiseptic for burns, sores, bruises, and wounds. Iroquois used leaf tea for colds, coughs, and asthma. The Chippewa inhaled fumes from resin on hot coals to treat headache. The oleoresin is pale yellow to greenish yellow; transparent and pleasantly scented. A primary commercial application has been as a sealing agent for mounted microscope slides. Newly discovered unique compounds, abibalsamins A and B, show research promise against several cancer cell lines. A diterpene, cis-abienol, is a major aroma component of the resin. **WARNING:** Resin may cause dermatitis in some individuals.

FRASER FIR, SHE BALSAM (NOT SHOWN) Resin

Abies fraseri (Pursh) Poiret Pine Family (Pinaceae)

Similar to *A. balsamea* (above), but needles and cones are generally smaller; *cone-scale margins toothed or jagged*. **WHERE FOUND:** Isolated to mountains. VA, NC, TN. Widely grown in the South on Christmas tree farms. **USES:** Cherokee used resin for chest ailments, coughs, sore throat, urinary tract infections, and wounds.

EASTERN HEMLOCK Leaves, bark

Tsuga canadensis (L.) Carr. Pine Family (Pinaceae)

Evergreen tree; 50–90 ft. Needles flat, 5/16–9/16 in. long, on short slender stalks. Needles bright green above, *silvery whitish beneath*. Cones drooping, to 1 in. long, with few scales; scales rounded. **WHERE FOUND:** Hills in rocky woods. NS to MD, GA mtns.; AL to KY, IN, e. MN. **USES:** The Iroquois used tea made from leafy twig tips in steam baths for rheumatism, colds, and coughs, and to induce sweating. Cherokee used leaf tips for kidney ailments; bark poulticed for itchy armpits. Inner-bark tea used for colds, fevers, diarrhea, coughs, "stomach troubles"; leafy tips eaten to treat scurvy (vitamin C deficiency). Tannins explain many of the indications. All green leaves, even Eastern Hemlock leaves, contain some vitamin C, which explains use in treating scurvy. Externally, used as a wash for rheumatism and to stop bleeding. Bark is very as-

LEFT: *Balsam Fir, a common Christmas tree in the Northeast.* ABOVE: *Eastern Hemlock needles are dark green above and silvery beneath, with small cones.*

tringent; formerly used as poultice for bleeding wounds, and in tanning leathers. The oleoresin derived from the bark is dark reddish brown, opaque, and has a characteristic turpentine-like fragrance.

NEEDLES MORE THAN 1 IN. LONG, IN CLUSTERS

TAMARACK, BLACK LARCH

Bark, gum, needles, buds

Larix laricina (Du Roi) K. Koch

Pine Family (Pinaceae)

Coniferous tree; to 100 ft. *Deciduous* needles, to 1 in. long, in *circular clusters*. Cones oval, to ¾ in. long; scales few, rounded. **WHERE FOUND:** Swamps, wet soils. NL to WV; n. IL; across s. Canada to AK. **USES:** Ojibwa used dried buds as a fumigant for coughs; bark used by various Native American groups for colds, coughs, and to treat wounds and inflammation. Bark tea traditionally used as a laxative, tonic, and diuretic for jaundice, rheumatism, and skin ailments. Gargled for sore throats. Poulticed on sores, swellings, and burns. Leaf tea astringent; used for hemorrhoids, diarrhea, dysentery, and dropsy; poulticed

Tamarack, Black Larch has deciduous leaves in whorls.

for burns and headaches. Gum chewed for indigestion. Used by Cree as an antidiabetic; recent research suggests it decreases glycemia and improves sensitivity to insulin; suggested as a possible treatment for managing obesity and diabetes. Contains antioxidant proanthocyanidins. Diterpenes from the bark are of research interest against colon cancer cell lines. **RELATED SPECIES:** The w. N. American *Larix occidentalis* Nutt. contains arabinogalactins, and products containing it are touted as immunostimulants and suggested for inhibition of liver and other cancers. **WARNING:** Sawdust can cause dermatitis.

SHORTLEAF PINE, YELLOW OR HARD PINE

Inner bark, buds, pitch

Pinus echinata Mill. — Pine Family (Pinaceae)

⚠ Straight evergreen tree; to 120 ft. Slender needles 3–5 in. long; *in 2s or 3s.* Cones oval; each *scale tipped with a short prickle.* **WHERE FOUND:** Dry woods. Se. NY, OH to FL; TX to s. IL. **USES:** Rappahannock used inner bark in tea to induce vomiting. Choctaw used cold tea of buds as a worm expellent. Pitch tea used as laxative and for tuberculosis; also for kidney ailments causing backaches. **WARNING:** Wood, sawdust, balsam, and turpentine of various pines may cause dermatitis in sensitive individuals.

LONGLEAF PINE

Pitch (resin), inner bark, buds, turpentine from wood

Pinus palustris Mill. — Pine Family (Pinaceae)

⚠ Evergreen tree; to 90 ft. Needles in 3s; *very long,* 7–12 (occasionally 18) in. Cones cylindrical, 6–10 in. long; each scale with a short, curved spine. **WHERE FOUND:** Sandy soil, coastal plains. Se. VA to FL; TX, AR. **USES:** The common Longleaf Pine of the Southeast was one of the most

ABOVE: *Shortleaf Pine, Yellow or Hard Pine is one of the most common pines of the South.* RIGHT: *Longleaf pine produces needles that grow to 1 ft. long.*

valuable natural resources of the American South. From the earliest sixteenth-century Spanish settlements to well into the twentieth century, "naval stores" such as pine pitch (crude resin), turpentine, and pine tar (rosin) were produced on a large scale. Turpentine is a volatile distillate produced from the resin within the wood. Pine tar or rosin is a solid by-product of turpentine production. All are used for a wide range of purposes, though have largely been replaced by petroleum products. Medicinal use of the tree and its products were important, especially during the Civil War. A tea or tincture (in gin) of the buds and inner bark were widely used on plantations as a cold and cough remedy. A decoction of the inner bark was used to treat diarrhea. Raw pine pitch (resin) rolled into pills were used as a diuretic. One-half part rosin (tar) and one part tallow melted in an iron pot were repeatedly brushed onto the boots of Confederate soldiers to help keep feet dry. A mix of turpentine with 5 percent beeswax was then rubbed on the boots as a polish and to provide more water resistance. Tar (rosin) mixed with plaster of Paris was sprinkled on wounds in Confederate hospitals to reduce odor. It was also suggested that the tree should be planted simply for its health-giving presence, as "ozone" from the tree would modify the atmosphere and improve control of malaria. Turpentine, distilled from the wood, formerly used for colic, chronic diarrhea, worms, and to arrest bleeding from tooth sockets; folk remedy for abdominal tumors. Externally it was applied as a rubefacient (local skin irritant or counterirritant) to increase blood circulation on rheumatic joints. **WARNING:** Considered potentially toxic.

WHITE PINE

Twigs, bark, leaves, pitch

Pinus strobus L.

Pine Family (Pinaceae)

Evergreen tree, to at least 150 ft. *Needles in 5s*; slender, pale green, glaucous. Cones cylindrical; to 8 in. long. **WHERE FOUND:** Common in East from Canada to GA mtns.; west to n. IL, cen. IA. **USES:** Algonquin groups used cold bark tea as a treatment for colds; the resin from bark wounds was boiled to make a decoction for sore throat, colds, and tuberculosis. The Iroquois used powdered, soft, dried, rotted wood as a kind of baby powder and for chafing caused by perspiration. Leaves used as incense in the home to prevent sickness of all kinds. Turpentine-rich knots were boiled and the decoction then sipped to treat tuberculosis. Used extensively by Native

White Pine, a majestic evergreen of the Northeast.

White Pine needles have a whitish cast.

Americans; pitch poulticed to "draw out" boils, abscesses; also used for rheumatism, broken bones, cuts, bruises, sores, abcess at fingertip, and inflammation. Twig tea used for kidney and lung ailments; emetic. Bark and/or leaf tea used for colds, coughs, grippe, sore throats, lung ailments; poulticed for headaches and backaches. Inner bark formerly used in cough syrups.

MISCELLANEOUS CONIFERS

EASTERN RED CEDAR

Fruits, leaves

Juniperus virginiana L. Cypress Family (Cupressaceae)

⚠ Spire-shaped; 10–50 ft. Leaves *scalelike, overlapping; twigs 4-sided.* Fruits hard, round, dry, blue-green. **WHERE FOUND:** Infertile soils, old pastures. Canada, ME to GA; TX to MN, MI. **USES:** The Kiowa used branches of the leaves as a prayer fumigant during the Peyote ceremony. The Cree used the berries as a diuretic. Other Natives Americans used fruit tea for colds, worms, rheumatism, coughs, and to induce sweating. They also chewed fruit for canker sores. Leaf smoke or steam inhaled for colds, bronchitis, rheumatism, and purification rituals. Berries once suggested as a diuretic substitute for Common Juniper (see p. 296). Berry tea once a popular folk remedy for dropsy (edema), also used for rheumatism, stiff joints, stimulant, and as an emmenagogue. Wood shavings (one bushel) boiled in an iron vessel produced about a cup of oil, used as an insect repellent. Essential oil aroma is familiar to most as the fragrance of the insect-repelling wood from which cedar chests and pencils are made. Said to contain the antitumor compound podophyllotoxin, best known from Mayapple (p. 64). **WARNING:** All parts may be toxic. Pollen is source of significant allergies in spring.

LEFT: *Eastern Red Cedar. Note its overlapping, scalelike leaves.* ABOVE: *Black Spruce needles are blunt tipped.*

BLACK SPRUCE

Inner bark, resin, twigs, cones, leafy tips

Picea mariana (Mill.) B. S. P. — Pine Family (Pinaceae)

⚠ Evergreen tree; 10–90 ft. Needles stiff, crowded, 4-angled, dark green, *mostly glaucous, blunt-tipped.* Cones are short-oval to rounded; dull gray-brown. **WHERE FOUND:** Woods. Canada to PA, VA mtns.; WI, IL. **USES:** Algonquins used the pitch as a poultice for boils, scabs, sores, and abscesses; also gum routinely chewed as a pastime. Twigs boiled to make a tea for coughs. Inner bark decoction for throat and lung problems. Various Native American groups made spruce beer by boiling twigs and cones, then adding maple syrup to catalyze fermentation. The beer, along with the leafy tips, was offered to the first explorers, particularly the French, as a treatment for scurvy. It is likely that spruce tips saved the ill-fated crew of Jacques Cartier from scurvy in 1536. In addition to vitamin C, preparations from the leaves and bark contain numerous amino acids that could have aided in recovery from scurvy. Inner-bark tea is a folk medicine for kidney stones, stomach problems, rheumatism. Resin was poulticed on sores to promote healing. Bark contains bioactive polyphenols such as anti-inflammatory and antioxidant resveratrol. **WARNING:** Sawdust, balsam (resin), and even the needles may produce dermatitis.

RED SPRUCE

Boughs, pitch

Picea rubens Sarg. — Pine Family (Pinaceae)

Evergreen tree; to 100 ft., with hairy branchlets. Needles slender; yellowish, *not glaucous.* Cones elongate-oval; brown to red-tinged brown. **WHERE FOUND:** Woods. Canada; New England to NC, TN, OH. **USES:** Cherokee used tea of boughs for colds and to "break out" measles. Pitch formerly poulticed on rheumatic joints, chest, and stomach to

Red Spruce produces hairy branchlets and yellowish needles.

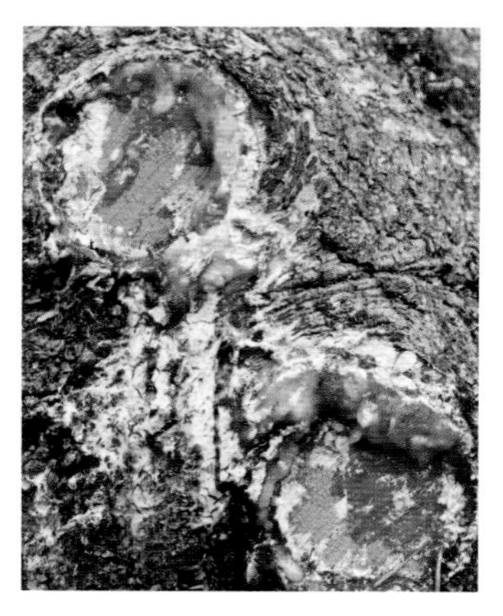

Red Spruce bark wounds exuding gum.

relieve congestion and pain. Used similarly to Black Spruce. **WARNING:** See under Black Spruce (above).

NORTHERN WHITE CEDAR

Leaves, inner bark, leaf oil

Thuja occidentalis L.

Cypress Family (Cupressaceae)

⚠ Evergreen tree; to 60 ft. Leaves in *flattened sprays*; small, appressed, overlapping. Cones *bell-shaped*, with *loose scales*. **WHERE FOUND:** Swamps; cool, rocky woods. NS to GA mtns.; n. IL to MN. By 1566 it was one of the first N. American trees introduced to Europe. **USES:** Native Americans used leaf tea for headaches, colds; also in cough syrups; in steam baths for rheumatism, arthritis, colds, congestion, headaches, gout; externally, as a wash for swollen feet and burns. Inner-bark tea used for congestion and coughs. Regarded as a general panacea and widely used in combination with other herbs. Physicians once used leaf tincture externally

Northern White Cedar leaves appear in flattened sprays.

on warts, venereal warts, hemorrhoids, ulcers, bed sores, and fungus infections. Internally, leaf tincture or decoction used for bronchitis, asthma, coughs, fevers, scurvy, pulmonary disease, rheumatism, enlarged prostate with urinary incontinence. Folk remedy for cancer. Experimentally, leaf oil is antiseptic, expectorant, counterirritant; extracts have antiviral properties. The leaves and the polysaccharides within stimulate the immune system and decrease inflammatory-inducing cytokines; experimentally shown to inhibit cancer cells. A combination product containing Wild Indigo, *Echinacea* root, and *Thuja* leaves is widely sold, with immunostimulating attributes. **WARNING:** Leaf oil is toxic, causing hypotension, convulsions. Fatalities have been reported.

BROAD LEAVES; EVERGREENS

AMERICAN HOLLY

Leaves, bark, berries

Ilex opaca Ait.

Holly Family (Aquifoliaceae)

⚠ Evergreen tree; to 90 ft. Leaves smooth, leathery; with few to many *spine-tipped* teeth. Fruits red or orange (rarely yellow); Sept.–Oct. Sprigs a familiar Christmas decoration. **WHERE FOUND:** Mixed woods. E. MA to FL; TX, OK to IL. **USES:** Alabama tribe used plant infusions for dysentery. Leaf infusion used as an eyewash and as a wash for toddlers too weak to walk. Choctaw also used leaf tea as eyewash for sore eyes; root for colic, fevers, and toothache remedy. Berries chewed for colic, indigestion. Leaf tea for measles, colds, flu, coughs, expectorant for pneumonia; externally a wash for sores, itching. Leaves boiled in beer, a frontier remedy for pleurisy. Thick syrup of berries formerly used to treat children's diarrhea. Chewing only 10–12 berries acts as strong laxative, emetic, and diuretic. Bark tea once used in malaria and epilepsy. **WARNING:** Fruits considered poisonous, inducing violent vomiting.

American Holly has distinctive, leathery, spine-tipped leaves. The holly of Christmas.

BULL-BAY, SOUTHERN MAGNOLIA

Bark, leaves, fruits, seeds

Magnolia grandiflora L.

Magnolia Family (Magnoliaceae)

Our largest magnolia, to 90 ft. Leaves leathery, evergreen, elliptical, to 10 in. long, 5 in. wide. Smooth above, rusty-hairy beneath. Flowers creamy white, cup-shaped, large, to 9 in. across; Apr.–July. **WHERE FOUND:** Rich woods, moist bottomlands. Se. VA, south to FL, west to TX.

LEFT: *Bull-bay, Southern Magnolia is our largest evergreen magnolia, with petals to 5 in. long.* ABOVE: *Sweetbay is evergreen with petals to 2 in. long.*

Widely grown for its beautiful evergreen foliage and dramatic flowers. **USES:** Choctaw used a wash of the bark to treat prickly-heat itching and sores. Crushed bark used in steam baths to treat water retention. In nineteenth-century America, bark used to treat malaria and rheumatism. Fruits used as a digestive tonic, for dyspepsia, and for general debility. Seeds used in Mexican traditions for antispasmodic activity. Also used for high blood pressure, heart problems, abdominal discomfort, muscle spasms, infertility, and epilepsy. Contains magnolol and honokiol, with antispasmodic activity. Leaves contain various alkaloids, including magnoflorine, lanuginosine, liriodenine, and anonaine; lignans from leaves and seeds have anti-inflammatory activity. Science confirms sedative activity of seed. Flower extracts have antioxidant activity; suggested as a possible skin-whitening agent in Asian cosmetics. Flowers contain a sesquiterpene, vulgarenol, which has coronary vasodilation activity. **WARNING:** Leaves have caused severe contact dermatitis.

SWEETBAY — Bark, leaves

Magnolia virginiana L. — Magnolia Family (Magnoliaceae)

⚠ Small tree or shrub; to 30 ft. Leaves leathery, evergreen (deciduous in north); 3–6 in. long. Flowers white, cup-shaped, *very fragrant*; petals to 2 in.; Apr.–July. **WHERE FOUND:** Low woods. MA; PA to FL; MS north to TN. **USES:** Native Americans used leaf tea to "warm blood," "cure" colds. Traditionally, bark was used like that of *M. grandiflora* (see above). Bark also used for rheumatism, malaria, epilepsy. Leaf essential oil with antioxidant activity; of research interest against several cancer cell lines. **WARNING:** Leaves may cause dermatitis.

DECIDUOUS TREES WITH OPPOSITE, COMPOUND LEAVES

OHIO BUCKEYE

Nuts

Aesculus glabra Willd.

Soapberry Family (Sapindaceae)

[Formerly in the Horsechestnut Family (Hippocastanaceae)]

⚠ Small tree; 20–40 ft. Leaflets 5 (rarely 4–7); toothed, 4–15 in. long. Twigs *foul-smelling* when broken. Buds not sticky; scales at tips strongly ridged. Bark rough-scaly. Flowers yellow; Apr.–May. Fruit husk with *weak prickles*; Sept.–Oct. **WHERE FOUND:** Rich, moist woods. W. PA, WV, e. TN, cen. AL, cen. OK to NE, IA. **USES:** Traditionally, powdered nut (minute dose) used for spasmodic cough, asthma (with tight chest), intestinal irritations. Externally, tea or ointment used for rheumatism and hemorrhoids. The Delaware carried a "buckeye" (seed) in pocket to ward off rheumatism. Native Americans put the pulverized nuts in streams to stupefy fish, which floated to the surface for easy harvest. Contains various toxic saponins. **WARNING:** Nuts toxic, causing severe gastric irritation. Still, Indians made food from them after elaborate processing.

HORSECHESTNUT

Nuts, leaves, flowers, bark

Aesculus hippocastanum L.

Soapberry Family (Sapindaceae)

[Formerly in the Horsechestnut Family (Hippocastanaceae)]

⚠ Large tree; to 100 ft. Leaflets 5–7; to 12 in. long; without stalks, toothed. Buds large, very sticky. Broken twigs *not foul-smelling* as in Ohio Buckeye (above). Flowers white (mottled red and yellow); May. Fruits *spiny or warty*; Sept.–Oct. **WHERE FOUND:** Planted in towns. Naturalized from

LEFT: *Ohio Buckeye has smooth leaves and yellow flowers.*
ABOVE: *Horsechestnut seed.*

Eurasia. **USES:** As in *A. glabra* (see above); also, peeled, roasted nuts were brewed for diarrhea, prostate ailments. Leaf tea tonic; used for fevers. Flower tincture used on rheumatic joints. Bark tea astringent, used in malaria, dysentery; externally, for lupus and skin ulcers. Horsechestnut seed extracts widely prescribed orally in European phytomedicine for edema with venous insufficiency, for varicose veins, to improve vascular tone, and to help strengthen weak veins and arteries in reducing leg edema, nighttime calf muscle spasms, thrombosis, and hemorrhoids; uses backed by clinical studies. Contains aescin, which reduces capillary wall permeability, lessening diameter and number of capillary wall openings, regulating the flow of fluids to surrounding tissue; increases blood circulation. Also used in gastritis and gastroenteritis. Topically, aescin-containing gels or creams widely used to allay swelling and pain in bruising, sprains, and contusions. Injectable forms of aescin used in European trauma centers to help stabilize brain-trauma patients. Only chemically well-defined products are used; not the crude drug. Antioxidant, helping to reduce oxidative stress caused by cellular free radicals. **WARNING:** Outer husks poisonous; all parts can be toxic. Fatalities reported. Seeds (nuts) contain 30–60 percent starch, but can be used as a foodstuff only after the toxins have been removed.

Horsechestnut is often grown as a shade tree for its showy flowers.

RED BUCKEYE

Nuts, bark

Aesculus pavia L.

Soapberry Family (Sapindaceae)
[Formerly in the Horsechestnut Family (Hippocastanaceae)]

⚠ Small tree or shrub; to 15 ft. Leaflets 5, oblong to oblanceolate, smooth above. Flowers *red* (rarely yellow) on reddish stalks; petals with glandular hairs on the margins; Apr.–May. Fruits with smooth, thick, leathery cover, nuts about an inch in diameter; Sept.–Oct. **WHERE FOUND:** Rich moist woods. IL to NC, south to FL, west to TX, s. MO. **USES:** Cherokee used the nuts as a talisman (carried in pocket) to ward off rheumatism, hemorrhoids, and for good luck. Externally, poultice of crushed nut used for cancers, sores, infections. Bark tea used to treat bleeding after childbirth. Pounded nuts used as a poultice or salve for tumors and sores. Contains various saponins of potential anticancer research interest. **WARNING:** All parts potentially toxic. Avoid use.

LEFT: *Red Buckeye produces red flowers.* ABOVE: *White or American Ash produces winged, maplelike seeds.*

WHITE OR AMERICAN ASH Bark, leaves, root, seeds
Fraxinus americana L. Olive Family (Oleaceae)

Large tree; to 100 ft. Twigs hairless. Leaves opposite, pinnate, with 5–9 leaflets oval, slightly toothed or entire; *white* or pale beneath. Flowers Apr.–June. Fruits narrow, winged; Oct.–Nov. **WHERE FOUND:** Woods. NS to FL; TX, NE to MN. **USES:** Iroquois used bark tea from small saplings as strong laxative; root used as poultice for snakebite; bark chewed as a cleansing emetic during deer hunt. Meskwaki used bark infusion as wash for sores, itch, scalp itch due to head lice; snakebite remedy. Inner bark chewed and applied as a poultice to sores. Seeds thought to be aphrodisiac. Inner bark a folk medicine as a bitter astringent for bleeding and fevers. Leaves poulticed for snakebites. Seed said to be aromatic, drying to moist tissue, and preventive for obesity. Burned leaf ashes thought to be diuretic. Little researched; various secoiridoids found in leaves. One study found that various phenolic compounds, such as secoiridoids and oleoside, occur in the medium-toasted heartwood; tannins were absent; wood developed a distinct vanillin fragrance, making it an interesting material for wine barrels.

LARGE HEART-SHAPED LEAVES; OPPOSITE, OR WITH 3 LEAVES AT EACH NODE

COMMON CATALPA Bark, leaves, seeds, pods
Catalpa bignonioides Walt. Bignonia Family (Bignoniaceae)

Ornamental tree; to 45 ft. Leaves opposite, or in 3s from each node, large—to 10 in. long and 7 in. wide; oval to heart-shaped, with *an abruptly pointed apex*; not toothed. Leaves foul-odored when bruised. Flowers whitish, marked with 2 orange stripes and numerous *purple spots* within; thimblelike, with 5 unequal, wavy-edged lobes; flowers in large, upright, showy clusters; June–July. Seedpods long, cigar-shaped; seeds have 2 papery wings. **WHERE FOUND:** Waste ground; a street tree. FL, AL, MS, LA; naturalized north to New England; NY, OH, and westward. **USES:** Bark tea formerly used as an antiseptic, snakebite antidote, laxative, sedative, worm expellent (a Chinese species is also used against worms). Leaves poulticed on wounds, abrasions. Seed tea used for asthma, bronchitis; externally, for wounds. Pods sedative; thought to possess cardioactive properties. *Catalpa* leaves contain iridoid glycosides such as catalpol and catalposide, the former of which is sequestered by Catalpa Sphinx caterpillars, a specialized feeder of the leaves, in order to deter predators. **RELATED SPECIES:** Northern or Hardy Catalpa (*Catalpa speciosa* Warder ex Engelm.) is a larger tree; leaves have a long, pointed tip, more pleasant scent when crushed; flowers have fewer spots. It is challenging even for botanists to distinguish the relatively subtle differences between *C. bignonioides* and *C. speciosa.* Original range is unclear—perhaps native from IN to e. AR. Now commonly naturalized in se. U.S. The Asian *C. ovata* G. Don is used in Traditional Chinese Medicine as a diuretic. **WARNING:** Honey from flowers is potentially toxic.

Catalpas are among our showiest spring-flowering trees.

PRINCESS-TREE, PAULOWNIA

All parts

Paulownia tomentosa Steud. — Paulownia Family (Paulowniaceae)
[Formerly in the Figwort Family (Scrophulariaceae)]

Medium-sized, thick-branched tree; 30–60 ft. Leaves *heart-shaped,* pointed at tip; large—*to 12 in. long and broad* (sometimes larger); velvety beneath; leafstalks to 8 in. long. Flowers fragrant, hairy, purple thimbles, to 2 in. long; with 5 flared, unequal lobes; in large, candelabra-like clusters; Apr.–May. Fruits upright; hollow hulls are filled with tiny winged seeds; hulls split in two, suggesting hickory fruits in shape; persist through winter. **WHERE FOUND:** Occurs from NY to FL and westward. Asian alien. Introduced to U.S. in 1843. The wood is highly valued by the Japanese and is exported at a high price. **USES:** In China, a wash of the leaves and capsules is used in daily applications to promote the growth of hair and prevent graying. Leaf tea is used as a foot bath for swollen feet. Inner-bark tincture (soaked in 2 parts whisky) given for fevers and delirium. Leaves or ground bark are fried in vinegar, poulticed on bruises. Flowers are mixed with other herbs to treat liver ailments. In Japan the leaf juice is used to treat warts. Flavonoids in plant are of potential research interest against enzymes associated with Alzheimer's disease; antioxidant, antibacterial, anti-viral. **WARNING:** Contains potentially toxic compounds. Safety undetermined.

Princess-tree, Paulownia is easily identified when flowering.

OPPOSITE LEAVES; MAPLES

BOX-ELDER, ASHLEAF MAPLE

Inner bark, sap

Acer negundo L.

Soapberry Family (Sapindaceae)

[Formerly in the Maple Family (Aceracae)]

Tree; 40–70 ft. Twigs *glossy green*. Leaflets 3–5 (occasionally 7); *similar to those of Poison Ivy*, but Box-elder leaves are *opposite*, not alternate; coarsely toothed (or without teeth); end leaflet often 3-lobed, broader than lateral leaflets. Fruits are paired, maple-type "keys"; seed itself is longer and narrower than in most maple species. **WHERE FOUND:** Riverbanks, fertile woods. NS to FL, TX; n. to cen. MB, s. AB; also in CA. **USES:** Native Americans used the inner-bark tea as an emetic (induces vomiting). Sap boiled down as a sugar source.

Box-elder, Ashleaf Maple leaves resemble those of Poison Ivy.

STRIPED MAPLE

Inner bark, leaves, twigs

Acer pensylvanicum L.

Soapberry Family (Sapindaceae)

[Formerly in the Maple Family (Aceracae)]

Slender tree; to 15 ft. Bark greenish, with *vertical white stripes*. Leaves 3-lobed, finely double-toothed; to 8 in. wide. Small greenish flowers, in long clusters; May–June. Fruits ("keys") with paired winged seeds, usually set widely apart; June–Sept. **WHERE FOUND:** Woods. NS and south through New England, mtns. of PA, OH to TN, NC, n. GA; west to MI. **USES:** Micmac used inner-bark tea for colds, coughs, kidney infections, and gonorrhea. Penobscot used for spitting-up of blood; wash used for swollen limbs and paralysis. Historically, bark tea was used as a folk remedy for skin eruptions, taken internally and applied as an external wash. Leaf and twig tea used both to allay or induce nausea, and induce vomiting, depending on dosage. **RELATED SPECIES:** Bark from a closely related Asian species has shown significant anti-inflammatory activity.

Striped Maple leaves have 3 large lobes on the upper half. Note the striped bark.

SUGAR MAPLE

Inner bark, sap

Acer saccharum Marsh.

Soapberry Family (Sapindaceae)

[Formerly in the Maple Family (Aceracae)]

Large tree; 60–130 ft. Leaves green on both sides. Leaves *5-lobed; lobes not drooping, notches between lobes rounded.* Twigs glossy. Fruits paired, maplelike "keys." **WHERE FOUND:** Rich, hilly woods, fields. NL to n. GA, e. TX; north to MN. **USES:** Potawatomi used inner bark in tea as an expectorant for congestion; also for coughs, diarrhea; diuretic, expectorant, blood purifier. Maple syrup said to be a liver tonic and kidney cleanser, and used in cough syrups. During the maple sap-gathering process in spring, New Englanders once drank the sap collected in buckets as a spring tonic. **RELATED SPECIES:** Red Maple (*A. rubrum* L.) has red flowers and reddish branches. The leaf lobes are sharply pointed rather than rounded. Range is similar.

Red Maple has V-shaped leaf lobe sinuses.

Sugar Maple. Note the rounded sinuses between lobes.

TREES WITH OPPOSITE LEAVES; SHOWY WHITE SPRING FLOWERS

FRINGETREE

Root bark, trunk bark, leaves

Chionanthus virginicus L.

Olive Family (Oleaceae)

Shrub or small tree; 6–20 ft. Leaves opposite, oval, 3–8 in. long; mostly smooth. Flowers white, in *drooping clusters*; May–June. Petals slender. Fruits bluish black, resembling small olives. **WHERE FOUND:** Dry slopes. NJ to FL; TX, e. OK north to MO, s. OH. **USES:** A folk medicine for fevers; bark applied externally to wounds and ulcers. Physicians formerly used 10 drops (every 3 hours) of tincture (1 part bark by weight in 5 parts 50 percent grain alcohol and water) for jaundice and hepatitis. In the late nineteenth century, Fringetree bark tincture was widely employed by physicians who thought it relieved congestion of glandular organs and the venous system. It was employed for hypertrophy of the liver, wounds, nephritis, and rheumatism. Once considered diuretic, blood purifier, cholagogue, and a useful tonic. Choctaw used the root-bark tea to wash inflammations, sores, cuts, and infections. Contains various lignans and secoiridoids that may contribute to claimed effects.

Fringetree produces beautiful strap-like flowers.

Leaves contain levels of oleuropein comparable to those found in Olive leaf (*Olea europaea* L.). Oleuropein is a polyphenol in Olive leaf extracts valued for antioxidant, anti-inflammatory, antimicrobial, antiviral, cholesterol-lowering, and blood sugar–reducing effects. Also found in olives and olive oil; responsible for extra virgin olive oil's bitter, pungent flavor tones. **RELATED SPECIES:** Leaves of the Chinese species *C. retusus* Lindl. et Paxton have been used in Asia as a tea substitute. **WARNING:** Overdoses cause vomiting, frontal headaches, slow pulse.

FLOWERING DOGWOOD

Inner bark, berries, twigs

Cornus florida L. — Dogwood Family (Cornaceae)

Our most showy deciduous tree; 10–30 ft. Leaves ovate; *latex threads appear at veins when leaves are split apart.* Flowers in clusters; *4 showy white (or pink) bracts* surround the true flowers; Apr.–May. Fruits scarlet, dry, inedible, very bitter. **WHERE FOUND:** Understory tree of dry woods. ME to FL; TX to KS. Widely cultivated in natural range and elsewhere as an ornamental. **USES:** Astringent root-bark tea or tincture widely used in South, especially during the Civil War, for malarial fevers (substitute for quinine); also for chronic diarrhea. Root bark also poulticed for external ulcers. Scarlet berries soaked in brandy as a bitter digestive tonic and for acid stomach. Flowers once suggested as a substitute for Chamomile tea. Twigs used as "chewing sticks"—forerunners of modern toothbrushes. An 1830 herbal reported that the Indians and captive Africans in Virginia were remarkable for the whiteness of their teeth, and attributed it to the use of dogwood chewing sticks. Once chewed for a few minutes, the tough fibers at the ends of twigs split into a fine soft "brush." Contains verbenalin, which has reported pain-reducing, anti-inflammatory, cough suppressant, uterotonic, and laxative qualities. A recent study found that components isolated from the

LEFT: *Flowering Dogwood produces extremely bitter, inedible red fruits.* ABOVE: *Flowering Dogwood, one of our most iconic spring-flowering trees.*

bark have a moderate antiplasmodial activity (against malaria-causing *Plasmodium falciparum*); therefore another mechanism of action may be responsible for historic antimalarial claims. **WARNING:** As with hard toothbrushes, Dogwood chewing sticks can cause receding gums.

MISCELLANEOUS TREES WITH COMPOUND LEAVES

TREE-OF-HEAVEN, STINKTREE

Bark, root bark

Ailanthus altissima (Mill.) Swingle

Quassia Family (Simaroubaceae)

⚠ Smooth-barked tree; 20–100 ft. Leaves compound, similar to those of sumacs; crushed leaves *smell like peanuts*. Each leaflet has *2 glandular-tipped teeth at base* (on underside). Flowers small, yellow; June–July. Male flowers foul-smelling. Fruits look like *winged "keys,"* persisting through winter. **WHERE FOUND:** Waste places. Throughout our area. This Chinese native was introduced in the late nineteenth century as an ornamental. It quickly established itself. In cities such as New York and Boston, it grows in harsh conditions where no other plants seem able to survive. Considered a weed tree in many American cities. **USES:** Two ounces bark infused in 1 quart water, given in teaspoonfuls for diarrhea, dysentery, leukorrhea, tapeworm; used in Traditional Chinese Medicine. In Korean traditional medicine, the bark has been used for fever, bleeding, infections, and inflammation. Components extracted from the bark have analgesic, anti-inflammatory, antimicrobial, and potential anti-asthmatic activity. National Cancer Institute researchers have reported several antimalarial compounds, 5 of which are more potent than the standard antimalarial drug, chloroquine. Contains various coumarins that are of research interest in age-related disorders. **WARNING:** Large doses potentially poisonous. Gardeners who cut the tree may suffer from rashes.

TOP: *Tree-of-heaven, Stinktree's winged fruits in large clumps.* ABOVE: *Tree-of-heaven, Stinktree. Note the gland at base of leaf lobe.*

Silk Tree, Mimosa flower close-up.

Silk Tree, Mimosa produces showy pink flowers and feathery leaves.

SILK TREE, MIMOSA Bark, flowers

Albizia julibrissin Durazz. Pea Family (Fabaceae or Leguminosae)

Fast-growing, broad-crowned, short-trunked tree; to 30 ft. Leaves with a graceful, feathery appearance, doubly compound, with 2–30 pairs of oblong leaflets (pinnules) to 1 in. long, *one side distinctly longer than the other*; sensitive to touch, folding up at night. Flowers pink (rarely white), *silky, fluffy blooms*; May–Aug. Flat, brown, dry, papery, pealike pod to 6 in. long. **WHERE FOUND:** Dry soils. Alien from tropical Asia, introduced by 1785. NY southward and west. Commonly naturalized throughout the South. **USES:** In Traditional Chinese Medicine, bark (*he huan pi*) and flowers (*he huan hua*) are still in use. Bark is used in tea (mostly in combination with other herbs) for depression, restlessness, and insomnia caused by anxiety. Externally, poultice is used for traumatic injuries. Flowers are used as sedative for insomnia. Flowers have components with a potential anti-obesity effect, inhibiting triglyceride accumulation; also an antidepressant effect; bark extracts have anxiolytic and antioxidant effects.

KENTUCKY COFFEE-TREE Bark, pods, leaf, seeds

Gymnocladus dioica (L.) K. Koch Pea Family (Fabaceae or Leguminosae)
(Spelling variant: *Gymnocladus dioicus*)

Medium-sized tree; to 50–60 ft. Compound leaves with 7–13 leaflets. Whitish flowers in axillary wands; May–June. *Hard, flat pods to 10 in.* (2½ in. wide), pulpy within; seeds large, hard. **WHERE FOUND:** Rich woods. Cen. NY to TN; AR to SD. This tree is becoming more rare in N. America, thanks to reproductive challenges. The large pods and seeds are dispersed only by water (today); researchers theorize the large seedpods were once the food of large-mammal herbivores (such as Woolly Mammoths) now extinct in N. America that may have dispersed seed. **USES:** Caramel-like pod pulp used by Meskwaki to treat "lunacy." Leaf and pulp tea formerly employed for reflex troubles, and as a laxative.

ABOVE: *Kentucky Coffee-tree. Note the compound leaves with 7–13 leaflets.* RIGHT: *Hoptree, Wafer Ash produces 3 leaflets and waferlike fruits.*

Various Native American groups of the Midwest used root-bark tea for coughs caused by inflamed mucous membranes; diuretic; given to aid childbirth in protracted labor, stops bleeding; used in enemas for constipation. Leaves considered laxative. Seeds once used as a coffee substitute. **WARNING:** Toxic to grazing animals. Leaves are a fly poison. Seeds contain toxic saponins.

HOPTREE, WAFER ASH Root, root bark, young leaves and shoots, fruits

Ptelea trifoliata L. Rue Family (Rutaceae)

Small tree; 10–20 ft. Leaves palmately divided into *3 leaflets*, mostly without teeth; black-dotted (use lens). Flowers small, greenish; May–July. Fruits round, 2-seeded "wafers"; July–Sept. Each seed pair is surrounded by a papery wing. **WHERE FOUND:** Rocky woods, outcrops. Sw. QC, ON, ME, NY to FL; TX (CO, NM) to n. KS, MN. **USES:** Meskwaki added root to strengthen other medicines; root used for lung problems. Menominee considered the tree a sacred medicine; used to make other medicine more potent. Historically root bark used by physicians as a tonic ("surpassed only by Goldenseal") for asthmatic breathing, fevers, poor appetite, gastroenteritis, irritated mucous membranes. A tea of the young leaves and shoots used as a worm expellant; aromatic leaves as poultice or tea to treat wounds. The bitter, slightly aromatic fruits were used as substitute for hops in the manufacture of beer. Contains antibacterial, antifungal, and anticandidal components. Little researched. **WARNING:** Do not confuse leaves of this small tree with Poison Ivy leaves. Coumarins in the leaves can induce photodermatitis.

TREES WITH COMPOUND LEAVES; TRUNK AND BRANCHES USUALLY THORNY

HONEY LOCUST

Pods, inner bark, seeds

Gleditsia triacanthos L.

Pea Family (Fabaceae or Leguminosae)

⚠ Deciduous tree; 80 ft. This tree is *usually armed with large, often compound thorns* (except in form *inermis,* which lacks thorns). Leaves feathery-compound; leaflets lance-shaped to oblong, barely toothed. Flowers greenish, clustered; May–July. Fruits flat, pods twisted; 8–18 in. **WHERE FOUND:** Dry woods, openings. NS to FL; TX, w. OK to SD. **USES:** The Meskwaki used bark tea to treat measles, fevers, and smallpox; also a cleanser for the kidneys and bowels; cure for a bad cold. Cherokee used bark tea for whooping cough, dyspepsia from overeating; tea of pods for measles; seed pulp to make a beverage, sweetens other medicines. Pod tea formerly used for indigestion, measles, catarrh of lungs. Inner-bark tea (with Sycamore bark) was once used for hoarseness, sore throats. Juice of pods is antiseptic. Bean gum researched as a potential binder in making tablets; polysaccharides from gum also studied as an edible coating for cheese. Seedpods historically used to make beer. **RELATED SPECIES:** The seedpods of a Chinese species, *G. sinensis* Lam., are used in Traditional Chinese Medicine for sore throats, asthmatic coughs, swellings, and stroke. Experimentally, these seedpods cause the breakdown of red blood cells, are strongly antibacterial, antifungal, and act as an expectorant, aiding in the expulsion of phlegm and secretions of the respiratory tract. Minute amounts of the seeds are taken in powder for constipation. The spines constitute another drug; they are used as a wash to reduce swelling and disperse toxic matter in the treatment of carbuncles and lesions. Early reports of cocaine in the

ABOVE: *Honey Locust leaves are feathery, compound.* RIGHT: *It is one of our most heavily armed trees.*

plant have been discredited. **WARNING:** All plant parts of both species contain potentially toxic compounds. Safety is undetermined.

BLACK LOCUST
Root, root bark, inner bark, flowers

Robinia pseudoacacia L.
Pea Family (Fabaceae or Leguminosae)

Large tree, to 70–90 ft.; armed with stout paired thorns, ½–1 in. long. Leaves *pinnately compound; 7–21 elliptic to oval leaflets.* Fragrant white to pink flowers in racemes; May–June. Pods smooth, flat; 2–6 in. **WHERE FOUND:** Dry woods. PA to GA; LA, OK to IA; planted as an ornamental worldwide since early 1600s. **USES:** Cherokee chewed root bark as an emetic; pounded root used to treat toothache. Native American groups made a declaration of love by presenting a branch with flowers to the object of their attachment. Seeds were received in France by Jean Robin (for whom the genus *Robinia* is named), herbalist to Henry IV of France; hence most early literature on the tree is European. An antispasmodic cough syrup was made of the aromatic flowers. Bark a folk tonic, purgative, emetic. Inner bark said to be sweet like licorice. Flower tea used for rheumatism. In China the root bark is also considered purgative and emetic, and flowers are considered diuretic. Flowers contain a glycoside, robinin, which is experimentally diuretic. **WARNING:** All parts are potentially toxic. The strong odor of the flowers historically reported to cause nausea and headaches in some persons.

Black Locust in flower. The strong-smelling flowers can induce nausea and headaches.

TREES WITH ALTERNATE COMPOUND LEAVES; WALNUTS, MOUNTAIN ASH

BUTTERNUT, WHITE WALNUT
Inner bark, nut oil, leaves, sap

Juglans cinerea L.
Walnut Family (Juglandaceae)

⚠ Large tree, to 80 ft. Stem pith *dark brown.* Leaves pinnate, with 7–17 leaflets; leaflets are *opposite, rounded at base, with minute clusters of downy hairs beneath.* Flowers Apr.–June. Fruits egg-shaped; sticky on outer surface. Nuts rough and deeply furrowed. **WHERE FOUND:** Rich woods. NB to GA; west to AR; north to ND. **USES:** Inner-bark tea or extract a popular early American laxative; thought to be effective in small doses, without causing griping (cramps). Inner bark of root considered best for use. When imported laxatives were scarce during the Revolutionary War, an extract of Butternut bark was widely used; also considered useful for dysentery. Various Native American groups used bark tea for rheumatism, headaches, toothaches; strong warm tea for wounds to stop bleeding, promote healing. Oil from nuts used for tapeworms, fungal infections. Leaves fragrant; when powdered, used for ringworm. Juglone, a component, is antiseptic and herbicidal; some antitumor activity has also been reported. Tapped in spring for its sap, boiled down to Butternut syrup and sugar, which is comparable to that from Sugar Maple. Contains plumbagin, a recent research subject as a novel chemopreventive agent against prostate cancer. **WARNING:** Fresh plant parts may cause irritation and blistering. Tree threatened throughout its range by invasion of an exotic fungus that causes Butternut canker.

Butternut. Note the terminal leaflet.

BLACK WALNUT — Inner bark, fruit husks, leaves, seed oil

Juglans nigra L. — Walnut Family (Juglandaceae)

⚠ Deciduous tree; 120 ft. Stem pith *light brown*. Leaves pinnate, with 12–23 leaflets; leaflets *slightly alternate, heart-shaped, or uneven at base*. Leafstalks and leaf undersides slightly hairy; hairs solitary or in pairs, not in clusters. Fruits rounded; Oct.–Nov. **WHERE FOUND:** Rich woods. W. MA to FL, TX to MN. **USES:** Cherokee used inner bark for smallpox; bark chewed for toothache; bark tea for sores; leaf tea for goiter. Iroquois used bark poultice for headache; inner-bark tea as an emetic, laxative; bark chewed for toothaches. Fruit-husk juice used on ringworm; husk chewed for colic, poulticed for inflammation. Leaf tea astringent, insecticidal against bedbugs. Walnut leaves and hulls have traditionally been used for their astringent activity against diarrhea. They have also been valued as a tonic and strong antifungal. The seed oil, dipped in a small piece of cotton, was applied for toothache. Leaves used for tuberculosis of lymph nodes (scrofula, King's evil). Recent scientific studies have found that the leaf extracts have strong antiviral activity against vesicular stomatitis, a protective effect on the vascular system, and an inhibitory effect on certain kinds of tumors. A recent animal study suggests that juglone, a growth inhibitor in the leaves, possesses sedative activity comparable with diazepam (the prescription drug Valium). Like Butternut, contains plumbagin, a recent research subject as a novel chemopreventive agent against prostate cancer. **WARNING:** Fruit husks and leaves can cause contact dermatitis.

Black Walnut. Note that there is not a terminal leaflet.

Black Walnut fruit husks drying on black plastic.

AMERICAN MOUNTAIN ASH

Fruits, bark, buds

Sorbus americana Marsh.

Rose Family (Rosaceae)

[*Aucuparia americana* (Marsh.) Nieuwl.; *Pyrus americana* (Marsh.) DC.]

Shrub or small tree; to 40 ft., with *gummy red buds*. Leaves compound, with 11–17 leaflets; leaflets toothed, long-pointed, narrow—3 times longer than broad. Flowers in clusters. Fruits in clusters, red, about ¼ in.; Aug.–Mar. **WHERE FOUND:** Woods, openings. NL to NC mtns.; IL to MB. **USES:** Native Americans used tea from ripe fruit for scurvy, worms; tea made from inner bark or buds for colds, debility, boils, diarrhea, tonsillitis; also as a blood purifier, appetite stimulant; astringent, tonic. Bark said to have the fragrance of cherry bark (commonly associated with "almond" fragrance). Fruits highly astringent and

European Mountain Ash, with showy fruit clusters.

acidic; when rotted once used to make a strong cider. Fruits with moderate antioxidant activity. **RELATED SPECIES:** Fruits of the European Mountain Ash (*S. aucuparia* L.) have been used similarly, for hemorrhoids, urinary difficulty, indigestion, gallbladder ailments, angina, and other coronary problems.

ALTERNATE LEAVES WITH ROUNDED LOBES OR ODD SHAPES

GINKGO

Seeds, leaves

Ginkgo biloba L. — Ginkgo Family (Ginkgoaceae)

⚠ Large deciduous tree; to 100 ft.; a living fossil surviving over 200 million years. Leaves alternate or in fascicles; *fan-shaped, 2-lobed, broader than long*; 1–3½ in. wide, with parallel veins. Male and female flowers on separate trees; females producing oval, fleshy, foul-smelling fruit with hard-coated, oval to elliptical seed. **WHERE FOUND:** Widely cultivated shade tree, adaptable to soil type, survives in cities where other trees do poorly. Alien (China). Throughout our range, near dwellings; though rarely naturalized, survives in cultivation. **USES:** Seeds (after removal of toxic flesh with obnoxious odor), cooked, are used in Traditional Chinese Medicine for treatment of lung ailments. Leaves rarely used historically for cough, asthma, and diarrhea. Externally, as a wash for skin sores and to remove freckles. Today, complex, highly processed, concentrated Ginkgo leaf extracts (calibrated to 24 percent flavonoids, 6 percent ginkgolides, with toxic components removed) are among the best-selling herbal preparations in Europe. The subject of hundreds of scientific studies, Ginkgo leaf extracts increase circulation and improve oxygen metabolism to the extremities and the brain; antioxidant. Clinically (though controversial) shown to improve short-term memory, attention span, and mood in early stages of Alzheimer's. Extract approved in Germany for memory deficits, poor concentration, peripheral arterial occlusive disease (improving pain-free walking distance), and for vertigo and ringing in the ears (tinnitus) caused by vascular disturbances. However, the crude leaf does not carry therapeutic claims, only the complex extract does. **WARNING:** Leaf extracts may cause relatively rare gastrointestinal upset, headaches, or skin allergies. Fleshy seed coat causes severe contact dermatitis (like Poison Ivy). Fruits and seeds are handled with rubber gloves.

TULIPTREE, TULIP POPLAR

Bark, leaves, buds, seeds

Liriodendron tulipifera L. — Magnolia Family (Magnoliaceae)

One of the largest native deciduous trees of e. N. America; to 150 ft. tall; 10 ft. in diameter. Leaves spicy; *4-lobed, apex notched*. Flowers to 2 in. long; tuliplike, green to greenish yellow or yellow-orange. Flowers Apr.–June. **WHERE FOUND:** Moist soil. MA to FL; LA, e. AR, IL to MI. Widely planted elsewhere. **USES:** Cherokee used root bark tea for fevers; inner bark in decoction for stomach upset, dysentery, rheuma-

Ginkgo produces unique fan-shaped leaves.

Tuliptree flowers resemble a tulip.

tism, menstrual problems, cough syrup ingredient, and for pinworms. Green bark chewed as an aphrodisiac, stimulant. Osage used root bark and green seeds for fevers and to treat worms in children. Historically, bark preparations considered tonic, diuretic, diaphoretic, and stomachic. Bark tea a folk remedy for worms and toothaches; ointment from buds used for burns, inflammation, chronic rheumatism, gout. The aromatic crushed leaves poulticed for headaches. Bark widely used as a substitute for Cinchona, often in combination with Dogwood bark as a treatment for malaria. Various aporphine alkaloids and sesquiterpenes isolated from the bark and leaves are reported to have significant antiplasmodial activity (killing *Plasmodium* spp. responsible for malaria), supporting historic use as an antimalarial remedy. **RELATED SPECIES:** The only other species in the genus is *L. chinense* (Hemsl.) Sarg., native to China with experimental cancer preventive activity.

WHITE OAK

Bark, acorns

Quercus alba L.

Beech Family (Fagaceae)

Tall tree; 60–120 ft. Bark light, flaky; flat-ridged. Leaves with *evenly rounded mostly deep lobes*, without bristle tips; glabrous and whitened beneath when mature. Bowl-shaped cup covers 1/3 or less of acorn. **WHERE FOUND:** Dry woods. ME to n. FL; e. TX to MN. **USES:** Penobscot ate acorns to induce thirst to promote the healthful practice of drinking plenty of water. Astringent inner-bark tea used for chronic diarrhea, dysentery, chronic mucous

White Oak. Oaks in this group have rounded lobes.

White Oak acorns.

Sassafras bark harvested by a beaver.

discharge, bleeding, anal prolapse, hemorrhoids; as a gargle for sore throat and a wash for skin eruptions, Poison Ivy rash, burns; hemostatic. Inner bark from younger tree parts considered best; gentler on the stomach than other oak barks. Folk remedy for cancer. Experimentally, tannins in bark are antiviral, antiseptic, cavity stabilizing, growth depressant, antitumor, and carcinogenic. In Germany, a related species, English Oak, *Quercus robur* L., is used externally for the treatment of inflammatory skin diseases. Wood famously used in cooperage for barrels to age alcoholic beverages.

SASSAFRAS Leaves, twig pith, root bark, flowers

Sassafras albidum (Nutt.) Nees Laurel Family (Lauraceae)

⚠ Deciduous tree; 10–100 ft. Leaves in 3 shapes: *oval, mitten-lobed, and 3-lobed*; fragrant, mucilaginous. Yellow flowers in clusters appear before leaves; Apr.–May. Fruits blue-black, 1-seeded. **WHERE FOUND:** Poor soils. S. ME to FL; TX to e. KS. **USES:** Delaware used root bark for fevers. Cherokee used bark tea as blood purifier, for colds, diarrhea, rheumatism. Choctaw drank root-bark tea to thin blood. Missouri tribes are said to have smoked the bark like tobacco. Earliest accounts are from the writings of Spanish physician Nicolaus Monardes in 1574, for fevers, "miasma," and an antidote to bad drinking water. Well known in European markets by the 1580s. In the early days of European settlement, Sassafras was a major export. The Plymouth colonies were in part funded on speculation of Sassafras exports. Root-bark and, more obscurely, flower tea was a famous spring blood tonic and blood pu-

LEFT: *Sassafras. Its aromatic leaves are mitten-shaped.* ABOVE: *Sassafras flowers before leafing out in spring.*

rifier; also a folk remedy for stomachache, gout, arthritis, high blood pressure, rheumatism, kidney ailments, colds, fevers, and skin eruptions. The mucilaginous twig pith has been used as a wash or poultice for eye ailments; also taken internally, in tea, for chest, bowel, kidney, and liver ailments. Leaves mucilaginous, once used to treat stomachache; widely used as a base for soup stocks (gumbo). **WARNING:** Safrole (found in essential oil of Sassafras) reportedly is carcinogenic; it is banned by the FDA. However, the safrole in a 12-ounce can of old-fashioned root beer is not as carcinogenic as the alcohol (ethanol) in a can of beer. Studies indicate that safrole is not soluble in water; furthermore, if safrole-containing plant parts are boiled in water for 1–5 minutes, some of the safrole may volatilize, but most of it transforms into degradation byproducts, suggesting the possibility that while toxic safrole and safrole-rich essential oil are dangerous, boiled plant parts containing safrole may not be significantly toxic. The question is, how much safrole is in a cup of boiled sassafras tea versus brewed sassafras tea?

LEAVES SHARP-LOBED AND TOOTHED

SWEETGUM — Inner bark, gum

Liquidambar styraciflua L. — Altingia Family (Altingiaceae)
[Formerly in the Witch-hazel Family (Hamamelidaceae)]

Large tree; to 125 ft. Outer branches often corky-winged. Leaves shiny; star-shaped or maplelike, with 5–7 lobes; lobes pointed; toothed leaves pine-scented when rubbed or crushed. Fruits *spherical* (to 1½ in.), with *projecting points*; autumn, persisting until following spring. **WHERE FOUND:** Moist woods, bottomland; usually along streams, riverbanks. Often invasive in old fields and in areas that have been recently logged. MO to IL, se. CT to FL, TX, Mex., Cen. America. Common in the South. The "gum" produced in pockets in the bark is used medicinally. The hard, spiny, 1-in.-diameter fruits make this tree a poor choice for lawn plantings, though the inherent ornamental value of the brilliant autumn leaves offsets this slight negative aspect. **USES:** Various Native American groups used the inner bark to treat dysentery and diarrhea; gum used as topical treatment for wounds, sores, cuts, bruises, and ulcers. Leaves used in some smoking mixtures. Gum or balsam (resin) was traditionally chewed for sore throats, asthma, bronchitis, coughs, colds, diarrhea, dysentery, ringworm; used externally for sores, skin ailments, wounds, hemorrhoids. An ingredient in "compound tincture of benzoin," it is available from pharmacies. Considered expectorant, antiseptic, antimicrobial, anti-inflammatory. Children sometimes chew the gum in place of commercial chewing gum. The mildly astringent inner bark was used as a folk remedy, boiled in milk for diarrhea and cholera infantum. The essential oil of the leaf contains similar components to Australian Tea Tree (*Melaleuca alternifolia*), well known for its antimicrobial activity.

Sweetgum produces spiny, globular fruit.

Sweetgum. Its leaves are aromatic and usually 5-lobed.

NORTHERN RED OAK

Inner bark

Quercus rubra L.

Beech Family (Fagaceae)

Large tree; to 60–120 ft. Bark dark, smoother than White Oak bark. Leaves hairless, thin, dull, with 7–11 *bristle-tipped lobes*; leaves 5–9 in. long, 3–6 in. wide. Cup covers ⅓ of acorn. **WHERE FOUND:** Woods. NS to n. GA; se. OK to MN. **USES:** Considered similar to but weaker than White Oak (see p. 363). Astringent inner-bark tea once used for chronic

Northern Red Oak. Oaks in the red or black oak group have pointed lobes.

diarrhea, dysentery, chronic mucous discharge, bleeding, anal prolapse, hemorrhoids; gargle for sore throats; wash for skin eruptions, Poison Ivy rash, burns; hemostatic. Folk remedy for cancer.

SYCAMORE

Inner bark

Platanus occidentalis L. Plane-tree Family (Platanaceae)

Large tree; to 150 ft. Bark mottled, multicolored, *peeling*. Leaves broadly oval, with 3–5 lobes; 5–8 in. long and wide, with round, shallow sinuses. Fruits globular; to 2 in. across. Without seeing leaf or fruit, it is often easy to identify this tree by its multicolored peeling bark, smooth and light-colored, especially on upper portions of the trunk. **WHERE FOUND:** Moist soils, swamps, lake edges, and streambanks. S. New England to FL; TX to NM, north to cen. IA, NE. **USES:** Cherokee used inner-bark tea mostly in combination with other herbs for dysentery, colds, lung ailments, measles, coughs; also as a blood purifier and emetic (to induce vomiting), laxative. Creek used wood chips and bark for treating tuberculosis. Bark once suggested for rheumatism and scurvy. A mixture of glycoside components from the bark is of recent research

Sycamore leaves are broadly oval with 3–5 toothed lobes.

Sycamore, with its multicolored, peeling bark.

interest as highly selective antibacterial agents against methicillin-resistant *Staphylococcus aureus* (MRSA).

LEAVES TOOTHLESS; INEDIBLE FRUIT

OSAGE-ORANGE
Root, fruit

Maclura pomifera (Raf.) Schneid. Mulberry Family (Moraceae)

⚠ Small tree; 30–60 ft. Branches armed with short spines. Leaves lustrous, oval or oblong to lance-shaped. Fruit large (to 6 in.), round, fleshy; surface *brainlike*; Oct.–Nov. **WHERE FOUND:** Roadsides, clearings. Mostly spread from cultivation; used as a living, impenetrable hedge. Originally, AR to TX. **USES:** Comanche used root tea as a wash for sore eyes. Fruit sections used in MD and PA as a cockroach repellent. Inedible fruits contain antioxidant and fungicidal compounds. Seed oil of promise as a biofuel diesel alternative. **WARNING:** Milky latex or sap may cause dermatitis.

CUCUMBER MAGNOLIA
Bark, fruits

Magnolia acuminata (L.) L. Magnolia Family (Magnoliaceae)

A deciduous magnolia; to 80 ft. Leaves large; oblong to lance-shaped. Greenish *cup-shaped flowers appear as leaves unfold*; Apr.–June. Fruits resemble small *cucumbers*. **WHERE FOUND:** Rich woods. W. NY to GA; AL, AR to s. IL; ON. **USES:** Iroquois and Cherokee used infusion of inner

Osage-orange. Its inedible fruits are softball-sized.

Cucumber Magnolia, a large tree with greenish flower petals.

bark held in mouth for toothaches. Bark tea historically used in place of Cinchona (source of quinine) for malarial and typhoid fevers, also for indigestion, rheumatism, worms, toothaches; tonic in debility, nervousness. Bark chewed to break tobacco habit. Fruit tea a tonic for general debility; formerly esteemed for stomach ailments. Extracts from plant have anti-inflammatory activity.

CAROLINA BUCKTHORN — Bark, wood

Frangula caroliniana (Walt.) Gray (*Rhamnus caroliniana* Walt.) — Buckthorn Family (Rhamnaceae)

Small tree; 10–30 ft. Leaves elliptic to oval, scarcely fine-toothed, usually smooth beneath when mature, but velvety in var. *mollis.* Flowers perfect (each one includes petals, sepals, and both male and female parts). Fruits black; *3-seeded, not grooved on back.* **WHERE FOUND:** Rich woods. VA to FL; TX to NE. **USES:** Creek used tea from wood to treat jaundice. Other Native American groups used bark tea to induce vomiting; also a strong laxative. Still used for constipation with nervous or muscular atony of intestines. **RELATED SPECIES:** The European species *R. cathartica* L., becoming increasingly naturalized in e. U.S., and the West Coast species *F. purshiana* (DC.) Cooper (*Rhamnus purshiana* DC., Cascara Sagrada) have been used similarly as official drugs for their laxative effects. Some botanists separate the 125 species of *Rhamnus* (5 petals and bisexual) from *Frangula* (4 petals, with male and female flower separate). **WARNING:** Fruits and bark of all 3 species will cause diarrhea, vomiting.

Carolina Buckthorn is the most common eastern buckthorn.

LEAVES TOOTHLESS; FRUITS EDIBLE OR A FLAT, INEDIBLE LEGUME

COMMON PAWPAW, CUSTARD APPLE — Fruits, leaves, seeds

Asimina triloba (L.) Dunal — Custard-apple Family (Annonaceae)

Small tree or shrub; 9–30 ft. Leaves oblong to lance-shaped (wider above); large—to 1 ft. long. Flowers *dull purple, drooping; petals curved backward*; Apr.–May. Fruits slightly curved, elongate; green to brown; edible (except seeds)—flavor and texture likened to that of bananas. Seeds toxic; large, lima bean–like. **WHERE FOUND:** Rich, moist woods. NJ to FL; TX; se. NE to MI. **USES:** Fruit edible, delicious; also a laxative. Leaves insecticidal, diuretic; applied to abscesses. Seeds emetic,

ABOVE: *Common Pawpaw, with deep maroon flowers.* RIGHT: *Common Pawpaw produces large, oblong leaves.*

narcotic (produce stupor). The powdered seeds, formerly applied to the heads of children to control lice, have insecticidal properties. Juice of unripe fruit as well as powdered seed formerly used as a vermifuge. Leaves once used to tenderize meat. Contains the acetogenin compound annonacin, which is toxic to neurons, raising the possibility that long-term ingestion of pawpaw fruit products could cause nerve degeneration. Acetogenins found in all parts of the plant show antitumor, antiviral, antimicrobial, and antimalarial effects among others, though could also pose significant health risks. Still, pawpaw extracts have been incorporated into head lice shampoos and flea and tick insecticidal sprays; also an ointment for oral herpes simplex virus. **WARNING:** Seeds toxic. Seeds and probably the leaves and bark contain potentially useful (anticancer), yet potentially toxic, acetogenins. More than one USDA chemist working with the acetogenins has experienced visual problems, which were corrected after they ceased working with the material. Leaves may cause rash.

REDBUD

Bark, flowers, root

Cercis canadensis L. — Pea Family (Fabaceae or Leguminosae)

Small tree with a rounded crown; to 40 ft. Leaves *heart-shaped, entire* (toothless); 3–6 in. long and wide. Flowers red-purple, pealike; on long stalks; *in showy clusters before leaves appear*; Mar.–May. Fruit a flat, pea pod–shaped, dry, inedible legume; Aug.–Nov. **WHERE FOUND:** Rich woods, roadsides. S. CT, s. NY to FL; TX to WI. Often planted as an ornamental. **USES:** Cherokee children (and authors of this book) eat the edible flowers; flowers also pickled. Bark tea for whooping cough. Alabama soaked root and inner bark in water, then used the solution for congestion and fever. Inner-bark tea highly astringent. An obscure

ABOVE: *The pink pealike flowers of the Redbud appear before the leaves do.* TOP RIGHT: *Redbud has heart-shaped leaves.* RIGHT: *Redbud flower close-up.*

medicinal agent once used for diarrhea and dysentery; also as a folk remedy for leukemia.

COMMON PERSIMMON — Bark, fruits

Diospyros virginiana L. — Ebony Family (Ebenaceae)

Deciduous fruit tree; 15–50 ft. Leaves shiny, elliptic; to 5 in. long. Flowers greenish yellow, thickish, lobed, urn-shaped; May–June. Fruits *globular, plumlike*, 1–2 in. across; with 6–8 compressed seeds; Sept.–Nov. **WHERE FOUND:** Dry woods. S. New England to FL; TX e. KS. **USES:** Cherokee used astringent inner bark for venereal disease, sore throat and mouth; chewed for heartburn; syrup for thrush; externally a wash for hemorrhoids. Native American groups ate fresh fruits, preserved by drying, forming into paste, or by making into a beer. For wine-making, astringent fruit skins were removed. Inner-bark tea highly astringent.

LEFT: *Ripening Persimmon fruit with leaves.* ABOVE: *Common Persimmon has plumlike fruits.*

In folk use, gargled for sore throats and thrush. Bark tea once used as a folk remedy for stomachaches, heartburn, diarrhea, dysentery, and uterine hemorrhage. The bark tea was used as a wash, or bark poulticed for warts and cancers. Fruits edible, but astringent before ripening; best after frost, ripening to a sweet, delicious fruit. Unripe bitter and astringent fruits formerly used in tincture, syrup, or tea for bowel complaints with bleeding. Seed oil is suggestive of peanut oil in flavor. Surprisingly little research. **WARNING:** Contains tannins; potentially toxic in large amounts.

MISCELLANEOUS TREES WITH ALTERNATE, TOOTHED LEAVES

SOURWOOD, SORREL-TREE — Leaves, twigs, bark

Oxydendrum arboreum (L.) DC. — Heath Family (Ericaceae)

(*Andromeda arborea* L. in older works)

Deciduous tree; to 80 ft. Leaves finely toothed, wide, lance-shaped; to 6 in. Leaf flavor acrid, sour—hence the common name. Flowers are *white urns, resembling Lily-of-the-Valley flowers, in drooping panicles* to 10 in. long; May–June. Fruits egg-shaped, upturned; about $^3/_8$ in. long. **WHERE FOUND:** Rich woods. PA to FL; LA to s. IN, OH. Cultivated as an ornamental elsewhere. **USES:** Catawba used cold infusion of tree (part not specified) to treat excessive menstrual flow; also for menopause symptoms. Cherokee used leaf tea as tonic for dyspepsia, asthma, and lung disease; chewed bark for mouth ulcers; bark exudate for treating itch. Leaf tea is a Kentucky folk remedy for kidney and bladder ailments (diuretic), fevers, diarrhea, and dysentery. Sour leaves chewed to allay thirst; decoction to treat fevers. Leaf tea described as lemonade-like. Decoction of leaves and bark considered tonic, especially for digestion. Flowers yield famous Sourwood honey.

WHITE WILLOW

Bark

Salix alba L.

Willow Family (Salicaceae)

Deciduous tree; to 90 ft. Branchlets pliable, *not brittle at base; silky.* Leaves lance-shaped, mostly without stipules; *white-hairy above and beneath* (use lens). **WHERE FOUND:** Naturalized; in moist woods, along stream edges. Throughout our range. Alien (Europe). **USES:** One of the best-known medicinal plants, the bark of this willow, and the very bitter and astringent bark of other willows, has traditionally been used for diarrhea, fevers, pain, arthritis, rheumatism; poultice or wash used for corns, cuts, cancers, ulcers, and Poison Ivy rashes. Salicylic acid, derived from salicin (found in bark), is a precursor to the most widely used semisynthetic drug, acetylsalicylic acid (aspirin), which reduces pain, inflammation, and fever. Aspirin reduces risk of heart disease in males; experimentally, delays cataract formation. In the intestines, compounds in the bark are transformed to saligenin, which is oxidized in the liver and blood to produce salicylic acid. Pain is reduced by inhibition of prostaglandin synthesis in sensory nerves. Bark preparations widely used in European phytomedicine for fever, rheumatic complaints, and headaches. In short, used similarly to aspirin as an antipyretic, antiphlogistic, and analgesic. Bark extracts shown to be anti-inflammatory, antioxidant, and immunomodulatory, among other activities. **RELATED SPECIES:** While many herb books list White Wil-

ABOVE: *Sourwood produces white urn-shaped flowers.* RIGHT: *White Willow. Leaves lance-shaped, mostly without stipules.*

low as the most common *Salix* species used, many *Salix* species are involved in the commercial supply of willow bark. In fact, other species contain 10 times as many active constituents as White Willow. Crack Willow (*S. fragilis* L.) and Basket Willow or Purple Osier (*S. purpurea* L.) are native to Europe and cultivated and escaped in our range. Both are higher in salicin than White Willow and are used as official sources of willow bark. Willows hybridize readily, making identification challenging; most are used similarly, effectiveness (and dose) depending upon amount of salicylic acid and other active phenolics.

WEEPING WILLOW

Bark, leaves

Salix babylonica L. — Willow Family (Salicaceae)

Perhaps the best known willow species, Weeping Willow is a medium-sized tree to 40 ft. with long, *pendulous (weeping)* branches. Leaves are lance-linear, to 6 in. Catkins appear at same time as leaves. **WHERE FOUND:** Moist soils. Alien. Native to China. Widely grown as an ornamental tree; occasionally naturalized. **USES:** Traditionally used in Europe for tonic, antiseptic, fever-reducing, and astringent qualities. Bark used for at least 2,000 years in China for rheumatoid arthritis, jaundice, and fevers. Leaves used in China to reduce heat (fevers), treat skin eruptions, regulate urination, and as a blood purifier. Used in the treatment of mastitis, toothache, scalds, and other conditions. Like most willows, contains salicin and tannins.

COASTAL PLAIN WILLOW, CAROLINA WILLOW

Bark

Salix caroliniana Michx. — Willow Family (Salicaceae)
[*S. amphibia* Michx.; *S. longipes* (Shuttlw. ex. Anderss.)]

Shrub or tree to 30 ft., trunk to 1 ft. in diameter. Leaves spreading, lanceolate, to 6 in. long (5 to 10 times longer than wide), gray-green,

The heavily drooping branches of the Weeping Willow.

Coastal Plain Willow, Carolina Willow is typically shrublike, on gravel bars. Note stipules at leaf base.

strongly glaucous (and gray powdery) beneath, with long, toothed, sharp tip. Stipules well developed and persistent, broadly kidney-shaped and toothed. **WHERE FOUND:** Flood plains, along rivers, creek beds, wet or low habitats. DE, MD, south to Cuba and Guatemala, west to e. KS and OK. **USES:** Widely used by Native Americans across its range. Root-bark tea used to thin blood, to alleviate fevers, stiff neck, backaches, rheumatism, headaches, diarrhea; to induce vomiting. Surprisingly little research considering its abundance in natural range.

BLACK WILLOW — Bark, leaves, root

Salix nigra Marshall — Willow Family (Salicaceae)

Tree to 100 ft. or more, sometimes shrublike; trunks often leaning; twigs yellowish brown, mostly hairless. Leaves often drooping, lance-linear to lanceolate, finely sharp-toothed, to 6 in. long, ¾ in. wide, dark green above, lighter *(not glaucous)* beneath. Male and female flowers on separate trees, with drooping catkins about 2 in. long. **WHERE FOUND:** Wet soils, flooded areas, along streams, ponds, and moist depressions. Throughout our range. Perhaps our most common willow. **USES:** Cherokee used the bark of this and other *Salix* species as a tonic, for diarrhea; inner-bark tea used for laryngitis; root chewed for hoarseness. Tea also used for fevers; as a hair wash to stimulate hair growth. Iroquois also used tea for throat and mouth sores and irritation; leaves poulticed on sprains, bruises, and sores. See White Willow (p. 373) for modern use. Used similarly.

Black willow, one of our largest willows.

AMERICAN BEECH — Nuts, bark, leaves

Fagus grandifolia Ehrh. — Beech Family (Fagaceae)

Large tree; to 80 ft. (occasionally 120 ft.). *Smooth gray bark*. Leaves oval, *sharp-toothed*, yellow-green, *persistent in winter*; veins silky beneath. Flowers Apr.–May. Fruits are edible; triangular nuts; Sept.–Oct. **WHERE FOUND:** Rich woods. PE to FL; TX to IL, ON. **USES:** Cherokee chewed nuts as a worm expellent. Dried leaves (persistent in winter), plucked from tree in winter, were used by Potawatomi as a restorative for frostbite, also for burns or scalding. Nuts widely used as food. Bark tea used for lung ailments. Leaf tea a wash for Poison Ivy rash (1 ounce to 1 pint of salt water). Beechnut oil used historically, harvested after first frost. Many nineteenth-century works considered American Beech the same

The American Beech has smooth, light gray bark. Leaves are persistent in winter.

Beech forest in early spring showing previous year's dried brown leaves still abundant on trees.

species as European Beech (*F. sylvatica* L.), hence uses were considered identical.

RED MULBERRY (NOT SHOWN) Root, fruit

Morus rubra L. Mulberry Family (Moraceae)

Small tree; 20–60 ft. Leaves heart-shaped, toothed, often lobed; *sandpapery above, downy beneath.* Flowers green in tight, drooping clusters; Apr.–May. Fruits like a thin blackberry; red, white, or black; June–July. **WHERE FOUND:** Rich woods. Sw. VT, NY to FL; TX, OK to SD. **USES:** Cherokee used an infusion of bark to treat tapeworms, relieve dysentery, and as a laxative. Creek used root tea as a restorative tonic; diuretic; larger doses as an emetic. Meskwaki considered root bark a useful tonic for any sickness. Bark decoction used as a wash for ringworm. Nutritious fruits used for lowering fever.

WHITE MULBERRY Leaves, inner bark, fruits, twigs, wood

Morus alba L. Mulberry Family (Moraceae)

Similar to Red Mulberry (above), but leaves *are less hairy, usually smooth and shiny* at fruiting time, and coarsely toothed, often with 3–5 lobes. Fruits whitish to purple; May–June. **WHERE FOUND:** Planted and naturalized in much of our range. Asian alien, introduced for silkworm production. **USES:** In China, leaf tea used for headaches, hyperemia (congestion of blood), thirst, coughs; "liver cleanser." Experimentally,

ABOVE: *Mulberry produces elongated fruits and smooth leaves.* TOP RIGHT: *Basswood, Linden, with an uneven, heart-shaped base.* RIGHT: *Basswood or Linden flowers, with typical straplike structure above sweetly fragrant flowers.*

leaf extracts are antibacterial. Young-twig tea used for arthralgia, edema. Fruits eaten for blood deficiency, to improve vision and circulation, and for diabetes. Inner-bark tea used for lung ailments, asthma, coughs, and edema. Bark experimentally produces a protective effect on kidneys in cases of high uric acid, leading to gout. Leaf extracts experimentally antidiabetic, antioxidant. Mulberry wood, high in oxyresveratrol, a potent antioxidant and free-radical scavenger, is experimentally beneficial and neuroprotective in brain trauma research models. Fruits are antioxidant. All plant parts widely used in Asian traditional medicine systems.

AMERICAN BASSWOOD, LINDEN — Flowers, bark, twigs, leaves

Tilia americana L. — Mallow Family (Malvaceae)

[Formerly in the Basswood Family (Tiliaceae)]

Variable deciduous tree, 60–80 (occasionally 120) ft., separated into at least 4 varieties. Leaves finely sharp-toothed; heart-shaped, base uneven; to 10 in. long. Flowers yellow, fragrant; from an *unusual winged stalk*; June–Aug. **WHERE FOUND:** Rich woods. NB to FL, TX to MB. **USES:** Cherokee used inner bark to treat dysentery; bark and twigs a treatment for heartburn in pregnant women; also for weak stomach and bowels; coughs and tuberculosis. Bark from tree struck by lightning chewed and applied as snakebite remedy. Iroquois used bark tea to treat urinary difficulties, tonic for injuries and feeling run down. Meskwaki used twig tea for lung troubles; decoction as wash to open boils. Tea or tincture of leaves, flower, and/or bud traditionally used for nervous headaches, restlessness, painful digestion, antispasmodic cough remedy. **RELATED SPECIES:** Small-leaved European Linden (*T. cordata* Mill.) and Large-leaved European Linden (*T. platyphyllos* Scop.), both cultivated and naturalized in N. America, are widely used in European herbal medicine. Various species also used in Traditional Chinese Medicine. Flowers used to treat colds, bronchitis, fever, inflammations, and influenza; experimentally immunostimulant. Flower tea used as a mild sedative for children. Flowers contain mildly sedative and anxiolytic flavonoids, and enhance pain-relieving response. In European phytomedicine, the flowers are approved for treatment of colds and cold-related coughs. Primarily used as a diaphoretic and mild sedative. Preparations of the leaves and wood also traditionally used as a folk medicine for fevers and cellulitis. **WARNING:** Samples of dried flowers in Europe were found to be a vector for infant botulism in rare instances.

TREES WITH ALTERNATE, TOOTHED LEAVES; ROSE FAMILY: CHERRIES, SERVICEBERRY

SERVICEBERRY — Root, bark

Amelanchier canadensis (L.) Medik. — Rose Family (Rosaceae)
[*Pyrus canadensis* (L.) Farw.]

Small tree; to 24 ft. Leaves fine-toothed, oblong, tip rounded; *veins in 10–15 main pairs, fading at edges.* Flowers white, in drooping clusters; Mar.–June. Fruits black. **WHERE FOUND:** In clumps. Moist thickets. S. QC, ME to GA, MS. **USES:** Chippewa used root-bark tea (with other herbs) as a tonic for excessive menstrual bleeding, "female tonic," and to treat diarrhea. Cherokee used in herb combinations as a digestive tonic. Bath of bark tea used for children with worms. Fruits contain antioxidant anthocyanins. **RELATED SPECIES:** Native

Serviceberry, one of our earliest spring-blooming trees.

Americans and Chinese used bark tea of other *Amelanchier* species to expel worms.

BLACK OR WILD CHERRY

Bark

Prunus serotina Ehrh.

Rose Family (Rosaceae)

⚠ Medium to large tree; 40–90 ft. Bark rough, dark; reddish beneath. Leaves oval to lance-shaped, blunt-toothed, often with asymmetrically paired glands near base; smooth above, pale beneath, with *whitish brown hairs on prominent midrib.* Flowers white in drooping slender racemes; Apr.–June. Fruits are small black cherries. **WHERE FOUND:** Dry woods. NS to FL; TX to ND. **USES:** Cherokee used inner bark decoction for fevers, warm tea at first signs of labor; coughs, gargle for laryngitis; externally a wash for old sores and ulcers. Delaware used inner bark as a wash for infections, ulcers, swollen lymph nodes. Iroquois used bark decoction for persistent coughs, headache, colds, fevers, soreness, and lung inflammations; bronchitis cough syrup made from bark. Ojibwa used bark tea for coughs and colds. Aromatic inner bark traditionally used in tea or syrup for coughs, blood tonic, fevers, colds, sore throats, diarrhea, lung ailments, bronchitis, pneumonia, inflammatory fever diseases, and bitter tonic dyspepsia. Experimentally, bark components anti-inflammatory. Described for coughs as "calming irritation and diminishing nervous excitability." Useful for general debility with persistent cough, poor circulation, lack of appetite; mild sedative, expectorant. Fruits used as "poor man's" cherry substitute. Bark extracts still widely used in natural cough syrups. **WARNING:** Bark, leaves, and seeds contain a cyanide-like glycoside, prunasin, which converts (when digested) to the highly toxic hydrocyanic acid. Toxins are most abundant in bark harvested in fall.

Black Cherry has drooping flower racemes.

CHOKECHERRY (NOT SHOWN)

Bark, fruits

Prunus virginiana L.
[*Padus virginiana* (L.) M. Roem.]

Rose Family (Rosaceae)

Shrub or small tree; to 20 ft. Smaller than Black Cherry. Leaves oval, sharp-toothed, *midrib hairless*. Flowers white in a thicker raceme; Apr.–July. Fruits reddish. **WHERE FOUND:** Thickets. NL to NC, MO, KS to SK. **USES:** Nonaromatic bark, similar to that of Black Cherry. Externally, used for wounds. Dried powdered berries once used to stimulate appetite and treat diarrhea and bloody discharge of bowels. **WARNING:** As with Black Cherry, seeds, bark, and leaves may cause cyanide poisoning.

TREES WITH ALTERNATE, TOOTHED LEAVES; STALKS MOSTLY FLATTENED; POPLARS

BALSAM POPLAR, BALM-OF-GILEAD, TACAMAHAC

Leaf buds, root, bark

Populus balsamifera L.

Willow Family (Salicaceae)

Medium to large tree; to 30–90 ft. Winter buds *yellowish, gummy, strongly fragrant*; end buds more than ½ in. long. Leaves broadly oval, with fine *wavy teeth*; leafstalks mostly rounded (rather than flat). **WHERE FOUND:** Moist soils. NL to AK; south to n. New England, WI, MN, IA to CO. **USES:** Cree used as treatment for diabetes. Buds boiled to separate resin, then dissolved in alcohol, once used as preservative in ointments. Historically, buds considered tonic, stimulant; good ingredient in ointments for rheumatism, gout, burns, sores, and various skin diseases. Folk remedy (balm) used for sores; tincture for toothaches, rheumatism, diarrhea, wounds; tea used as a wash for inflammation, frostbite, sprains, and muscle strain. Bud tea used internally for cough, lung ailments (expectorant). Innerbark tea used for scurvy, also as eyewash, blood tonic. Root tea used as a wash for headaches. Probably contains salicin, explaining its aspirinlike qualities. Dried, unopened leaf buds used in European phytomedicine for treatment of skin injuries, hemorrhoids, frostbite, and sunburn. Extracts are antibacterial and stimulate wound healing. Recently found to inhibit fat cell accumulation, decrease sugar levels in blood, and reduce liver inflammation, suggesting possibility for future research in treating diabetes and obesity.

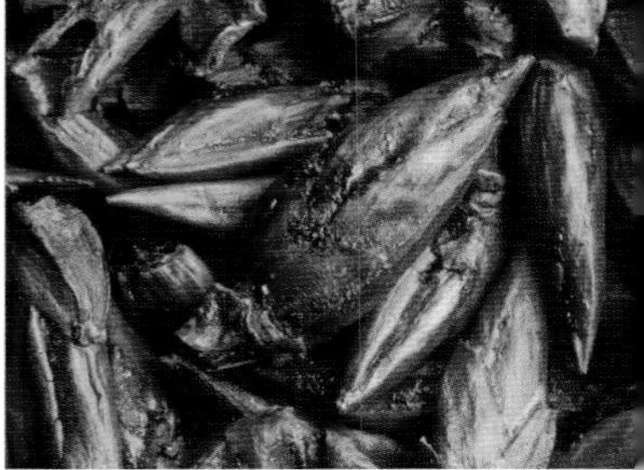

Balsam Poplar, Balm-of-Gilead, Tacamahac. Sticky leaf buds are used in herbal medicine.

TOP LEFT: *Leaves of Cottonwood, broadly oval to triangular in shape.* BOTTOM LEFT: *Quaking Aspen often occurs in populations where it is the dominant deciduous tree.* TOP RIGHT: *Quaking Aspen has flattened leaf stalks, causing leaves to twist in a breeze.*

COTTONWOOD Bark

Populus deltoides Bartr. ex Marsh. Willow Family (Salicaceae)

Large tree; to 150 ft. Leaves broadly oval, coarsely toothed; stalks *flattened* with *2–3 glands* at top of each stalk (use lens). Seeds dispersed by cottony "parachutes." **WHERE FOUND:** Along streams and rivers. S. QC, w. New England to FL; TX to MB and westward. **USES:** Inner-bark tea used for scurvy and as a female tonic. Tree held sacred by Native Americans of the prairies. Bark contains the aspirinlike compound salicin. Contains antimicrobial flavonoids.

QUAKING ASPEN Root, bark, leaves, leaf buds

Populus tremuloides Michx. Willow Family (Salicaceae)

Tall, straight tree; to 60 ft. Bark smooth, greenish to gray-white. Leaves roundish to broadly oval, smooth, *fine-toothed*; leafstalks *flattened*. Branches and leaves *sway restlessly in breeze*. **WHERE FOUND:** Widely distributed in n. U.S.; absent south of n. MO, TN; found again in highlands of w. TX and westward. **USES:** Algonquin groups used bark tea for colds and to produce sweating in fevers; shredded roots applied externally for rheumatism. Iroquois used bark for pleurisy (applied externally as a poultice); leaf tea used for cramps; bark tea for worms. Meskwaki used buds in ointment, applied to nose to treat coughs and colds. Various Native American groups used root-bark tea for excessive menstrual bleeding; poulticed root for cuts, wounds. Inner-bark tea used for stomach pain, venereal disease, urinary ailments, and as an appetite stimulant. Bark tincture a folk remedy used for fevers, rheumatism, arthritis, colds, worms, urinary infections, and diarrhea. Bark contains aspirinlike salicin, which is anti-inflammatory, analgesic; reduces fevers.

ALTERNATE, DOUBLE-TOOTHED LEAVES

SWEET OR BLACK BIRCH

Betula lenta L.

Bark, twigs, leaves, essential oil

Birch Family (Betulaceae)

Deciduous tree; to 50–70 ft. Nonpeeling, *sweet, aromatic, black bark*, often smooth, like that of our more familiar white birches, but black and not papery. Leaves oval, toothed; to 6 in. long. Buds and leaves *hairless*. Broken twigs and, to a lesser extent, the leaves have a *strong wintergreen fragrance*. Inconspicuous, separate male and female flowers in catkins, early spring. Fruits are oblong, upright, ¾–1¼ in. long. **WHERE FOUND:** Rich woods. S. QC, sw. ME to n. GA, AL; north to e. OH. **USES:** Our most fragrant birch was widely used by Native Americans and early settlers; bark tea for fevers, stomachaches, lung ailments; twig tea for fever. In folk medicine, bark tea used as a blood purifier for rheumatism, gout, and skin diseases. Tea of leaves used as a diaphoretic for fevers. Essential oil (methyl salicylate)

TOP RIGHT: *The nonpeeling bark of Black Birch has a strong wintergreen fragrance.* BOTTOM RIGHT: *Broken twigs and leaves of Sweet or Black Birch have a wintergreen fragrance.*

distilled from bark was used for rheumatism, gout, scrofula, bladder infection, neuralgia; anti-inflammatory, analgesic. To alleviate pain or sore muscles, the oil has been applied as a counterirritant. Essential oil was formerly produced commercially in Appalachia. But now, methyl salicylate is produced synthetically, using menthol as the precursor. **WARNING:** Essential oil toxic. Easily absorbed through skin. Fatalities reported.

PAPER (AMERICAN WHITE) BIRCH

Bark, twigs, leaves, sap

Betula papyrifera Marsh. Birch Family (Betulaceae)

Small to medium tree; 50–70 ft. Bark *white, peeling*; one of the easiest trees to identify from bark. Leaves are oval with acute tips, 2–3 in. long, ½–2 in. wide; base wedge- or heart-shaped; margins doubly serrated; prominent raised midrib is yellow; marked by black glandular dots. Above, the leaves are dark green and shiny; beneath, yellowish. Male flowers appear before leaves in early spring in long clusters or pairs; 3–4 in. long. Female catkins are 1–1½ in. long, with pale green lance-shaped scales. **WHERE FOUND:** Rich, moist woods, streams, and lake banks from NL to Long Island and n. PA; IA to MI; NE, SD, MT to w. WA. **USES:** Famous as the material of birch-bark canoes. Menominee used inner-bark tea to treat dysentery. Various Native American groups used twigs or bark as flavoring for other medicines. An infusion of the inner bark as an enema for constipation; with maple sugar to make a syrup for stomach cramps. Inner bark used as survival food. Sap suggested by colonists against scurvy, as a diuretic and laxative. Externally, the leaves and bark poulticed for "hard tumors." Betulinic acid from many species, including birches, is a promising anticancer compound (against melanomas); also anti-inflammatory and antiviral. Betulin, contained in bark, has anticancer, antiviral, and anti-inflammatory activity. A glycoside from the bark, platyphylloside, is of promise for potential anticancer activity. Chaga (Inonotus), a black fungus on White Birch, has a folk reputation against skin and internal cancers, and is of scientific interest against melanoma. It is said to take up betulin from birches.

Paper (American White) Birch is easily recognized by its peeling white bark.

The American Elm has been decimated by Dutch elm disease.

American Elm. Close-up of flowers in early spring before leaves appear.

AMERICAN ELM Inner bark

Ulmus americana L. Elm Family (Ulmaceae)

Our largest elm, decimated by Dutch elm disease, often with drooping or arching branches; to 120 ft. tall. Leaves oval; 3–6 in. long, to 2½ in. wide; *smooth or only somewhat rough* above, sharply double-toothed. Fruits papery, winged, to ½ in. long; tips incurved (often overlapping), with prominent hairs along margins; Mar.–May. **WHERE FOUND:** Moist to dry soils. NS to FL, west to TX, north to SK. **USES:** Mohegan used an infusion of the bark for colds and coughs. Potawatomi used inner-bark tea for cramps and diarrhea. Ojibwa used as treatment for diarrhea. Seldom used except by Native Americans, who also used the bark tea for severe coughs, menstrual cramps, internal hemorrhage, hemorrhoids, and as a folk remedy for cancer. Bark mostly used in combinations with other remedies; sometimes an adulterant to Slippery Elm bark.

SLIPPERY ELM Inner bark

Ulmus rubra Muhl. Elm Family (Ulmaceae)
(*Ulmus fulva* Michx.)

Deciduous tree; to 40–60 ft., with large, *rust-hairy* buds. White, mildly scented inner bark is very mucilaginous (slippery). Leaves oval; 3–7 in. long, to 3 in. wide; sides of base distinctly unequal. Leaves *sandpapery* above, soft-hairy below, *sharply double*-toothed. Papery, winged, yellowish green, 1-seeded fruits, about ½ in. wide, *without hairs on margins*; Mar.–May. **WHERE FOUND:** Moist woods. ME to FL; TX to ND. **USES:** One of the most important, yet poorly researched, American medicinal plants. Widely used by Native Americans and colonists throughout the tree's range; still important today. The Osage applied bark poultices to extract thorns and gunshot balls. American Revolution surgeons used bark poultice as primary treatment for gunshot wounds, easing extraction of the musket ball. Three tablespoons of inner bark in a cup of hot water makes a thick, mucilaginous tea, traditionally used for sore throats,

ABOVE: *Slippery Elm leaves are sandpapery above.* TOP RIGHT: *Slippery Elm. Fruits mature before leaves appear in spring.* CENTER RIGHT: *Inner bark is highly mucilaginous in spring, hence the name "Slippery" Elm.* BOTTOM RIGHT: *Close-up of Slippery Elm flowers in early spring.*

upset stomach, indigestion, digestive irritation, stomach ulcers, coughs, pleurisy; said to help in diarrhea and dysentery. Inner bark considered edible. Once used as a nutritive broth for children, the elderly, and convalescing patients who had difficulty consuming or digesting food. Externally, the thick tea, made from powdered inner bark, was applied to fresh wounds, ulcers, burns, scalds. Inner-bark decoction taken two months before childbirth was well known among Native American women throughout the tree's range; a practice subsequently adopted as a folk practice by colonists. Used in early America as a substitute for Marshmallow root. Science confirms tea is soothing to mucous membranes and softens hardened tissue. Despite few scientific studies, approved by the FDA as a nonprescription demulcent for use in throat lozenges. Bark once used as an antioxidant to prevent rancidity of fat. Slivers of inner bark once used—dangerously—as a mechanical abortifacient.

BARK GREEN; LEAVES SIMPLE, SHINY; VINES AND MISTLETOE

SUPPLEJACK, RATTAN VINE — Leaves, bark, stems, roots

Berchemia scandens (Hill) K. Koch — Buckthorn Family (Rhamnaceae)

High-climbing, twining woody vine; stems smooth, green. Alternate oval leaves with *conspicuous parallel veins*; no teeth. Tiny white flowers in panicles; May. Dark blue fruit in clusters; Sept.–Oct. **WHERE FOUND:** Moist woods, thickets. Se. VA to FL, TX; north to s. IL. **USES:** Choctaw used bark or leaf tea as a blood purifier. The Houma of Louisiana used bark and leaf decoction to restore youthful vigor and sexual vitality in men and women. Kosati infused burned stems in water to treat coughs. Roots once prescribed as a remedy for syphilis.

MISTLETOE — Leafy branches

Phoradendron leucarpum (Raf.) Reveal & M. C. Johnston — Sandalwood Family (Santalaceae) [Formerly in the Mistletoe Family (Viscaceae)]

[*P. serotinum* (Raf.) M. C. Johnston; *P. flavescens* (Pursh) Nutt.]

Parasitic, thick-branched perennial; semi-evergreen. Leaves oblong to obovate; to 3 in. long. Flowers small, inconspicuous. Fruits *translucent white;* Oct.–Dec. **WHERE FOUND:** On trees. NJ to FL; MO north to OH, MN. **USES:** Cherokee used dried, crushed plant for fits or epilepsy; hot water infusion a medicine for pregnant woman; tea a wash for headache, high blood pressure; and to cure lovesickness. Mistletoe from oaks considered best. Creek used green branches and leaves as an ingredient in prescriptions for lung ailments and tuberculosis. Formerly used to stop bleeding after childbirth; folk medicine for epilepsy, convulsion, vertigo, pleurisy, and dysentery. Once considered equivalent to the European Mistletoe (*Viscum album* L.). Used as a folk anticancer remedy in Mexico. A recent study found that extracts of the plant induce immunomodulatory activity, producing immunity-related cytokines, and exert-

Supplejack, Rattan Vine bark is typically smooth and olive green.

Mistletoe, a parasite, is easy to spot after leaves drop from trees.

ing a moderate toxicity to cancer cells. European Mistletoe controversially used as a potentially toxic cancer treatment in Europe. **WARNING:** Often considered poisonous. Unconfirmed reports of deaths have been attributed to eating berries. May cause dermatitis.

SAWBRIER, SAW GREENBRIER

Root, stems, leaves

Smilax bona-nox L.

Greenbrier Family (Smilacaceae)
[Formerly in the Lily Family (Liliaceae)]

Highly variable clambering, climbing vine. Stems 4-angled, often *zigzagging*; prickles absent or present; if present, black-tipped. Leaves usually evergreen, alternate; *pale green with white spots*, oval to lance-shaped, base flattish or heart-shaped, with or without prickles along margins. Flowers not showy. Berries black; July–winter. **WHERE FOUND:** Thickets. MD to FL; TX to OK, IN, OH. **USES:** Choctaw used stem decoction as a general tonic. Creek moistened leaves and rubbed on face to make skin appear young. Root decoction used as a diuretic. **RELATED SPECIES:** Of 350 *Smilax* spp. worldwide, most are tropical or subtropical, extending into temperate regions. The 17 species of *Smilax* in our range are notoriously highly variable and confusing to identify. Tropical American species are the source of sarsaparilla; also difficult to identify; for practical purposes, most species are treated in a generic sense. Some *Smilax* spp. contain diosgenin, which can be converted (by chemists—not in the human body) to testosterone and other steroidal compounds. Science confirms anti-inflammatory, antioxidant, estrogenic, cholesterol-lowering, and antistress activity of various *Smilax* species.

Sawbrier is one of 17 Smilax *spp. in our range, many of which are highly variable; tropical species is the source of Sarsaparilla.*

GREENBRIER, CATBRIER

Leaves, stems, roots

Smilax rotundifolia L.

Greenbrier Family (Smilacaceae)
[Formerly in the Lily Family (Liliaceae)]

Green, stout-thorny, climbing shrub. Stems angular. Leaves leathery or papery, round, shiny; base *mostly heart-shaped*, but variable; prickles slightly curved, usually *dark-tipped*. Flowers Apr.–June. Fruits blue-black; July–Nov. **WHERE FOUND:** Thickets. Weedy native vine. NS to FL, e. TX to s. IL. **USES:** Cherokee used root decoction for rheumatism; the spiny stems of this and other *Smilax* spp. were used to irritate skin (increasing blood circulation at the site of application), then other medicines applied for treatment of localized pain. Leaves parched and pow-

TOP LEFT: *Greenbrier, with blue-black fruits.* ABOVE: *Greenbrier leaves. Rounded base is mostly heart-shaped.* BOTTOM LEFT: *Greenbrier flowers in spring.*

dered, applied as a treatment for scalds. Recently wilted leaves are a treatment for boils; leaf tea for stomachache. Roots used as food. Most of our *Smilax* species are used similarly.

LEAVES SIMPLE; BARK NOT GREEN

AMERICAN BITTERSWEET

Bark, root bark, leaves

Celastrus scandens L.

Staff-tree Family (Celastraceae)

Highly variable climbing, twining shrub; to 50 ft. Leaves ovate to oblong to lance-shaped, sharp-pointed, fine-toothed. Flowers greenish, in clusters, *mostly terminal*; May–June. Fruit in terminal clusters; capsules, scarlet to orange, split to reveal scarlet seeds. **WHERE FOUND:** Rich thickets. QC to GA, AL, OK to ND. **USES:** Root-bark tea induces sweating; diuretic; emetic. Folk remedy for chronic liver and skin ailments, rheumatism, leukorrhea, suppressed menses. Externally, bark used in ointment for burns, scrapes, skin eruptions. Native Americans used this plant as above, also used astringent leaf tea for diarrhea, dysentery. Root-bark tea used for pain of childbirth. Bark extracts thought to be cardioactive. **RELATED SPECIES:** Oriental Bittersweet (*C. orbiculatus* Thunb.), an Asian species naturalized in many areas, differs in that the flowers occur in groups of 1–3, in *axillary cymes*; leaves *mostly*

LEFT: *Oriental Bittersweet often has rounded leaves.* ABOVE: *Oriental Bittersweet fruits in axils. American Bittersweet fruits mostly in terminal clusters. Highly variable.*

rounded, but often highly variable in shape. Difficult to tell the American and Asian species from one another. A serious weed in some areas, locally abundant. Uses in Asia similar to those for American species. **WARNING:** Fruit toxic. All parts potentially toxic.

YELLOW JESSAMINE

Roots

Gelsemium sempervirens (L.) J. St.-Hil. Gelsemium Family (Gelsemiaceae)
[Formerly in the Logania Family (Loganiaceae)]

Twining, tangling evergreen shrub. Leaves lance-shaped to oval; shiny. Flowers yellow, shaped like an open trumpet with 5 *rounded petal lobes* that are *notched* at end; sweetly fragrant; one of the earliest blooming late-winter flowers in the South; Feb.–June. **WHERE FOUND:** Dry or wet woods, thickets. Se. VA to FL; TX to AR. **USES:** Root preparations were once used as a powerful CNS-depressant, deadening pain, reducing spasms. Externally, a folk remedy for cancer. Contains the highly toxic alkaloid gelsemine, which, as an isolated compound, is of interest for possible development of pain-

Highly toxic Yellow Jessamine; fatalities have been attributed to consuming a single flower.

relieving and anxiolytic drugs; researched (but not developed) for anticancer potential. Used in homeopathic preparations in highly diluted dosage (with no detectable alkaloids). **RELATED SPECIES:** There are 2 N. American species and 1 Asian species. *Gelsemium elegans* Juss. from e. Asia, with similar components. *Gelsemium rankinii* Small, Rankin's Trumpetflower, is limited to southeastern coastal states. **WARNING:** Deadly poison. Eating a single flower has resulted in death; nectar within is toxic. Can also cause contact dermatitis.

JAPANESE HONEYSUCKLE

Flowers, leaves, stems

Lonicera japonica Thunb. Honeysuckle Family (Caprifoliaceae)

Evergreen trailing, twining vine. Leaves oval, entire (not toothed). Flowers white or buff; lobes strongly spreading from throat, stamens protruding; Apr.–Oct. The flowers are *white, but quickly fade yellow*, earning it the name Gold and Silver Flower in China (*jin yin hua*). **WHERE FOUND:** Noxious weed in much of the South; north to MA, IN. Alien (China and Japan). **USES:** Leaves and flowers a beverage tea (Japan). Flowers traditionally used (in e. Asia) in tea for bacterial dysentery, enteritis, laryngitis, colds, fevers, flu; externally, as a wash for rheumatism, sores, tumors (especially breast cancer), infected boils, scabies, swelling. Stem tea is weaker. Experimentally, flower extracts lower cholesterol; also antiviral, antibacterial, antioxidant, tuberculostatic. Widely used in prescriptions and patent medicines in Traditional Chinese Medicine to treat colds and flu; available as over-the-counter preparations in nearly every Chinese pharmacy. Pills are made from floral concentrates. Both authors have used such preparations for bronchitis, colds, and flu. When *Echinacea* and Garlic have failed against flu, Jim Duke has used this plant as a last resort. Flowers contain at least a dozen antiviral compounds. With the rapid evolution of viruses, synergistic combinations of phytochemicals, such as those found in Japanese Honeysuckle, are less liable to lead to resistant viral strains than solitary chemical compounds. This serious weed might be managed by using it for proven medicinal purposes.

Japanese Honeysuckle is called "Gold and Silver Flower" in China.

ENGLISH IVY, IVY Leaves

Hedera helix L. Ginseng Family (Araliaceae)

⚠ Evergreen, sprawling, climbing shrub, often to great height in trees. Leaves *leathery evergreen*, dark above, *whitish beneath*, mostly alternate, distinctly 3- to 5-lobed, without teeth, tips rounded. Flowers in round simple umbel; May–Aug. Fruit black globular berries; Aug.–winter. **WHERE FOUND:** Widely naturalized as escape from cultivation; especially in thickets near human habitats. Alien (Europe). **USES:** Leaves and inedible berries bitter, emetic, purgative. Historically used to induce sweating in fevers, gout, rheumatism; prevent drunkenness; dissipate effects of wine. During the great plague in London it was mixed with vinegar as a futile treatment. Leaves once used externally for itch; to kill head lice; folk remedy for cancer with glandular enlargements. Leaf extracts used in European phytomedicine for coughs and upper respiratory tract infections, chronic inflammatory bronchial diseases; often in combination with Thyme; expectorant, antispasmodic. Various clinical studies have suggested good effectiveness, but some studies suffer from various flaws that leave the science in question. Contains various saponins, including hederacoside C and alpha-hederin, which may contribute to medicinal effects; anti-inflammatory. Carefully manufactured extracts are almost always used rather than leaf tea. **WARNING:** Fresh leaves cause contact dermatitis and allergic reactions; some toxicologists consider ivy contact dermatitis underdiagnosed thanks to lack of commercially available patch-test allergens for this extremely common and widespread plant.

Alien, but common, English Ivy in flower.

English Ivy leaves are evergreen.

A seriously invasive alien, Kudzu covers millions of acres in the South.

Kudzu flowers are scented like grape bubblegum.

KUDZU

Root, flowers, seeds, stems, root starch

Pueraria montana var. *lobata* (Willd.) Maesen & S. Almeida; *P. montana* (Lour.) Merr. [*Pueraria lobata* (Willd.) Ohwi]

Pea Family (Fabaceae or Leguminosae)

Noxious, robust, trailing, climbing vine. Leaves *palmate, 3-parted*; leaflets entire or palmately lobed. Flowers reddish purple, *grape bubblegum-scented*; in a loose raceme; July–Sept. **WHERE FOUND:** Waste ground. PA to FL; TX to KS. Asian alien. Perhaps this pernicious invasive weed of the South could best be controlled by harvesting its economic and medicinal potential. **USES:** In China, root tea used for headaches, diarrhea, dysentery, acute intestinal obstruction, gastroenteritis, deafness; to promote measles eruptions, induce sweating. Experimentally, lowers blood sugar and blood pressure. Flower tea

Kudzu leaves superficially resemble those of Poison Ivy.

used for stomach acidity; "awakens the spleen," "expels drunkenness." Seeds used for dysentery and also to expel drunkenness. Stem poulticed for sores, swellings, mastitis; tea gargled for sore throats. Root starch (used to stimulate production of body fluids) eaten as food. See also p. 288.

POISON IVY

Leaf preparations

Toxicodendron radicans (L.) Kuntze (*Rhus radicans* L.)

Cashew Family (Anacardiaceae)

⚠ Highly variable—grows as a trailing or climbing vine or an erect shrub. Leaves on long, glossy to hairy stalks; *3 highly variable leaflets, outer one on a longer stalk*. Flowers whitish. Berries *white*; Aug.–Nov. **WHERE FOUND:** Woods, thickets. Most of our area. **USES:** Identify to avoid. Contains an oily mixture of compounds known as urushiol, found in all plant parts, which is responsible for the well-known contact dermatitis from the plant; can also lead to internal poisoning if ingested or from exposure to smoke from burning Poison Ivy. Internal exposure may occur, for example, when wood of a tree with Poison Ivy stems on the outside is placed in a campfire. Once used by physicians for paralytic and liver disorders. Fighting fire with fire, Native Americans rubbed leaves on Poison Ivy rash as a treatment. Jim Duke does not recommend this approach, nor eating tiny amounts of the leaves to prevent Poison Ivy rash, nor drinking milk from goats who have eaten Poison Ivy (did not work for his goat-tending sister-in-law). Micro doses are used homeopathically to treat Poison Ivy rash. Duke has received numerous unsolicited favorable comments on using homeopathic "Rhus tox" to immunize

Poison Ivy leaves are highly variable.

Poison Ivy in flower.

oneself against Poison Ivy. Smoke from burning weeds (which included Poison Ivy) next door put Duke's father-in-law in the hospital with an internal case of Poison Ivy. Crushed Jewelweed, or Touch-me-not, is also rubbed on skin to prevent or relieve outbreak of rash. **WARNING:** If human skin comes in contact with Poison Ivy, severe dermatitis often results. Internal consumption of Poison Ivy may cause severe effects, necessitating steroid or other therapies. Dried plant specimens more than 100 years old can still cause dermatitis, as can the smoke from a burning plant. Ironically, the active ingredient, urushiol, inhibits prostaglandin synthesis.

DOMESTIC GRAPE

Leaves, berries, seeds

Vitis vinifera L. (and related species)

Grape Family (Vitaceae)

A high-climbing woody vine. At least 10 native species of grapes are found in our range. Leaves simple, lobed, or rounded in outline, with tendrils opposite leaves (rarely absent). Flowers in cymose panicles; fruits a juicy berry, with 4 or fewer ovoid seeds. A familiar species is the common domestic grape, *Vitis vinifera.* Many thousands of grape varieties have been produced since antiquity from the 65 northern temperate species of *Vitis,* with fermented grape juices dating to 7,000 years ago in the Republic of Georgia in the Caucasus, which, rightly or wrongly, claims to be the place where wine originated. More than 8,000 varieties have been described, 20 percent of which are still grown. Fruits are purple-black, amber, white, or red grapes in a cluster; Sept.–Oct.

Fox Grape, a widespread wild grape, has much smaller fruits than do cultivated grapes.

Grapes have been cultivated for at least 7 millennia.

WHERE FOUND: Cultivated throughout our range. Native *Vitis* species found in thickets, woods, fence rows. **USES:** In the past decade the health benefits of antioxidant resveratrol has famously made red wine a health food, benefiting cardiovascular and brain health, with claims to helping prevent stroke, among many other health benefits. Vines, when cut in summer, yield potable water, possibly purer than today's acid rainwater. Seeds contain oligomeric procyanidins (OPCs), from which commercial extracts are made; valued for antioxidant activity. Ten to a hundred times more of the compounds have been found in the leaves compared to fruits or wine. Grape-seed extracts have been scientifically evaluated (with positive results) for microcirculatory disorders, such as a tendency toward bruising (particularly in the elderly) and varicose veins, as well as circulatory problems. **RELATED SPECIES:** *Vitis labrusca* L., Fox Grape, indigenous to N. America and found in wooded thickets from the Canadian Maritimes to GA, AR, n. to ON, is the source of the eleventh-century Norse name for N. America—Vinland. It produces small wild grapes and was used to develop cultivated varieties in colonial America. Cherokee used leaf tea for diarrhea, hepatitis, stomachaches, thrush; externally, poulticed wilted leaves for sore breasts; also poulticed leaves for rheumatism, headaches, fevers. Components in fruits and seeds considered protective to the liver, kidneys, and cardiovascular system; antioxidant. Other *Vitis* species have been used similarly. **WARNING:** Do not confuse this with Canada Moonseed, which is considered toxic.

MISCELLANEOUS WOODY VINES

DUTCHMAN'S-PIPE

Leaves

Aristolochia tomentosa Sims

Birthwort Family (Aristolochiaceae)

⚠ Climbing woody vine. Leaves heart-shaped, blunt-tipped; *lower surface densely covered with soft white hairs*. Flowers *pipelike*; calyx yellowish; May–June. **WHERE FOUND:** Rich riverbanks. NC, FL, TX; north to e. KS, MO, s. IL, s. IN. **USES:** Similar to but much weaker in effect than Virginia Snakeroot (*A. serpentaria,* p. 294). Aromatic weak tea promotes sweating, appetite; expectorant. Used for fevers, stomachaches, indigestion, suppressed menses, snakebites. Little used. **RELATED SPECIES:** *A. macrophylla* Lam. (not shown) differs in that it produces nearly smooth, sharp-pointed leaves. Flowers brown-purple. Ironically, Virginia farmers spray *A. macrophylla* with herbicides as a weed. It contains the antiseptic antitumor compound aristolochic

Dutchman's-pipe is a woody, twining vine. Flowers are shaped like a pipe.

acid, which is also a strong toxin to kidney function. **WARNING:** Potentially irritating in large doses. Aristolochic acid is considered an insidious toxin, inducing mutations.

CANADA MOONSEED, YELLOW PARILLA — Leaves, root

Menispermum canadense L. — Moonseed Family (Menispermaceae)

Climbing woody vine; 8–12 ft. Root bright yellow within. Leaves smooth, *with 3–7 angles or lobes; stalk attached above base.* Flowers small, whitish, in loose clusters; June–Aug. **WHERE FOUND:** Rich, moist thickets. QC, w. New England to GA; AR, OK. **USES:** Cherokee reported to use root tea for indigestion, arthritis, bowel disorders, also as a blood cleanser and "female tonic"; externally, salve used for chronic sores. Historically, physicians used root (tincture) as a laxative, diuretic; for indigestion, rheumatism, arthritis, syphilis, general debility, and chronic skin infections. Most research has been conducted on the Asian species *M. dahuricum* DC., which contains alkaloids such as daurisoline, with an experimental antiarrhythmic effect, and dauricine, among others; preparations widely used in China for cardiovascular disorders and thrombosis; also for cervical and esophageal cancers. **WARNING:** Poisonous. Fatalities have been reported from children eating seeds and fruits. Some people reportedly confuse this plant with edible wild grapes.

Canada Moonseed stalk is attached just above leaf base.

VIRGINIA CREEPER, WOODBINE — Root, leaves

Parthenocissus quinquefolia (L.) Planchon — Grape Family (Vitaceae)

Climbing (or creeping) vine with *adhesive disks on much-branched tendrils.* Rather than coiling around supports, the tendrils adhere to objects, producing a flattened disk that secretes an adhesive compound to firmly anchor it. The adhesive is a weather-resistant mix of chemicals; perhaps a research lead for the next "super glue." Leaves divided into *5 leaflets, radiating from a central point*; elliptical to oval, sharply toothed. Small flowers in terminal groups; June. Nearly black small fruit 3/16 in. in diameter on loose terminal clusters; Sept.–Nov. **WHERE FOUND:** Thickets. Weedy. ME to FL; TX to KS, MN. **USES:** Cherokee used plant tea for jaundice. Creek used root tea for gonorrhea, diarrhea. Iroquois used leaf tea to wash swellings and Poison Sumac rash; mixed with vinegar for wounds and lockjaw; astringent and diuretic. **WARNING:** Berries reportedly toxic. Leaves toxic; touching autumn foliage may cause dermatitis on human skin.

Virginia Creeper. Note the 5 leaflets radiating from a single point.

FERN ALLIES: HORSETAILS, CLUBMOSS

FIELD HORSETAIL — Whole plant

Equisetum arvense L. — Horsetail Family (Equisetaceae)

⚠ Stiff-stemmed, apparently leafless herb; to 1 ft. Internodes elongate; sheaths of nodes with 8–12 distinct teeth (the leaves). *Branchlets radiating upward from nodes.* Fertile stalks without branches to 18 in. Variable. **WHERE FOUND:** Damp sandy soil. Most of our area. **USES:** Cherokee used plant tea for kidney and bladder ailments, constipation. Chippewa used leaf tea for dysuria. Iroquois and Potawatomi used plant tea to treat lumbago. Asian Indians consider the Field Horsetail diuretic, hemostatic. Root given to teething babies. Folk remedy for bloody urine, gout, gonorrhea, stomach disorders. Poultice used for wounds. High silica content. Also once used in tea for tubercular lung lesions. Shown to be valuable against inflammation, though scientific validity in question. Used in European phytomedicine for treatment of posttraumatic edema; irrigation therapy for bacterial and inflammatory diseases of the lower urinary tract and kidney and bladder gravel; externally for wounds, burns. **WARNING:** Toxic to livestock; questionable for humans—may disturb thiamine metabolism.

Field Horsetail grows in moist soils. Note its branched stems.

SCOURING RUSH, GREATER HORSETAIL — Whole plant

Equisetum hyemale L. — Horsetail Family (Equisetaceae)

(*E. hiemale* L. is a common alternate spelling)

Evergreen, hollow-stemmed, rough-surfaced, jointed primitive perennial; to 5 ft. Variable. The jointed, apparently leafless, rough, finely ribbed, nonbranching stems make it easy to distinguish from other species of horsetail in our range. **WHERE FOUND:** Moist sandy soils, along streambanks, moist depressions, and pond edges. Throughout our range and beyond. **USES:** Essentially the same as for *E. arvense* (above), though this species is considered stronger by some authors. A folk remedy used throughout the Northern Hemisphere. Rough stems are used like sandpaper to give a very fine, satiny finish to wood. Once widely used to polish cabinet work and turned wooden articles and to polish metal goods. Early settlers used the stems to scour pots and pans. Homeopathically used for cystitis, bladder ailments, urinary incontinence, urethritis. Contains up to 25 percent silica (dry weight), consisting of silicic

Scouring Rush, Greater Horsetail has ribbed stems, apparently leafless.

Scouring Rush, Greater Horsetail. Close-up of ribbed stems.

acids and silicates, available in tea as water-soluble components (up to 80 percent). A number of flavonoids in the plant may contribute to diuretic activity. Externally, strengthens and regenerates connective tissue (thanks to the silica content). Recent studies have focused on how these primitive plants concentrate silica through a process called biosilicification, which may involve a protein and polysaccharides such as callose. Science confirms antibacterial and antioxidant activity. **WARNING:** Toxic to livestock; questionable for humans—may disturb thiamine metabolism; case reports relate to potential liver toxicity.

Common or Running Clubmoss. Note the linear leaves tipped with soft, hairlike bristles.

COMMON OR RUNNING CLUBMOSS, GROUND PINE

Leaves, spores

Lycopodium clavatum L. Clubmoss Family (Lycopodiceae)

Mosslike evergreen; 3–15 in., with long, creeping runners. Tiny linear leaves, tipped with *soft, hairlike bristles*. Spores on leafy-bracted stalk, with 1–6 strobiles. **WHERE FOUND:** Dry woods. Canada south to NY, NC mtns.; west to WI, WA. **USES:** Aleut used plant tea for postpartum pains. Montagnais used for fever, weakness. Potawatomi used the spores to stop bleeding and coagulate blood. In folk medicine, spores used for diarrhea, dysentery, rheumatism; also as diuretic, gastric sedative, aphrodisiac, styptic; externally, in powders for a baby's chafing, tangled or matted hair with vermin, herpes, eczema, dermatitis in folds of skin, erysipelas. Spores, called vegetable sulphur, formerly used to coat pills and suppositories. A related Chinese species in the Clubmoss family [*Huperzia serrata* (Thunb.) Rothm.] and its potentially toxic alkaloid, huperzine A, are being researched as a potential treatment for Alzheimer's disease. Recent research suggests potential antidiabetic, anticancer, and anti-inflammatory potential. **WARNING:** Contains a toxic alkaloid.

EVERGREEN FERNS

WALKING FERN

Whole plant

Asplenium rhizophyllum L. — Asplenium Family (Aspleniaceae)
[*Camptosorus rhizophyllus* (L.) Link]

Small, usually evergreen fern with erect stems and characteristic elongated arrow-shaped leaves, with *long narrow tip* that comes into contact with the ground and vegetatively forms a new frond, hence the name Walking Fern. Base is *heart-shaped*, with rounded lobes. Sterile fronds usually smaller than fertile fronds. **WHERE FOUND:** Moist, shaded limestone outcrops. QC to GA mtns.; west to e. OK, KS, north to MN, ON. **USES:** Cherokee used fronds in decoction as a wash for swollen breasts; also as an emetic to induce vomiting.

ROCK POLYPODY

Root, whole fern

Polypodium virginianum L. — Fern Family (Polypodiaceae)

⚠ Vigorous, evergreen, mat-forming fern; to 12 in. Leaves leathery; deep green above, *smooth below*; 10–20 leaflet pairs; veins obvious, variable. **WHERE FOUND:** Shaded, rich, shallow soils or among rocks. NL to GA mtns.; AR to e. SD, MN. **USES:** Algonquin used plant tea for heart disease. Other Native Americans used root tea for pleurisy, hives, sore throats, stomachaches; poulticed root for inflammations. Historically, root steeped in milk as laxative for children. Once considered valuable for lung ailments and liver disease. Tea or syrup of whole plant used for liver ailments, pleurisy, worms. Like Male Fern (p. 405), this fern was believed to be toxic to tapeworms. The root has a unique, rather unpleasant odor, and a sweet (cloying) flavor at first, but then quickly becomes nauseating. Root contains fructose, glucose, and sucrose, plus methyl salicylate (wintergreen flavor—see p. 36). Root contains up to 2 percent insect-regulating "hormones" (steroidal prohormones), including ecdysterone and ecdysone (the former offered in lean muscle–

Walking Fern. When long leaf tips touch the ground, they take root to produce a new frond, hence the name Walking Fern.

Walking Fern. Note elongated, undivided fronds with sori beneath.

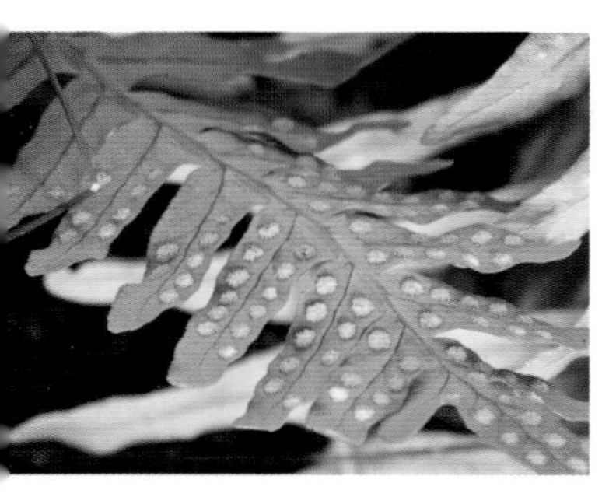

Common Polypody sori in middle to edge of leaf margin.

Common Polypody is an evergreen, mat-forming fern.

mass body-building supplements). They are from a class of compounds called phytoecdysteroids, also found in Cordyceps, the adaptogenic fungi of Nepalese fame, and in Common Polypody (*Polypodium vulgare* L). Resins active against worms. **WARNING:** Of unknown toxicity.

CHRISTMAS FERN — Root, fiddleheads, leaves

Polystichum acrostichoides (Michx.) Schott. — Dryopteris Family (Dryopteridaceae)

Shiny evergreen fern; 1–3 ft. Leaves lustrous, leathery, tapering, leaflets *bristle-tipped, strongly eared*; teeth incurved. **WHERE FOUND:** Rich wooded slopes. NB to n. FL; e. TX to KS, WI. **USES:** Cherokee used cold root tea to treat stomachache and bowel complaints, rheumatism, fever, and pneumonia. Ate curled spring fiddleheads as food. Iroquois used root as an emetic for sick stomach. Small leaves were carefully picked and decocted, taken in teaspoonful doses for fever and cramps in children. Root tea used as a footbath to treat rheumatism in back and legs. Micmac used root tea to treat sore throat. **WARNING:** Of unknown safety.

Christmas Fern. Note lobe at base of leaflet.

MAIDENHAIRS AND BRACKEN FERN

VENUS MAIDENHAIR FERN — Whole fern

Adiantum capillus-veneris L. — Maidenhair Fern Family (Adiantaceae)

Fronds to 20 in.; oblong in outline, mostly twice-compound. Leaflets *very thin,* papery; wedge-shaped, lobed at apex. Stalks delicate, brittle; shiny, dark, scaly at base. **WHERE FOUND:** Wet limestone rocks, waterfalls, bluffs, usually in shaded areas. VA to FL, TX, and CA; north through UT, CO to MO, SD. Introduced as a curio farther north. Found throughout many parts of the world. **USES:** In folk tradition, a handful of dried leaves are steeped to make a tea drink as an expectorant, astringent, and tonic for coughs, throat afflictions, and bronchitis. Used as a hair wash for dandruff and to promote hair growth. In colonial America, a syrup of Venus Maidenhair Fern (*Sirop de Capillaire*) was imported from France as a cough medicine. Leaf tea suggested as a pleasing summer drink. In Traditional Chinese Medicine, the leaves are similarly used for bronchial diseases and as an expectorant. This fern has also been used as a worm expellent, an emetic, and an agent to reduce fevers. Externally, it has been poulticed on snakebites and used as a treatment for impetigo. Science confirms significant antimicrobial, anti-inflammatory, and analgesic effects.

LEFT: *Venus Maidenhair Fern. Note black stems and triangular leaf pattern.* ABOVE: *Maidenhair Fern leaves grow in a horseshoe pattern.*

MAIDENHAIR FERN — Whole fern

Adiantum pedatum L. — Maidenhair Fern Family (Adiantaceae)

One of the easiest ferns to recognize, given the arrangement of the wedge-shaped leaflets in a *horseshoelike frond atop shiny ebony-colored* stems. A distinctive fern; to 1 ft. Leaflets long, fan-shaped, lobed on upper side; alternate. **WHERE FOUND:** Rich woods, moist limestone ravines. ME south to GA, LA; west to OK, north to MN and westward. **USES:** Cherokee used whole-plant tea as an emetic for fevers, rheumatism, chills. Powdered leaves used as snuff for asthma. Considered expectorant, cooling, and antirheumatic. Tea or syrup used for nasal congestion, asthma, sore throats, hoarseness, colds, fevers, flu, and pleurisy. This fern was highly valued as a medicinal plant by some nineteenth-century medical practitioners, suggesting that its efficacy should be investigated by science. Native Americans throughout N. America used stems as a hair wash to make hair shiny.

BRACKEN FERN — Root

Pteridium aquilinum (L.) Kuhn — Cup Fern Family (Dennstaedtiaceae)
(*Pteris aquilina* L.)

⚠ Our most common fern; 3–6 ft. tall, forming large colonies. Leaves *triangular, divided into 3 parts*; leaflets blunt-tipped; upper ones not cut to midrib. Variable. **WHERE FOUND:** Barren soils. Much of our area; mostly absent from Great Plains. Occurs in much of the world; considered the fifth most widespread common weed species worldwide. **USES:** Cherokee used root as a tonic, antiseptic, to relieve nausea, and to treat cholera morbus (a nineteenth-century term for non-epidemic cholera). Other Native American groups used root tea for stomach cramps, diarrhea; smoke for headaches; poulticed root for burns and sores, caked breasts; wash to promote hair growth; astringent, tonic. Historically, root tea used for worms. **WARNING:** Poisonous—disturbs thiamine metabolism. Contains at least 3 carcinogens. One of the few plants known to induce cancer in grazing animals; contains the carcinogenic toxin ptaquiloside (among others), which damages DNA. Human exposure to this carcinogen can come from consuming milk or meat. Nineteenth-century writers warned that it has an injurious effect on the mucous membranes of the bladder and causes inflammation of the urinary organs.

Bracken Fern. Note the lateral, thrice-divided leaves.

MISCELLANEOUS FERNS

LADY FERN

Root (rhizome), stem

Athyrium filix-femina (L.) Roth. — Cliff Fern Family (Woodsiaceae)

"Discouragingly variable" (*Gray's Manual,* 5th ed.). Fern; to 3 ft. Stems *smooth,* with a few pale scales. Grows in circular clumps from horizontal rootstock. Leaves (fronds) broad, lance-shaped; mostly twice-divided, lacy-cut; *tips drooping. Leaflets toothed.* **WHERE FOUND:** Moist, shaded areas. Much of our area. **USES:** Iroquois used root tea as a diuretic, to stop breast pains caused by childbirth, induce milk in caked breasts. Stem tea taken to ease labor. Like many ferns, this one was traditionally used to eliminate worms. Dried powdered root used externally for sores. **RELATED SPECIES:** Japanese researchers found anti-gout potential in the related fern *A. mesosorum* (Makino) Makino.

RATTLESNAKE FERN

Root

Botrychium virginianum (L.) Swartz — Adder's-tongue Family (Ophioglossaceae)

The largest *Botrychium* in our area. Delicate, lacy, nonleathery, *broadly triangular leaf* (sterile frond); to 10 in. long, 12 in. wide. Fertile frond on a much longer stalk, bearing bright yellow spores. **WHERE FOUND:** Rich, moist, or dry woods. PE, MN to FL; CA, BC. Unfurls before other *Botrychium* species. **USES:** The Cherokee boiled the root down to an extract the consistency of syrup and applied topically to treat snakebites; also poulticed for bruises, cuts, and sores. In folk medicine, root tea emetic, induces sweating; also an expectorant, used for lung ailments. In

Lady Fern has twice-divided, lacy leaves.

Rattlesnake Fern unfurls in late winter or early spring.

Appalachia and the Ozarks, this fern is called "seng pointer" as its presence is an indicator plant for the habitat that American Ginseng favors.

MALE FERN

Roots

Dryopteris filix-mas (L.) Schott

Dryopteris Family (Dryopteridaceae)

⚠ Yellow-green, leathery, semi-evergreen fern; 7–20 in., with blackish thick-wiry roots. Leaves divided into about 20 lance-shaped, pointed leaflets; leaflets narrow, oblong, cut nearly to midrib, with rounded lobes or subleaflets, or slightly toothed. **WHERE FOUND:** Rocky woods. ME, VT, NY to MI. **USES:** Used in classical times and mentioned by Theophrastus, Dioscorides, and Pliny; in the Middle Ages it was used as a folk remedy. It emerged again in medical practice in the sixteenth century, revived as a secret remedy for tapeworm. By the mid-nineteenth century it was widely used by physicians for tapeworm. An oleoresin extracted from the roots has been used as a worm expellent. It is toxic to tapeworms. Components in root are of potential research interest as chemopreventives. **WARNING:** Toxic poison and skin irritant.

Male Fern root was formerly used to expel worms.

LARGE GRASSES OR GRASSLIKE PLANTS

AMERICAN SWEETFLAG; AMERICAN CALAMUS
Rhizome

Acorus americanus (Raf.) Raf. Sweet Flag Family (Acoraceae)
[*Acorus calamus* var. *americanus* (Raf.) H. Wolf.] [Formerly in Arum Family (Araceae)]

Now recognized as a separate species, *A. americanus* differs from *A. calamus* by leaves with prominent *single midrib with 1–5 additional raised midribs* above the leaf surface. This is a fertile diploid species, producing viable seeds. **WHERE FOUND:** Northeastern states and Canadian provinces. See p. 124.

SWEETFLAG, CALAMUS (NOT SHOWN)
Rootstock

Acorus calamus L. Sweet Flag Family (Acoraceae)
[Formerly in Arum Family (Araceae)]

A sterile Old World triploid with a *single vertical midrib*. **WHERE FOUND:** Much wider distribution in N. America compared with *A. americanus. A. calamus* was spread by humans. See p. 124.

GIANT CANE
Root

Arundinaria gigantea (Walt.) Muhl. Grass Family (Poaceae or Gramineae)

⚠ *Bamboolike*, woody-stemmed grass; to 10 ft. Lance-shaped leaves, in fanlike clusters. Flowers in racemes, on leafy branches. **WHERE FOUND:** Among our largest native grasses, this distinctly bamboolike perennial forms large thickets along rivers, creeks, and moist soils south of DE. S. DE to FL; TX to IL. **USES:** Houma Indians used root decoction to stimulate kidneys. **WARNING:** Ergot, a highly toxic fungus, occasionally replaces the large seeds of Giant Cane. Do not collect or use any specimens from areas with diseased plants.

COMMON CATTAIL
Root, flowerheads, seed down

Typha latifolia L. Cattail Family (Typhaceae)

Perennial; 4–8 ft., forming thick stands. Leaves swordlike. Stiff, erect flowering stalks, topped with pollen-laden yellow male flowers above *hot dog–shaped brown female flowerheads*; May–July. **WHERE FOUND:** Fresh marshes, ponds. Throughout our area. **USES:** Native Americans poulticed jellylike pounded roots on wounds, sores, boils, carbuncles, inflammations, burns, and scalds. Fuzz from mature female flowerheads applied to scalds and burns and used to prevent chafing in babies. Young flowerheads eaten for diarrhea. Root is infused in milk to cure dysentery and diarrhea. Anthers, pollen, rhizomes, and shoots have all served as human food. Pulp can be converted to rayon. Used for thousands of years for the treatment of skin disorders, burns, and as an absorbent wound dressing; recent research confirms a scientific basis for traditional use. Polysaccharides in the cottony fluff of fruiting

ABOVE: *American Sweetflag, Calamus has aromatic, grasslike leaves with 2–5 raised midribs. Sweetflag from Europe has a single midrib.* TOP RIGHT: *Giant Cane has a bamboolike stalk.* RIGHT: *Common Cattail has an easily recognizable seed head.*

bodies have a strong stimulatory effect in the proliferation of skin cells and their early differentiation in skin tissue, aiding in the wound-healing process. **RELATED SPECIES:** Root tea of Narrowleaf Cattail (*T. angustifolia* L.) has been used for "gravel" (kidney stones). **WARNING:** Though it is widely eaten by human foragers, Cattail is suspected of being poisonous to grazing animals.

ABOVE: *Corn is one of the most important food plants from the Americas.* RIGHT: *Corn silk from Corn is used as an herbal diuretic.*

CORN Whole plant

Zea mays L. Grass Family (Poaceae or Gramineae)

So well known that a description is unnecessary. **WHERE FOUND:** Introduced to the U.S. by Native Americans centuries ago; cultivated throughout our area. **USES:** Corn "silk," a well-known herbal diuretic, was once used in tea for cystitis, gonorrhea, gout, and rheumatism. Recently, water-soluble components of corn silk (chitinases and glucanases) have been found to be antifungal, inhibiting infection from aflatoxin produced by *Aspergillus flavus*; seeds contain allantoin (best known from Comfrey, p. 243), a cell-proliferant, wound-healing substance. Science has confirmed diuretic, hypoglycemic, and hypotensive activity in animal experiments with corn extracts.

MISCELLANEOUS GRASSES

QUACK GRASS Whole plant

Elytrigia repens (L.) Desv. Grass Family (Poaceae or Gramineae)

[*Agropyron repens* (L.) Beauvois; *Elymus repens* (L.) Gould]

Grass; to 3 ft. Spreads on creeping *yellow* rhizomes. Leaves soft, flat, somewhat drooping; *crowded with fine ribs*. Flower spike is not square, as in most *Agropyron* species; 2–9 flowered spikelets, bract below spikelets not stiff, with slender keel and ribs. **WHERE FOUND:** Fields, gardens. Troublesome weed throughout our area. Eurasian alien. **USES:** Native Americans used tea as a diuretic for "gravel" (kidney stones) and urinary incontinence; worm expellent; wash for swollen limbs. Ap-

Quack Grass is a troublesome weed throughout our area.

proved in Germany in irrigation therapy for inflammatory conditions of the urinary tract and prevention (not treatment) of kidney gravel. Also used for bronchitis, cold, cough, fever, infections, pharyngitis, and stomatitis. In times of famine, Native Americans used the rhizomes to make breadstuffs; also scorched as a coffee substitute. Roots sometimes chewed like licorice. Considered an antidote to arrow poisons in Africa. Essential oil in root is antimicrobial.

BROOMSEDGE (NOT SHOWN)
Leaves, root

Andropogon virginicus L. — Grass Family (Poaceae or Gramineae)

Highly variable grass; 28–55 in. Bluish or green stem. Leaf sheath overlapping, keeled, and strongly compressed. Small flowers emerge from envelopes that are not inflated; racemes usually in pairs, with silvery white hairs. **WHERE FOUND:** Dry soil, open woods. MA, NY south to FL; west to TX, KS; north to IL, IN, OH. **USES:** Catawbas used root decoction for backaches. Cherokee used leaf tea for diarrhea; externally, as a wash for frostbite, sores, and itching. Also used for hemorrhoids and Poison Ivy rash.

WILD OATS
Stems, flowering plant

Avena fatua L. — Grass Family (Poaceae or Gramineae)

Annual; coarse grass; 2–4 ft. tall. Leaves rough to soft-hairy. Flowers in a panicle with spreading branches. Spikelets mostly nodding, usu-

Also called Wild Oats, Avena sativa *has milky juice within its stems.*

Wild Oats produce nodding spikelets.

ally longer than 1¼ in. Bracts (glumes) *nearly as long as the flower spikelets*; rounded at back, with numerous veins. **WHERE FOUND:** Waste places, cultivated ground. Scattered throughout our range. Alien (Europe). **USES:** A folk medicine used as a nerve tonic, diuretic, and antispasmodic. In modern herbal medicine, oat straw is commonly used for treatment of skin conditions characterized by inflammation and pustules. A tincture prepared from the fresh flowering plant is used as a sedative, primarily for sleeplessness and nervous exhaustion, anxiety, and related conditions. Oat straw tea is considered diuretic, used in the treatment of gout, rheumatism, and skin disorders. An oat straw bath is traditionally used to relax patients with hypertension. Oats themselves are considered nutritive and demulcent; they lower cholesterol. **RELATED SPECIES:** Oats *Avena sativa* L., the common oats of commerce, derives from *A. fatua*. They differ on small technical characteristics. *Avena sativa* has *milky juice* in aboveground parts.

SWEET GRASS

Leaves

Hierochloe odorata (L.) P. Beauv. Grass Family (Poaceae or Gramineae)
[*Anthoxanthum nitens* (G. H. Weber) Y. Schouten & Veldkamp]

Vanilla-scented grass; 10–24 in. Spreads on slender, creeping rhizomes. Leaf clumps arise from dead foliage of previous year and wither soon after flowering. Flowers in pyramid-shaped clusters. **WHERE FOUND:** Meadows. NS to PA, OH, IA, SD. Sweet Grass is threatened

Sweet Grass is growing increasingly rare because of overharvest.

because of overcollection. **USES:** Native Americans widely used Sweet Grass as incense for ceremonies. Tea used for coughs, sore throats, chafing, venereal infections; to stop vaginal bleeding, expel afterbirth. Antioxidant compounds have been isolated from the leaves. **WARNING:** Roots contain a coumarin, sometimes considered carcinogenic.